Automotive Engine Repair

A Text-Lab Manual

Glencoe Automotive Technology Series

Ivan D. Hinerman
Automotive Technology Instructor
Mount San Jacinto College

Glencoe Publishing Co., Inc.
Encino, California
Collier Macmillan Publishers
London

Glencoe Automotive Technology Series

Automotive Suspension and Steering: A Text-Lab Manual
Automotive Brakes: A Text-Lab Manual
Automotive Power Trains: A Text-Lab Manual
Automotive Engine Repair: A Text-Lab Manual
Automotive Transmissions: A Text-Lab Manual

Printed in the United States of America

Glencoe Publishing Co., Inc.
17337 Ventura Boulevard
Encino, California 91316
Collier Macmillan Canada, Ltd.

Library of Congress Catalog Card Number: 77-81911

ISBN 0-02-818600-1

4 5 6 7 8 9 83 82

Contents

Preface

The automotive service technician of today must know the fundamentals of the trade and be capable of building a storehouse of knowledge to keep up with changing vehicle service requirements. A balance of fundamental information and current service practices is presented in this text.

This course of instruction will serve equally well in a program of individualized instruction or in traditional automotive training at the secondary and post-secondary level. The material has been used successfully at Mount San Jacinto College, San Jacinto, California in an open-entry, open-exit, individualized instruction program. This text-lab manual is written for the vocational automotive course where the emphasis is on skill development.

Important concepts are explained in simple, everyday language. Very often, comparisons are made to ideas or functions that are found elsewhere in the automobile or in the home. When a new technical term is introduced in the text, an explanation follows.

A large number of illustrations accompanies the text to both further explain and to lessen the time required in order to fully understand the most difficult topics. The illustrations appear as close as possible to the written explanation, with each figure keyed to the text.

General and specific safety hazards will be discussed throughout the text at a time when the student has the greatest need for learning about them.

This book is organized in unit form. Each unit begins with a short statement concerning the major topics of discussion. This is followed by a written test objective which spells out exactly what the student is expected to achieve. These are labeled with a unit number and a letter indicating position in the unit for example: 3-A. Below the objective are the text pages which the student needs to study. This is the preparation assignment. The preparation assignment provides all the general and specific information required to fully complete the objective.

A manipulative objective, or objectives, follows the text material. These are also labeled with the unit number, location within the unit, and a number to maintain consecutive order of completion. The manipulative objectives are stated in specific performance terms. Step-by-step procedures of a general nature follow the manipulative objective with specific directions for completing a service operation typical of those performed by today's automotive technician.

At the end of each unit or at the end of each major division within a unit is a written test which covers material and service procedures from the written and manipulative objectives. These tests are multiple choice and are labeled the same as the written objective found earlier in each unit.

Seldom does one find two schools whose automotive programs teach exactly the same subjects in the same manner. Each instructor has individual preferences on assignments to be completed. This text, hopefully, includes the majority of assignment preferences. It is the author's experience that the most usable learning will take place when this text is used as a guide in completing the lab assignments, and as a reference when completing the tests and reading assignments.

Unlike many other automotive texts, this program is a reflection of practical teaching at the community college level whose object is to provide the most relevant materials and the most conducive climate for young men and women to gain the skills necessary to be competent service persons. It is with these same goals that this text is presented.

unit 1

Engine Basics

Most people have driven an automobile at some time or another and have a casual acquaintance with its engine. All drivers understand that the engine is the source of power that takes them where they want to go. Usually they know where the engine and the battery are, how to lift the hood, check the coolant level, and check the oil level. Beyond these basic operations there are wide differences in people's knowledge of the automotive engine. The explanations in Unit 1 are sufficiently simple and elementary to serve as a beginning point on the journey to understanding automotive engines and becoming competent in their repair.

OBJECTIVE 1-A (written test)

After completing the preparation assignment, you will correctly answer a minimum of seven out of ten multiple choice questions in a written test, without using reference material, on the following topics:

1. Engine definition
2. Engine types
3. Internal combustion
4. Engine classifications
5. Engine operation
6. Engine systems
7. Basic engine parts

Preparation:

Study pages 1 through 12.

ENGINE DEFINITION

What is an engine? Briefly stated, it is any machine capable of converting one form of energy to another. The automobile engine converts gasoline into heat energy, which is turned into mechanical energy (rotational power) at the crankshaft (Fig. 1-1). There are many kinds of engines for many different uses. This text will discuss only the type used in nearly all of the automobiles of today. The technical name for this engine is the gasoline-fueled, four-stroke cycle, internal combustion engine. This name is very specific but much too long to be convenient. Therefore, in this text, the single word *engine* will be used to mean this gasoline-fueled, four-stroke cycle, internal combustion engine.

ENGINE TYPES

Many other types of engines have been developed which have possibilities as automobile power plants; however, most are currently in the experimental or developmental stage. Some of these are the turbine, commonly called the jet engine from its applications in aircraft; the steam engine; the Sterling engine; and a variety of noncombustion power plants such as the fuel cell or the electric combination (small engine with generator). All of the alternative types of engines have some drawbacks which have discouraged their widespread use in automobiles up to this time. The only alternative automobile engine which is currently in production is the Wankel Rotary, used in import vehicles including the Mazda from Japan. Scientific breakthroughs or ongoing development may lessen the automobile's dependence on the four-stroke, internal combustion engine.

The only explanation or repair procedures discussed in this text will be for the four-stroke cycle engine.

INTERNAL COMBUSTION

To better understand the functions of the automobile engine, an explanation of each part of its long, complicated name may be helpful. Gasoline-fueled means that its primary source of heat energy is the common hydrocarbon compound called gasoline. In the early years of automobile transportation, the most convenient fuel with the greatest power-producing capacity was gasoline. Even today, gasoline is the overall best fuel choice.

This fuel is mixed with air and ignited "internally" in a closed chamber within the engine framework—hence the name *internal combustion engine* (Fig. 1-2).

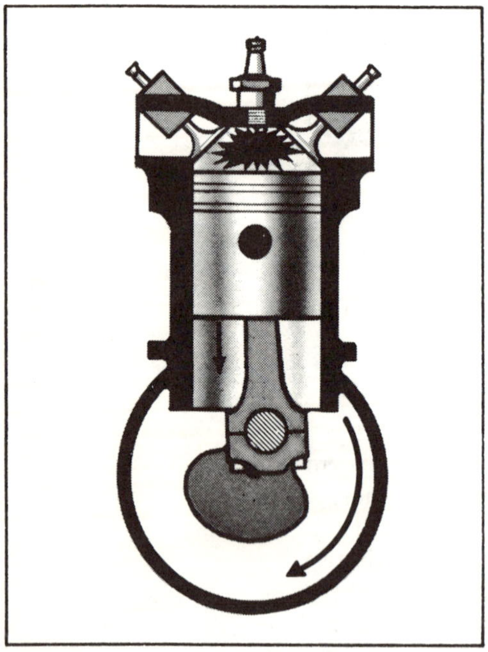

FIG 1-1 Mechanical energy, created by igniting the fuel, which forces the piston downward.

(Federal-Mogul Corporation)

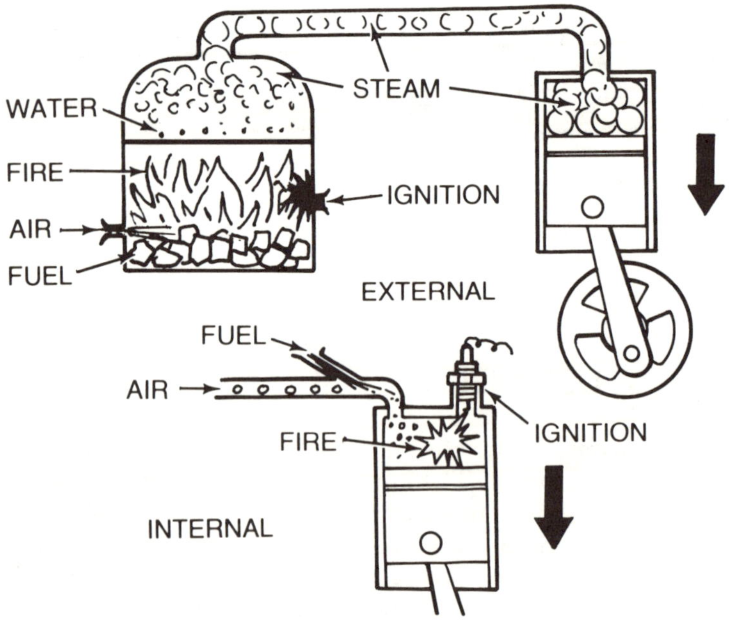

FIG 1-2 Internal combustion takes place within a confined space such as the combustion chamber.

(General Motors Corporation)

The term *cycle* indicates that the engine parts go through a series of events which recur at regular intervals and in the same event sequence each time. *Four-stroke* refers to the four up-and-down or back-and-forth movements the piston makes to complete one cycle. A more detailed explanation of the four-stroke cycle will appear later in this unit.

The objective of any automotive engine is to propel the vehicle by converting heat energy into mechanical energy. Heat energy from gasoline is released as it begins to burn. The heat causes the air-fuel mixture to expand, forcing the piston downward from the closed end of the combustion chamber and causing the crankshaft to rotate. This greater-than-original volume of gas, along with the heat formed by combustion, expands the carbon dioxide and water formed as a result of the burning fuel. All of this action helps the engine complete its objective.

As the heat of combustion is converted to mechanical energy, large amounts of heat are lost (Fig. 1-3). Part of the heat loss occurs during the exhaust stroke which passes hot burned gases out through the exhaust system to prepare the cylinder for a fresh incoming charge. Heat is also lost in the heat transfer of the combustion chamber gases to the engine metals, which have to be cooled by air or liquid from the cooling system (Fig. 1-4). The last major source of heat loss is in the radiation of engine heat energy that builds up under the automobile hood.

A rough estimate of the heat conversion efficiency of an automotive engine is 25–30 percent. This means only 25–30 percent of the potential heat energy (gasoline) put into the fuel tank is actually converted to mechanical energy (vehicle movement). Besides the heat loss, friction in the mechanical parts of the engine and power train of the car reduces efficiency. Wind resistance and rolling resistance of the tires are also major factors in the reduction of a vehicle's efficiency.

The basic function of the engine is to produce continuous and controllable rotating motion with enough power to move the vehicle.

FIG 1-3 Areas within an engine where heat energy is lost.
(Chrysler Corporation)

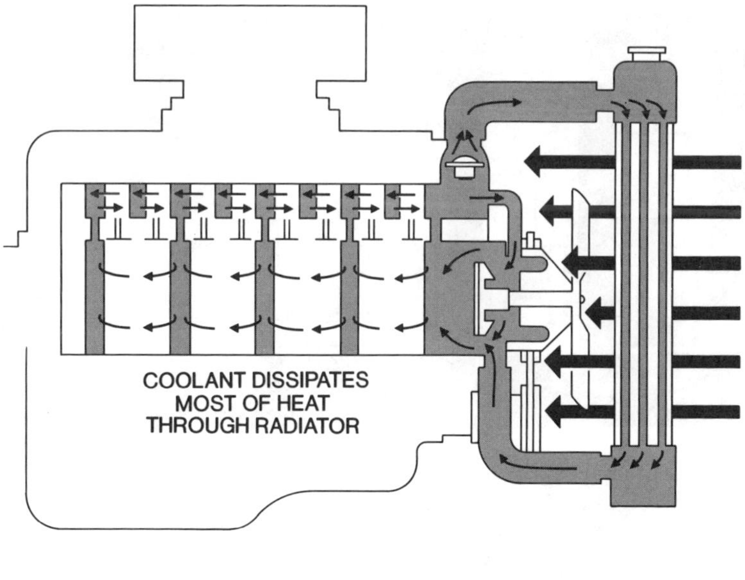

FIG 1-4 Air passing through the radiator dissipates (takes away) much of the engine's unusable heat.
(Chrysler Corporation)

ENGINE CLASSIFICATIONS

Gasoline engines are classified in different ways to give information about the design of the engine. The four common gasoline engines are classified by number of cylinders and their arrangements; valve and camshaft arrangements; number of cycles; and the type of cooling employed.

Oftentimes a slightly different means of classification or identification is necessary to obtain specifications or parts for an engine to be repaired. Both classification information and identification processes will be discussed in this text.

One important point to remember is that, regardless of the differences in classifications, there is very little difference in the fundamental operations of automotive engines.

CYLINDER CLASSIFICATIONS

The most common way of classifying an engine is by the number of cylinders. This tells very little about the performance or capabilities of an engine other than that, generally, the more cylinders an engine has, the smoother it runs and the greater power it can develop. The most common engine designs used in automobiles today have four, six, or eight cylinders.

When people purchase automobiles or light trucks, they often consider the number of cylinders in the engine as a guide to economy of performance. A four- or six-cylinder engine is the most economical, while the eight-cylinder engine is the one most people would choose for power and performance.

Cylinder arrangement is mostly an engineering design decision. The common arrangements in both foreign and domestic automobiles are the in-line and the V-types (Figs. 1-5 and 1-6).

In the in-line design, cylinders are arranged in a straight line and, generally positioned from front to rear in the engine compartment. In-line engines of some European manufacturers have the cylinders mounted across the frame lines. In-line cylinder arrangements usually have four or six cylinders. Eight-cylinder in-line

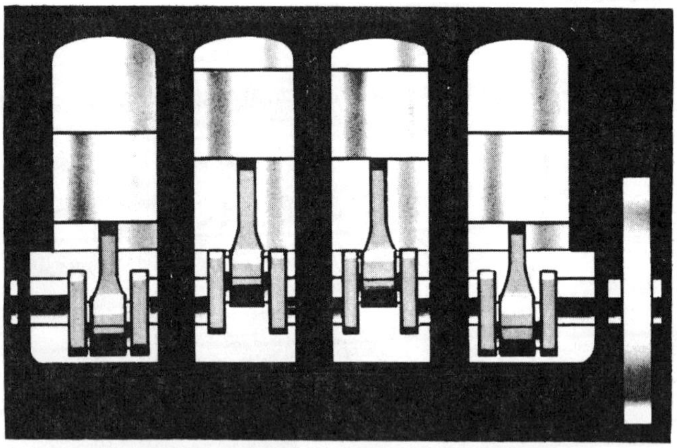

FIG 1-5 In-line cylinder arrangement.
(Federal-Mogul Corporation)

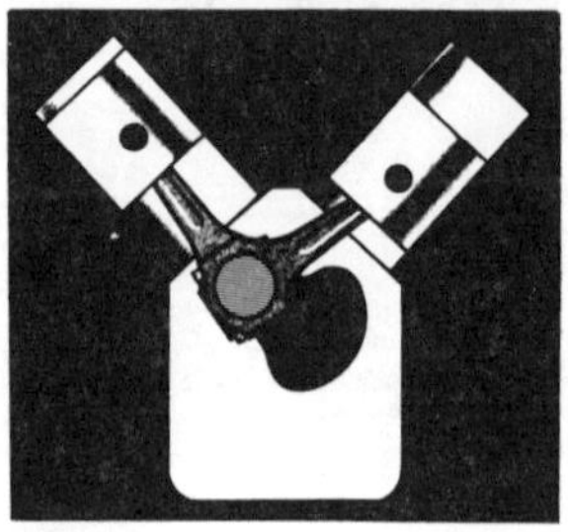

FIG 1-6 V-type arrangement with 90° between rows of cylinders.
(Federal-Mogul Corporation)

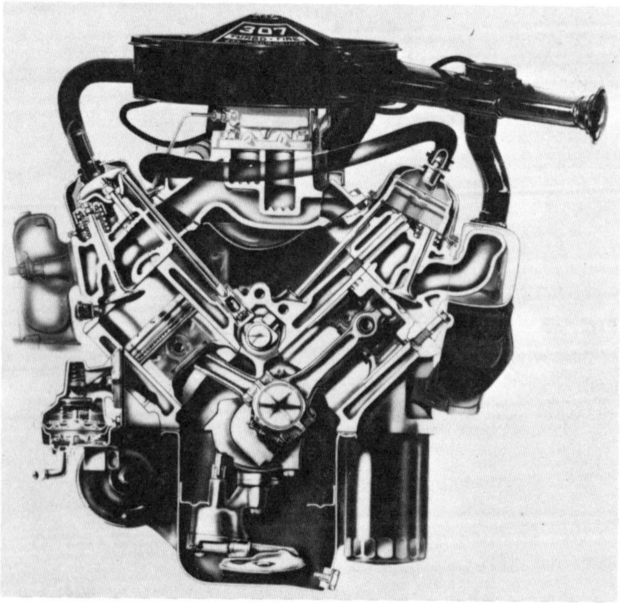

FIG 1-7 V-8 engine. (Oldsmobile Division, GMC)

engines are no longer produced for automobiles because of problems of excessive weight and inefficient operation.

The V arrangement of cylinders gets its name from the shape of the letter *V*. With two rows of cylinders arranged beside one another at approximately a 90° angle, roughly twice as many cylinders can be placed in the same area that one row would otherwise hold (Fig. 1-7). This arrangement is very popular today, especially for eight-cylinder engines, but there are also V-6 and V-4 arrangements.

In the opposed cylinder design (sometimes called flat or pancake) the cylinders are on opposite sides of the crankshaft. This design is very popular for air-cooled automobile engines and those where very small engine space was allotted by body designers. The most popular application of this design is the four-cylinder Volkswagen manufactured in Germany (Fig. 1-8).

VALVE CLASSIFICATIONS

Three types of valve arrangements are found in automotive engines of today: the I-head, the L-head, and the F-head. The L-head and F-head are not currently being manufactured for automotive use, but service work on older models is still being performed.

The most popular design by far is the I-head or, as it is referred to more commonly, the overhead or OHV, meaning overhead valve. The letter *I* is used to describe this design because placement of the cylinder block and cylinder heads resembles that letter. Occasionally, the term *valve-in-head* is still used to describe this design. Figure 1-9 shows an engine block (V-8) with cylinder heads installed. Notice how the valves are located within the cylinder head, which is mounted above the cylinders.

A number of reasons may be advanced for the popularity of this valve arrangement. Ease of service is one major reason why many people prefer this design. The cylinder heads can be removed, serviced, and replaced with relative ease compared to other designs. From the engineering standpoint, this design allows for better cooling of the combustion chamber and valve area. Performance enthusiasts find

FIG 1-9 Valves placed directly above the piston in the cylinder head.

(Cadillac Motor Car Division, GMC)

10. Intake air distributor
11. Injector
12. Valve
13. Oil cooler
14. Oil bath air cleaner
15. Cylinder head
16. Spark plug
17. Heat exchanger
18. Piston
19. Ignition distributor
20. Connecting rod
21. Thermostat
22. Cylinder

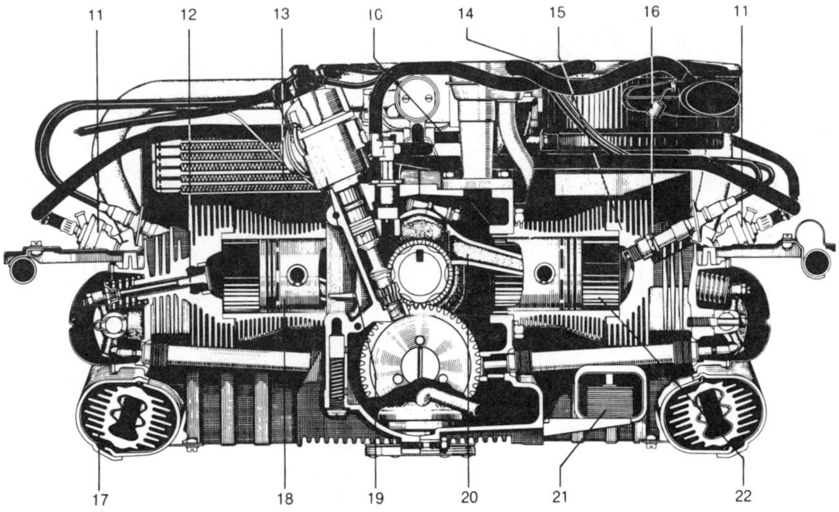

FIG 1-8 An opposed cylinder arrangement.

(Volkswagen)

this valve arrangement adaptable to changes that bring increased horsepower and torque through higher compression ratios and better breathing qualities. Higher compression ratios mean more driving force to the crankshaft, while better breathing allows more air-fuel mixture into the combustion chamber. This produces higher gas pressures pushing down on the piston during combustion on the power stroke. The disadvantage of this design is that it requires extra parts, such as pushrods and rocker arms, which means more costly production, extra engine friction, and higher total weight.

The L-head arrangement of valves is usually referred to as a flat-head because of the relative flatness of the cylinder heads. A cross-sectional view (Fig. 1-10) shows how the valves are placed within the engine block, in a pattern resembling an upside-down letter *L*. This design has disappeared from current automobile production because of basic inefficiencies and difficulties in meeting exhaust emission standards presently in force. Industrial and small horsepower engines continue to use the flat-head design because of its less complicated and less costly valve train in applications where it does not have to meet any stringent exhaust emission standards.

Another classification of valve arrangements is based on the placement of the camshaft. During the early years of American automotive engines, the camshaft was located within the block or between the two halves of the air-cooled block. The major reasons for locating the camshaft inside the block were to provide the most rigid housing for the support of the camshaft and to position the camshaft gear as close as possible to the driving crankshaft gear.

The vast majority of automobile engines today have their camshafts placed within the engine block near the crankshaft. This eliminates the complicated and expensive gear groups, belts, or lengthy chains necessary to drive the overhead camshaft.

Engines whose camshaft is in or on top of the cylinder head are usually labeled OHC for overhead camshaft. Further designations are sometimes used to distinguish between a single overhead camshaft (SOHC) and a dual overhead camshaft (DOHC) (Fig. 1-11).

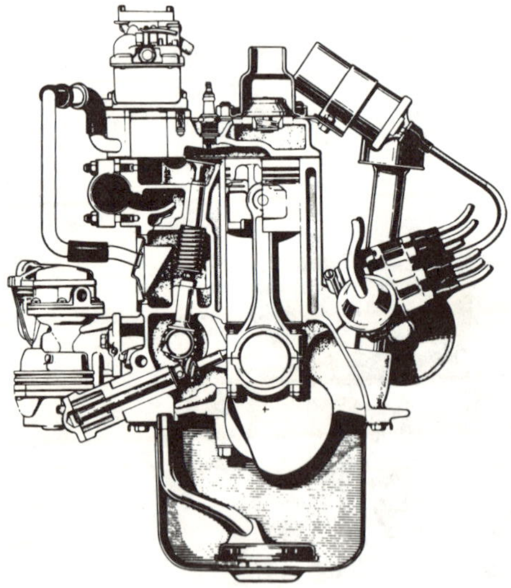

FIG 1-10 An L head or flat head engine.

(American Motors Corporation)

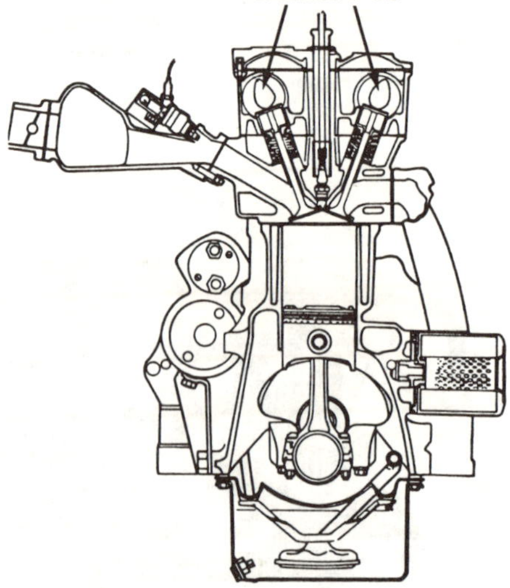

FIG 1-11 A dual overhead camshaft (DOHC) valve arrangement.

(Chevrolet Motor Division, GMC)

CYCLING CLASSIFICATIONS

Internal combustion engines are also classified by the number of strokes per cycle. There are two-stroke and four-stroke engines (Fig. 1-12). Two-stroke engines are common power sources for motorcycles or lawnmowers.

With few exceptions, automotive engines in use today are of the four-stroke design. Many times the term *four-cycle* will be used to describe the four-stroke cycle engine. In this abbreviation, the word *stroke* is eliminated.

All internal combustion engines considered in this text are of the four-stroke cycle design. This design, sometimes called the Otto cycle, was originally named for its inventor, Nikolaus Otto, who built and patented the first practical engine of this type in 1877.

COOLING SYSTEM CLASSIFICATIONS

Engines used for automotive power are also classified by the type of cooling system they employ. The systems in use today are the liquid (water) cooled and air cooled.

During normal engine operation, a great deal of the heat of combustion is wasted and must be constantly removed from the engine. In the liquid system, the coolant circulates through the engine passages (water jackets) soaking up the excess heat. This coolant then goes to the radiator where the heat is taken away by the passing air.

The air-cooled system has fins designed into hot areas of the engine. These fins provide increased surface area, allowing greater heat transfer. A fan or blower pushes cooler air over the fins, which takes the heat away (Fig. 1-13).

Varied ways of classifying engines are necessary for different situations. The most common classifications have been discussed in order to give an understanding of how many automotive engine variations are in use today.

When the actual service work begins, some means of basic engine classification is needed so specifications and replacement parts can be obtained. The additional service information for identification of an engine—such as size, year, make, etc.—will be discussed in Unit 3.

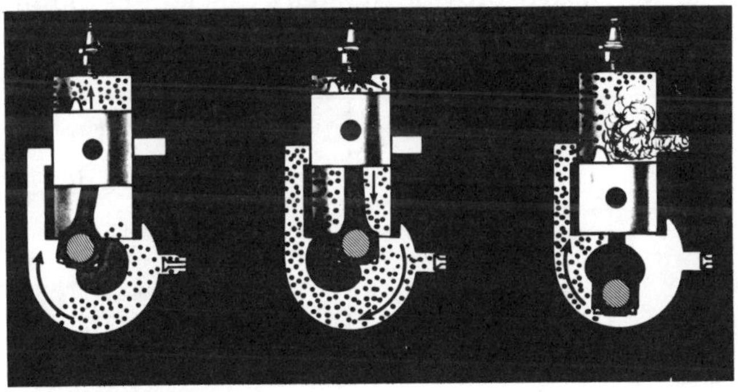

FIG 1-12 Stroke sequence of a two-cycle engine.
(Federal-Mogul Corporation)

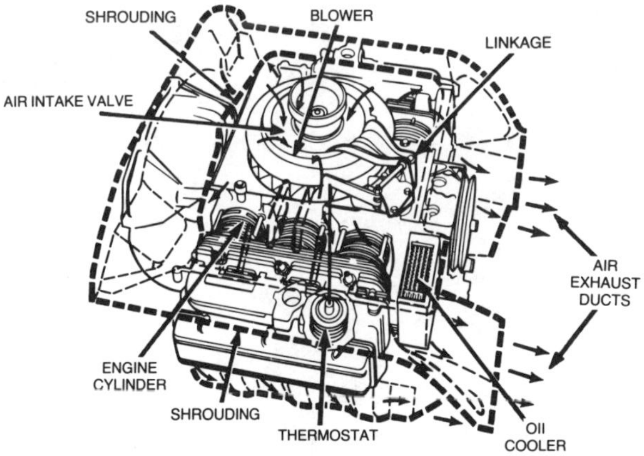

FIG 1-13 Air-cooled cooling system
(Chevrolet Motor Division, GMC)

ENGINE OPERATION

Three things are basic to the operation of all internal combustion engines: air (oxygen), fuel (gasoline), and ignition (spark and compression). All three must be present before any engine will function. If one of these is absent, the engine will not operate.

AIR

Obtaining the air required to operate an engine is not usually a problem since air surrounds us all of the time and can be easily moved into the engine. The only part of the air which is really needed for combustion is the oxygen. The other gases in the atmosphere (Fig. 1-14) serve no purpose during combustion. The oxygen is necessary to make the fuel burn within the interior area of the engine, called the combustion chamber.

A great amount of air is used to make an engine run. As an example, the engine of an average car running at 55 miles per hour (88.514 km) on a level road can use 10,000 gallons (37 854.27 L) of air for every gallon of gasoline it burns.

FUEL

The fuel used in most automotive engines is gasoline. Some diesel fuel engines are in use today, but they will not be covered in this text. Gasoline is a blend of refined hydrocarbon materials found in the underground areas of the earth in the form of raw petroleum.

Gasoline is an excellent source of energy, since it is easily stored, dispensed, and controlled for use within an engine. Its greatest energy value lies in the huge amount of expansion or pressure buildup it has when ignited. This expansion takes place within the closed area of the engine (combustion chamber), driving the piston downward. Additional engine and power train parts make the vehicle move.

IGNITION

The basic function of the ignition system is to provide the spark that ignites the air-fuel mixture within the engine through the use of a spark plug in the combustion

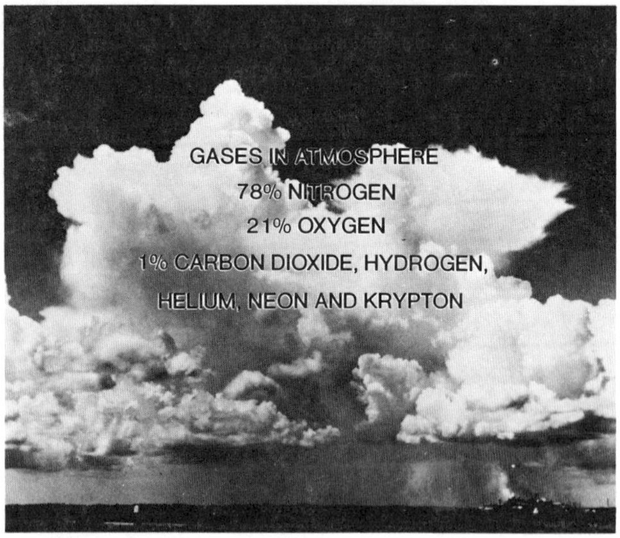

FIG 1-14 Gases in the air we breathe and use for combustion.

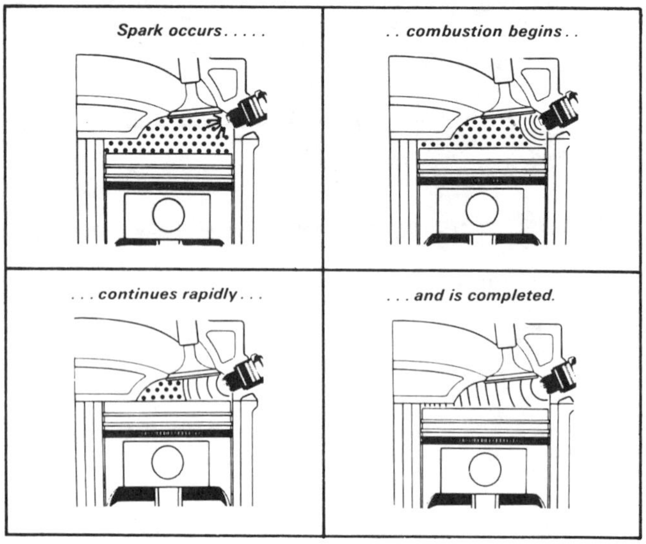

FIG 1-15 In normal combustion, the spark ignites the fuel, which burns smoothly across the combustion chamber. (Champion Spark Plug Company)

chamber. If the mixture is ignited, it expands, thereby producing usable power (Fig. 1-15).

Ignition is usually considered to be the electrical impulse jumping the gap at the spark plug. The electrical function of the ignition system is absolutely necessary; however, the air-fuel mixture must be compressed within the combustion chamber before it will ignite. The more the mixture is compressed, the more power the engine can develop. When compression pressures become excessive, abnormal combustion can take place within the combustion chamber. This causes the engine to detonate or knock, which is harmful to internal engine parts and causes power loss.

ENGINE SYSTEMS

Every engine has support systems which are required to make it run or keep it running. Those listed below are the most basic support systems:

1. Fuel system
2. Ignition system
3. Lubrication system
4. Cooling system
5. Starting system
6. Charging system

These systems often rely upon each other for assistance to perform properly. As an example, for the lubricating system to do its job, the cooling system must keep the engine oil within normal operating temperatures. When the oil is too cool, it does not lubricate as well as it should; when the oil becomes overheated, it burns (oxidizes), which causes it to lose much of its slippery quality.

FUEL SYSTEM

The fuel system comprises the fuel tank, lines or tubing, fuel pump, fuel filter, and carburetor or injectors. Its prime function is to deliver the correct amount of fuel to the intake manifold or cylinder under all conditions of engine speed and load requirements.

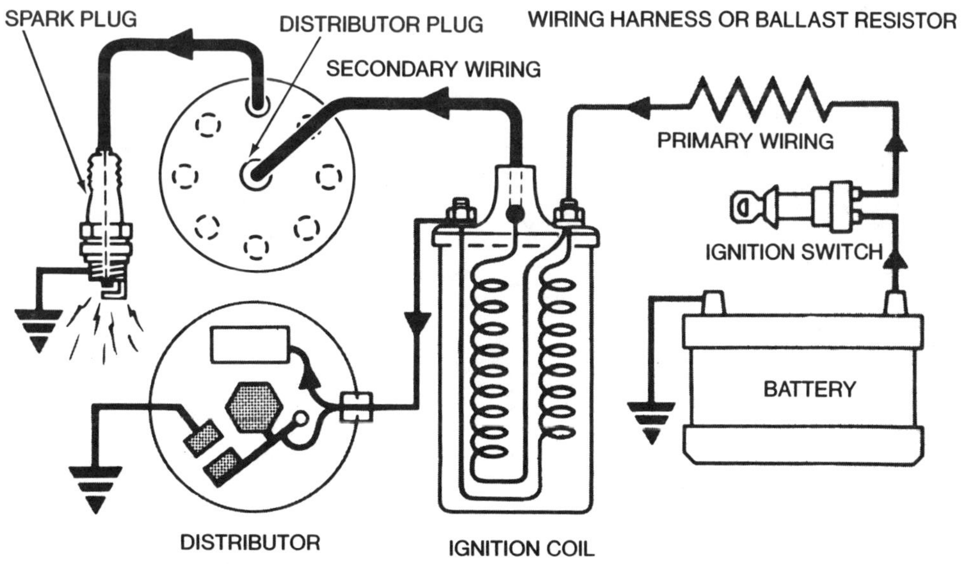

FIG 1-16 Typical ignition system.

IGNITION SYSTEM

The ignition system (Fig. 1-16) consists of the ignition switch, wiring harness, resistance or ballast resistor, ignition coil, distributor, high tension (secondary) wiring, spark plugs, and low tension (primary) wiring. Electrical energy to operate the ignition system is obtained from the battery while cranking, and from the charging system while running.

LUBRICATION SYSTEM

The lubrication system (Fig. 1-17) consists of the oil pan, oil pump, oil filter, oil galleries (oil passages in the engine block), and the oil itself. When an engine is run, great internal friction develops. The natural characteristics of oil reduce friction and the resulting heat. The pump picks up oil from the pan and forces it through the filter. From that point, the oil goes to each of the bearings and moving parts requiring lubrication. Oil is splashed on those surfaces that are not directly fed with pressure oil.

COOLING SYSTEM

The parts of a liquid cooling system (Fig. 1-18) include the radiator and its hoses, water pump, water jackets (passages within the block where coolant flows), and thermostat. As an engine is run, the heat that was not used in creating power in the cylinder must be taken away. This job is performed by the cooling system. The water pump takes cooled liquid from the radiator and pushes it through the engine block. A thermostat placed near the block outlet regulates the flow and temperature of the coolant. The radiator serves to reduce the temperature of the coolant. In the air-cooled engine a fan pushes cool air across fins on the cylinders and cylinder heads to lower the temperature to acceptable levels. Some air-cooled engines have a thermostat to regulate the air flow which brings the temperature to the correct level.

STARTING SYSTEM

The parts that make up the starting (cranking) system are the battery, battery cables, starter motor, solenoid and/or relay, and starting switch (usually part of the ignition switch).

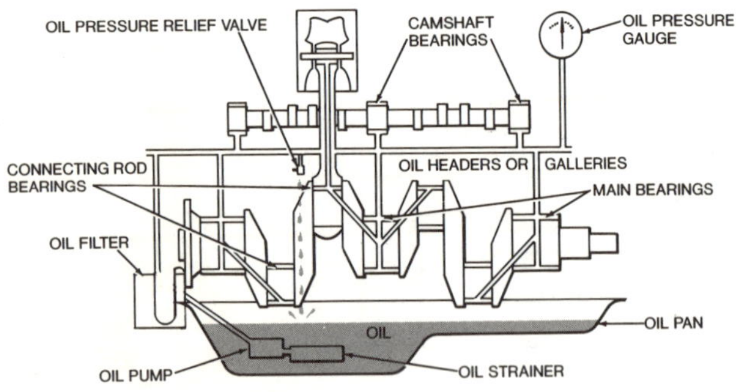

FIG 1-17 Typical lubricating system. (Federal-Mogul Corporation)

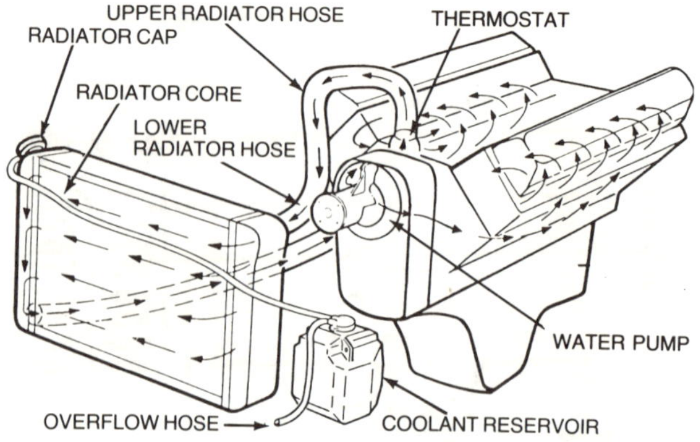

FIG 1-18 Typical cooling system. (Cadillac Motor Car Division, GMC)

The battery is the source of electricity used to crank the engine. A chemical reaction within the battery produces electrical energy. The starting system will operate when the starting switch is turned on and the solenoid or relay is activated, causing the motor to revolve. As the motor turns, it engages the flywheel of the engine to make it rotate. The cables provide the path of electricity from the battery to the starter motor.

CHARGING SYSTEM

The charging system consists of the battery, the alternator and its regulator, and the connecting wiring. The functions of the charging system are to recharge the battery and to provide electrical energy for all purposes when the engine is running.

To charge the battery properly as the engine runs, a regulator is needed to sense the state of charge. The regulator also senses the amount of electricity required by the rest of the vehicle while the battery is being charged.

Designs and types of charging systems vary widely, but their functions remain the same.

BASIC ENGINE PARTS

To understand how the engine operates, a knowledge of the names of its basic parts is first needed. The following discussion and illustrations show a one-cylinder engine for the sake of explanation. Since most automotive engines have at least four cylinders, simply expand the explanation given here to fit a specific engine.

The simplest of engines requires at least four basic parts: a cylinder, a piston, a connecting rod, and a crankshaft (Fig. 1-19). The cylinder is a round, hollow tube closed off at one end and open at the other end. It is the stationary part of the engine; all other parts move in one way or another. The cylinder in the automobile engine is part of the basic framework known as the cylinder block. The block, usually made of cast iron or aluminum, contains four or more cylinders.

The piston is a round piece of aluminum designed so that it will fit into a cylinder very snugly, yet be able to slide up and down with some freedom. The piston is commonly flat or domed-shaped at one end and open at the other. A large, hollow

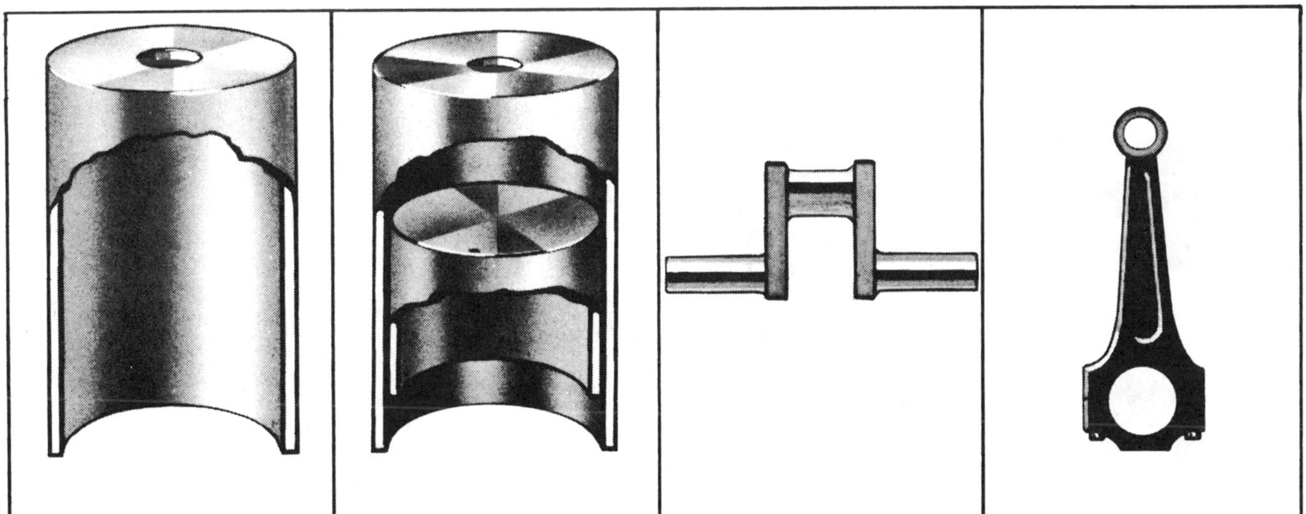

Simple cylinder. Piston and cylinder. Crankshaft. Connecting rod.

FIG 1-19 Basic parts of an engine.

steel pin running from side to side in the piston allows for the attachment of the connecting rod (Fig. 1-20). When the piston moves within the cylinder until it reaches the upper limit of its travel, the space remaining is known as the combustion chamber. Although they are not shown in Figure 1-20, pistons always have grooves to hold a group of sealing rings. One of the primary functions of the piston and its rings is to seal the expanding gases during combustion.

We have said that heat energy from combustion is converted into mechanical reciprocating, up-and-down, motion of the piston. To do this, air and fuel are brought into the cylinder and ignited, normally by a spark. This air-fuel mixture burns very rapidly. As this burning occurs, the gases expand, which increases the pressure within the cylinder and drives the piston downward. The reaction of the burning air and fuel in an engine can be compared to the firing of a rifle. The powder in the cartridge burns after being ignited, causing a pressure buildup of gas. This expanding gas drives the bullet, the only piece that can move, out of the barrel. In the engine, the piston is the movable piece.

The connecting rod is the link between the piston and the crankshaft. It usually has a small and a large end. The piston pin passes through the small end, and the large end is attached to the connecting rod journal of the crankshaft. The connecting rod itself is made from steel forged into an I-beam shape. This method of manufacture and shape give it the strength which it needs to transmit all the downward force from the piston to the crankshaft. The crankshaft converts reciprocating piston movement into rotary motion.

To show how an engine operates, it can be compared to the action of a bicycle rider. The knee of the rider moves almost straight up and down, like the piston (Fig. 1-21). The rider's lower leg is like the connecting rod. The foot follows the pedal, which goes around in a circle. As the piston is forced downward by the pressure of the expanding gases, the knee of the rider pushes downward on the leg. The lower end of the connecting rod must also move, and the only path it can travel is the circular one prescribed by the crankshaft. Thus, the job of the crankshaft is to turn up-and-down motion to rotary motion. This is the basic operation of most automotive engines in use today.

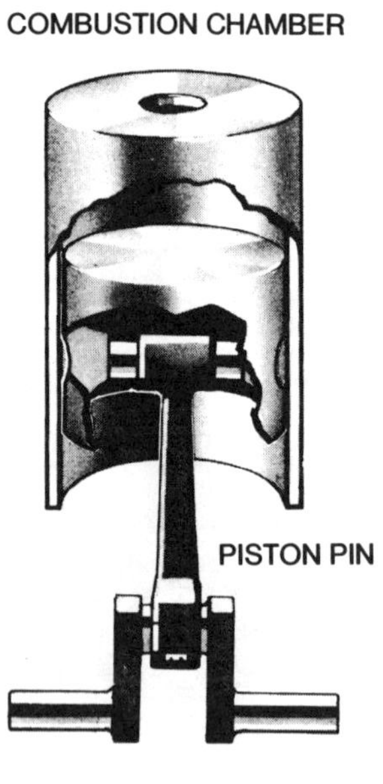

FIG 1-20 Basic engine parts assembled.

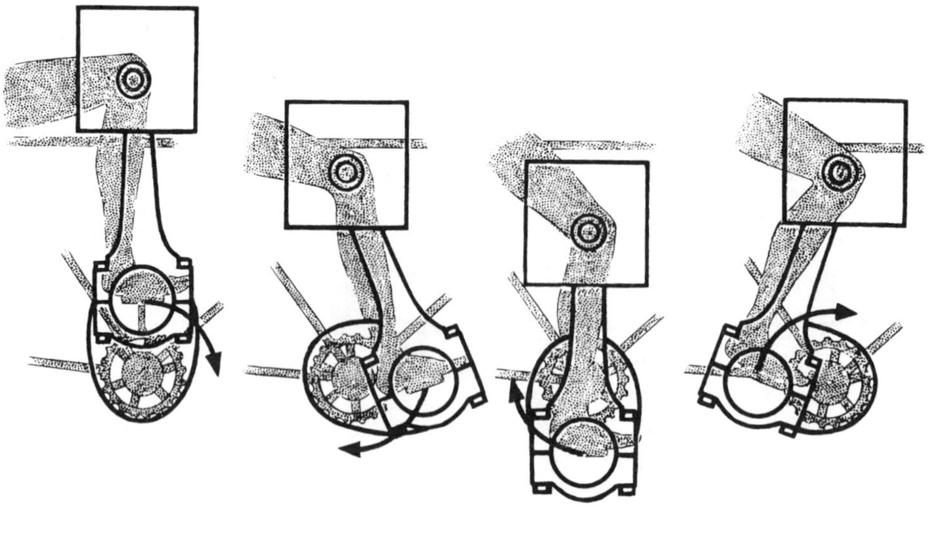

FIG 1-21 Piston connecting rod and crankshaft movements compared to a bicycle rider. (General Motors Corporation)

Name Dewis Kirby Date

DIRECTIONS

Select the best answer to each question and circle the letter at the left.

1. In the basic operation of an engine,
   I. expanding gases push the piston to rotate the crankshaft.
   II. air is being compressed in the cylinder to turn the crankshaft.
   A. I only
   B. II only
   C. Both I and II
   D. Neither I nor II

2. The automotive engine is driven by
   I. internal combustion.
   II. external combustion.
   A. I only
   B. II only
   C. Both I and II
   D. Neither I nor II

3. Engine efficiency is
   I. the ability to convert heat energy to mechanical energy.
   II. determined by the amount of heat energy lost during its operation.
   A. I only
   B. II only
   C. Both I and II
   D. Neither I nor II

4. The I-head valve arrangement is the most common because
   I. service work can be performed more easily than with other types.
   II. it is less costly to produce.
   A. I only
   B. II only
   C. Both I and II
   D. Neither I nor II

5. The L-head engine has
   I. the intake valves in the head and exhaust valves in the block.
   II. both intake and exhaust valves in the head.
   A. I only
   B. II only
   C. Both I and II
   D. Neither I nor II

6. The term *flat head* is used to describe
   I. an opposed cylinder arrangement.
   II. an L-head valve arrangement.
      A. I only
      B. II only
      C. Both I and II
      D. Neither I nor II

7. One cycling classification of an automotive engine is the
   I. Otto.
   II. four-stroke.
      A. I only
      B. II only
      C. Both I and II
      D. Neither I nor II

8. The part of the air necessary for engine combustion is
   I. hydrogen.
   II. nitrogen.
      A. I only
      B. II only
      C. Both I and II
      D. Neither I nor II

9. Excessive compression pressures can cause
   I. abnormal combustion.
   II. internal engine damage.
      A. I only
      B. II only
      C. Both I and II
      D. Neither I nor II

10. Electricity to operate the ignition system comes from the
   I. battery
   II. charging system.
      A. I only
      B. II only
      C. Both I and II
      D. Neither I nor II

DATE COMPLETED ______________________ INSTRUCTOR'S SIGNATURE ______________________

OBJECTIVE 1-B (written test)

After completing the preparation assignment, you will correctly answer a minimum of 14 out of 20 multiple choice questions in a written test, without reference material, on the following topics:

1. Four-stroke cycle
2. Multi-cylinder engines
3. Piston displacement
4. Compression pressure
5. Compression ratio
6. Cylinder numbering
7. Firing order
8. Companion cylinders
9. Performance

Preparation:

Study pages 15 through 26.

FOUR-STROKE CYCLE

If the common engine parts and basic requirements just discussed are assembled and put into operation, the four-stroke cycle can be further explained. Remember, the final result of these strokes should be to convert heat energy into rotative power at the crankshaft. Also keep in mind that one stroke is the travel of the piston from one end of the cylinder to the other; this can be from top to bottom or bottom to top. A four-stroke engine has two strokes that move the piston from the top to the bottom of the cylinder and two other strokes that move the piston from the bottom to the top.

Very often the term *four-cycle* is used to describe automotive engines which have one cycle with four strokes. The cycle, or repeated series of events, which occurs inside a typical automotive engine can be explained more simply by describing a one-cylinder engine as an example. The four strokes of its cycle are intake, compression, power, and exhaust.

INTAKE STROKE

The first stroke of a four-stroke cycle engine is the downward movement of the piston. This creates a vacuum in the combustion chamber area which draws the burnable air-fuel mixture from the carburetor (Fig. 1-22). A more technical explanation is that the downward-moving piston creates an area of less than atmospheric pressure in the combustion chamber. Since a fluid (liquid or gas) will travel from an area of high pressure to one of lower pressure, the higher pressure above the carburetor causes the air to mix with the fuel at the carburetor on its way to the combustion chamber and cylinder.

Figure 1-22 also shows two valves at the top of the cylinder, the intake and exhaust valves, which regulate the flow of gases that enter and leave the cylinder. The valve movements are controlled by the camshaft, which, in turn, is controlled by the crankshaft movement. During the intake stroke (piston moving downward), the intake valve must be open to allow entry of the air-fuel mixture. During the intake stroke, the crankshaft has traveled 180 degrees, halfway through one complete 360° revolution; one stroke equals one-half of a crankshaft revolution.

COMPRESSION STROKE

After the piston has traveled to the bottom of the cylinder (known as Bottom Dead Center, or BDC), the spring on the intake valve closes the opening to the carburetor, sealing the cylinder. As the crankshaft continues to turn, the piston begins its travel to the top of the cylinder. The area above the piston becomes smaller and smaller as the piston continues to move upward (Fig. 1-23), squeezing (compressing) the air-fuel mixture into a smaller space. The particles of air and fuel move closer together, which increases the temperature of the mixture.

When the piston reaches its extreme top limit of travel (called Top Dead Center, or TDC), the mixture of air and fuel is confined to a fraction of 1/8 or less of the space it originally occupied at Bottom Dead Center. The pressure of gases at this point is often 120–180 psi (pounds per square inch) or, in Metrics, 825-1240 kPa (kilopascal). This completes the compression stroke, which has moved the crankshaft another one-half turn or 180 degrees.

POWER STROKE

Just before the piston reaches TDC, an electric spark from the coil jumps from the center of the spark plug to its ground electrode. This spark ignites the compressed air-fuel mixture, causing it to burn very rapidly (about 3/1000ths of a second). The burning of this mixture produces gases whose pressure to expand (approximately four times the original pressure) forces the piston downward (Fig. 1-24). The crankshaft changes the gas expansion piston movement into rotative motion. The amount of force produced at the crankshaft is known as torque. The crankshaft has now turned another 180 degrees, a total of 540 degrees since the beginning of the intake stroke.

EXHAUST STROKE

At the end of the power stroke, the camshaft begins to open the exhaust valve and the piston moves upward in this last of the four strokes (Fig. 1-25). The main job of this stroke is to force out all burned gases from the cylinder so a fresh charge can be brought in during the next cycle's intake stroke. The burned gases are forced out past the open exhaust valve, through the exhaust manifold, and eventually into the atmosphere. When the piston reaches TDC the exhaust valve closes as the intake valve opens to begin a new cycle. When the exhaust stroke is completed, a total of two complete crankshaft revolutions (720 degrees) have taken place.

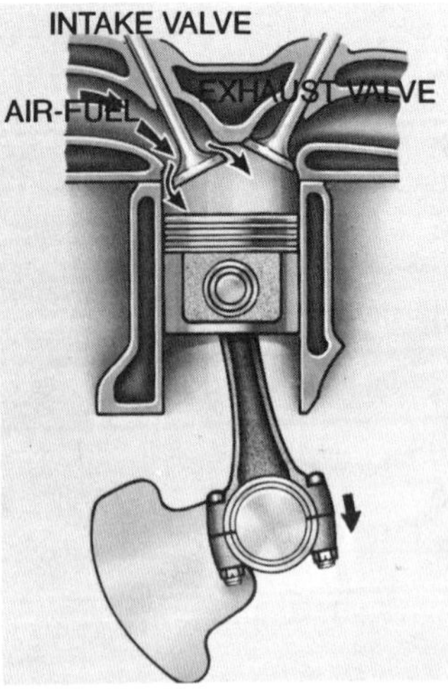

FIG 1-22 Incoming air-fuel mixture during the intake stroke. (Chevrolet Motor Division, GMC)

FIG 1-23 Compression stroke with both valves closed. (Chevrolet Motor Division, GMC)

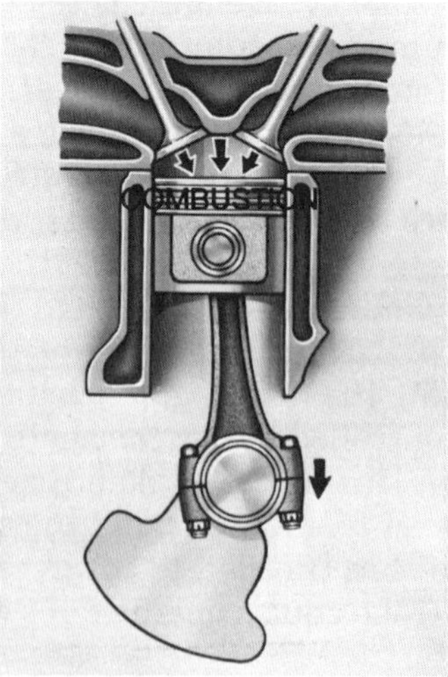

FIG 1-24 Power stroke with the burning gases expanded to push the piston. (Chevrolet Motor Division, GMC)

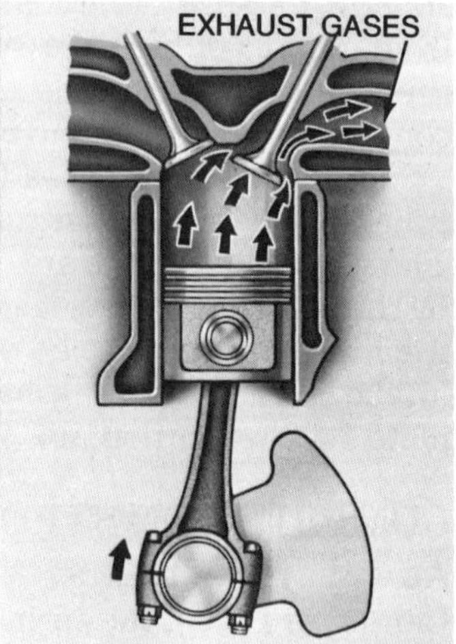

FIG 1-25 Piston pushing the exhaust gases past the exhaust valve. (Chevrolet Motor Division, GMC)

MULTI-CYLINDER ENGINES

The above explanation of the four-stroke cycle has described an engine of only one cylinder for the sake of example. Automobiles of today need engines of four, six, or eight cylinders to provide power output and smoothness of operation desired by the owners.

When more cylinders are added to an engine, the crankshaft, block, camshaft, connecting rods, and piston must be increased in numbers or size accordingly. Regardless of the number of cylinders, however, a four-stroke cycle engine completes all four strokes of all cylinders in two complete crankshaft revolutions or 720 degrees.

In the typical four-cylinder, in-line engine, the crankshaft is like four single-cylinder crankshafts fastened together end to end. The crankshaft is cast or forged so as to allow each cylinder to operate in different strokes at the same time. One piston will be drawing in the air-fuel charge, one compressing the air-fuel mixture, one pushing the connecting rod downward during the power stroke, and one pushing the exhaust gases past the exhaust valve. Figure 1-26 shows a cutaway view of this process. This engine will have four power strokes in 720 degrees of crankshaft movement (two revolutions): one power stroke every 180 degrees.

If an eight-cylinder engine could be viewed inside as it runs, we would see that it has eight power strokes or impulses as all cylinders complete their four strokes. This means it has a power impulse every 90 degrees of crankshaft rotation, or every quarter of a revolution. Generally speaking, the more cylinders an engine has, the smoother it will run.

To further explain how the engine operates, the accuracy and design of its internal parts must be discussed. These are factors that engineers must consider before an engine is placed in a particular vehicle.

To obtain the maximum power output from an engine, the valve timing (when and how long the valves are open) must be exact. In the explanation of the four strokes, for example, the intake valve was said to open *as* the piston moved downward in the intake stroke. This is not strictly accurate. In actual engine operation, the intake valve must open *before* the piston starts its downward movement in the intake stroke (Fig. 1-27). This means the intake valve is actually opening as the

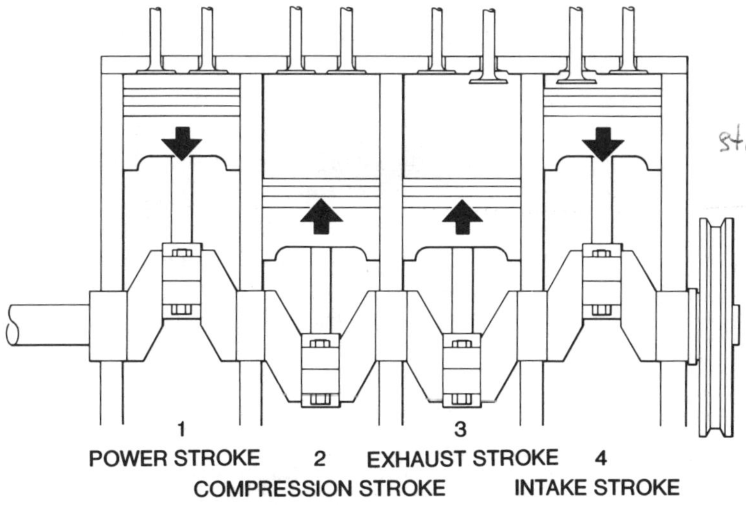

FIG 1-26 Four-cylinder engine in different stroke positions.

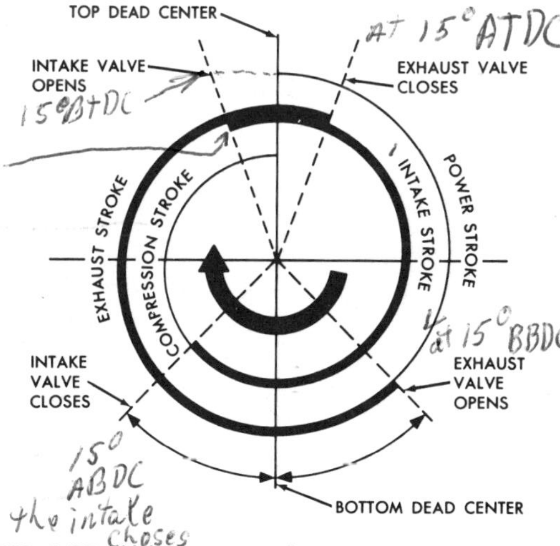

FIG 1-27 Valve movements during 720° of crankshaft rotation —one complete cycle of four strokes. (Chevrolet Motor Division, GMC)

piston is finishing the exhaust stroke. The intake valve also stays open beyond the point when the piston reaches the bottom of its downward travel in the intake stroke. This early opening and late closing is necessary because the valve cannot be opened and closed instantly. Up to a point, the longer the intake valve is open, the more air-fuel mixture will enter the cylinder. Figure 1-27 shows the valve timing in degrees of crankshaft movement.

In like manner, the exhaust valve opens before the piston reaches BDC on the power stroke and closes later than TDC of the exhaust stroke. Both the incoming air-fuel mixture and the burned exhaust gases are fluids which have weight. The early opening and late closing of the valves is necessary to allow for the time it takes to make these fluids move.

Notice in Figure 1-27 that the exhaust valve is still open during the last of the exhaust stroke while the intake valve is already opening to begin the intake stroke. This period of valve timing is called *valve overlap.* Excessive valve overlap creates poor low-speed performance while not enough overlap causes poor high-speed performance.

Ignition timing (the moment the spark ignites the air-fuel mixture) is another operation that must be exact if the engine is to operate efficiently. In explaining the power stroke, sparking was said to occur "just before the piston reaches TDC." The exact time the spark occurs is a design factor studied at length by engineers. It may vary for the sake of different engine speeds and loads and, more recently, for reduced exhaust emissions.

The design factors mentioned above are but a few considerations made before an engine is ever manufactured. Further explanations will accompany the service portions of each engine part.

PISTON DISPLACEMENT

The most common designation of engine size in automobiles today is cubic inch or cubic centimetre displacement, usually abbreviated to read CID (CI) or cc. The engine sizes of some luxury cars are stated in litres, which is a multiple of cubic centimetres.

Piston displacement is a major consideration in determining how well an engine can convert heat energy to mechanical energy. One of the basic physical principles governing engine operation is that the larger the cylinder diameter (bore) and/or length of piston movement within the cylinder (stroke), the greater the potential twisting movement of the crankshaft. Hence, engines with large displacement are usually more powerful.

Piston displacement is a measurement of volume. This may be calculated by measurement of cylinder cross section by the length of the piston stroke. That total would then be multiplied by the number of cylinders in a multi-cylinder engine to give the total piston displacement of that engine. For example, an eight-cylinder engine with cylinders four inches in diameter and a four-inch movement (stroke) would have a piston displacement of 401.92 cu. in. The formula used to arrive at this figure is as follows:

$$\text{Piston displacement} = \pi r^2 \times L \times \text{No. of cylinders}$$

π (pi) is the constant factor that represents the ratio of the circumference of a circle to its diameter; its value is 3.14. A circle with a diameter of 1 inch (25.4 mm) is 3.14 inches (79.756 mm) around; if its diameter is 2 inches (50.8 mm), it is 6.28 inches (159.212 mm) around (Fig. 1-28).

The next part of the formula multiplies pi by the radius squared (r^2). This is often awkward because bore is always given as the piston diameter, which is double the radius. Let us change the first part of the formula by substituting *D* (diameter) for *r* (radius):

$$\frac{\pi \text{ (or 3.14)} \times (2r)^2 \text{ (or } D^2) \times L \text{ (stroke)} \times \text{No. of cylinders}}{(2)^2 \text{ (or 4)}}$$

$$= \frac{3.14 \times D^2 \times L \times \text{No. of cylinders}}{4}$$

Diameter can now be used instead of radius because the upper and lower halves of the formula were doubled, then squared. Reducing this to one simple formula gives us

$$\text{CID} = 0.7854 \times \text{Bore}^2 \times \text{Stroke} \times \text{No. of cylinders}$$

The earlier example was figured on a calculator by this procedure:

$$0.7854 \times (4)^2 \times 4 \text{ (stroke)} \times 8 \text{ (no. of cylinders)}$$

	Enter	Press	Display Shows	
Bore	4	☒	4	
Squaring the bore	4	☒	16	
π factor	0.7854	☒	12.566	
Stroke	4	☒	50.265	
No. of cylinders	8	⊟	402.12	cubic inches

This example was figured in cubic inches. An example in cubic centimetres would be figured in essentially the same way. A 1 600 cc (1.6 litre)* Volkswagen engine's piston displacement is calculated in the following manner:

$$\text{cc} = .7\,854 \times (85.5 \text{ mm})^2 \times 69 \text{ mm} \times 4$$

*To convert cubic centimetres to litres, simply move the decimal point to the left 3 places.

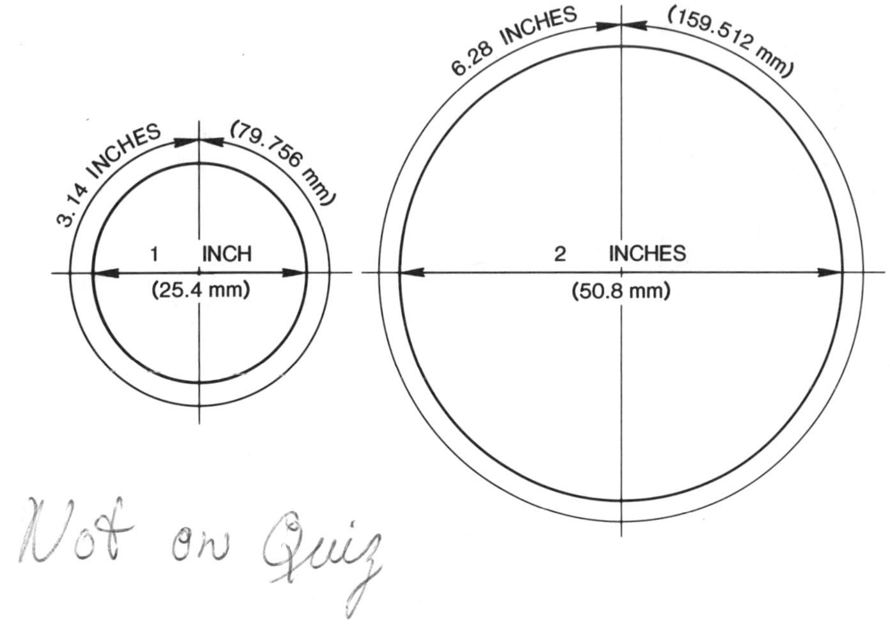

FIG 1-28 Relationship of diameter to circumference: pi (π)

	Enter	Press	Display Shows	
Bore	8.55	☒	8.55	(85.5 mm = 8.55 cm)
Squaring the bore	8.55	☒	73.1025	
π factor	0.7854	☒	57.414	
Stroke	6.9	☒	396.161	
No. of cylinders	4	⊟	1584.6458, or, 1600	

The only way the displacement can be changed is to enlarge the bore or lengthen the stroke. The precedure for enlarging the bore, called *boring,* is frequently done in small amounts such as + 0.030 in., + 0.040 in., or + 0.060 in. to recondition the cylinder walls of a worn engine block. This will be fully explained in Unit 9, Cylinder Block Preparation. Lengthening the stroke, called *stroking,* is seldom done except on high performance engines.

COMPRESSION PRESSURE

Automotive service work is an attempt to improve the engine's or vehicle's efficiency through adjustments or repairs. As the engine wears, it becomes less efficient, which means the customer is not gaining full advantage from the gasoline purchased. In gasoline engines, the air-fuel mixture is compressed at approximately 120–180 psi (825–1240 kPa) if the engine is in good mechanical condition. At that compression, the mixture can be ignited quickly and there will be good gas expansion in the combustion chamber. Generally speaking, the more the air-fuel is compressed on the compression stroke, the greater the expansion of gases. This in turn means more force driving the piston downward.

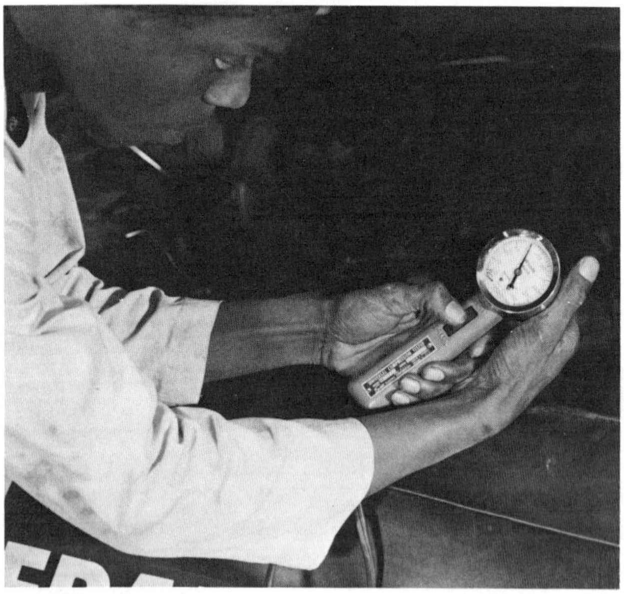

FIG 1-29 Pressure buildup within the combustion chamber measured by a compression tester.

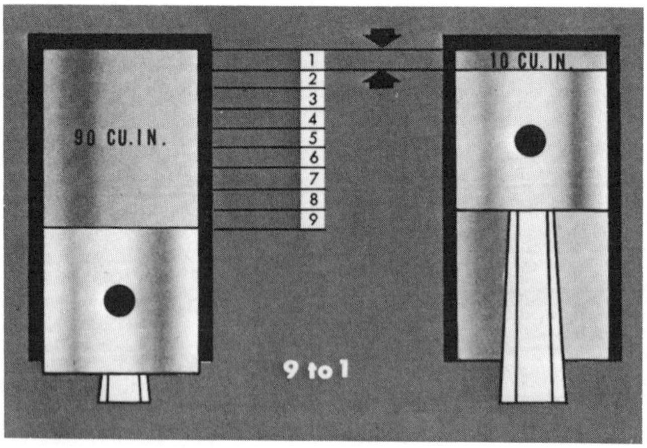

FIG 1-30 Air-fuel mixture being compressed to 1/9 of its original volume.
(Federal-Mogul Corporation)

Compression pressures, which are gauge readings of the amount of pressure buildup, measured at the spark plug hole, are designed into engines after consideration of many factors (Fig. 1-29). Some of the factors engineers consider are: What type of operating conditions will the engine be required to meet? How much horsepower is required to do the proper job? What is the quality of fuel available? Usually, the service person only needs to know the compression pressure specification and is not concerned about the design factors. The recommended compression pressure is commonly found in tuneup specifications under the make, year, and cubic inch displacement of the engine. The pressure specification may be a number or a percentage variation of high to low gauge readings. Psi indicates pounds per square inch, or kPa for kilopascal, while kg/cm indicates kilograms per square centimetre (as listed by some import car manufacturers).

The compression pressure of an engine is roughly the product of the amount of atmospheric pressure (14.7 psi) times the compression ratio. An engine with 8:1 compression ratio will have approximately 118 pounds compression pressure (8 x 14.7 = 118). This is only a general, rough means of figuring compression pressure and should be used only if specifications are not available.

COMPRESSION RATIO

Engine descriptions usually include the figure known as compression ratio, which is 8:1, 8.5:1, or 9:1 for the great majority of passenger car engines produced today. When read aloud, these figures would be stated as "8 to 1, 8½ to 1," etc.

A ratio is a comparison of two things. An engine compression ratio is a comparison of the volume or space within the combustion chamber and cylinder when the piston is at two different positions. Engine compression ratio may be defined as the total volume of gas (air-fuel) in the cylinder when the piston is at BDC compared to its volume when the piston is at TDC. Since compression ratio is a mathematical calculation, the formula CR = CV at BDC: CV at TDC can be used for any engine. (CR = compression ratio; CV = total cylinder and combustion chamber volume.) Remember to keep units of measurement consistent (cubic inches *or* cubic centimetres).

Figure 1-30 shows an engine with 9:1 compression ratio. The total volume of the cylinder including the combustion chamber with the piston at BDC is 90 cubic inches or 1 475 cubic centimetres. When the piston is moved to TDC, the air-fuel mixture is compressed to only 10 cubic inches or 164 cubic centimetres. CR therefore, equals 90: 10 or 9:1.

In this engine, the air-fuel mixture would be compressed to 1/9 of its original volume. This would make it much more easily ignited, since the more the air-fuel mixture is compressed (up to a point, approximately 11:1 for passenger cars), the hotter it becomes, making it burn faster. More pressure also makes the burning gas expand with much greater force, which, in turn, develops more crankshaft timing force, torque, and horsepower.

Lower compression ratios (less than 7:1) are not considered efficient for passenger cars because they cannot produce adequate combustion pressures. Compression ratios above 11:1 are quite efficient and produce high horsepower output. However, the quality of gasoline generally available will cause abnormal combustion with compression ratios above 11:1. The results of abnormal combustion will be discussed in Unit 6, Measurements and Inspection. High compression ratios also cause excessive amounts of NO_x, oxides of nitrogen, which are among the undesirable chemical compounds in exhaust emissions associated with smog.

CYLINDER NUMBERING

Earlier in this unit different cylinder arrangements were discussed, including the in-line, V-type and opposed designs. To simplify the discussion of individual cylinders, every manufacturer has a cylinder numbering system for each engine. For in-line engines, it is simply a matter of designating the cylinder closest to the front as number one, and continuing to the back of the engine with consecutive numbers.

Numbering systems for V-type engines can vary greatly. Number one cylinder is always up front, but the side on which it is located will vary among manufacturers. The most common cylinder numbering patterns are shown in Figure 1-31.

FIRING ORDER

Every multi-cylinder engine has a designated firing order. This is the sequence or order in which the air/fuel mixture is fired (ignited) in the cylinders. The most common firing order for a four-cylinder in-line engine is 1–3–4–2. This means the spark ignites the mixture in those cylinders in that order every two revolutions of the crankshaft.

The firing order is determined by the arrangement of the crankpins on the crankshaft and the camshaft design. A four-cylinder engine with a firing order of 1–3–4–2 will have a crankshaft similar to that pictured in Figure 1-32. Notice how crankpins are spaced opposite one another. This means two crankpins will move to a vertical (TDC) position every 180° (½ turn) of rotation. Figure 1-32 shows #1 and #4 crankpins at TDC. If both valves were closed in cylinder #1, it would be completing the compression stroke and the #4 cylinder would be at the end of the exhaust stroke. After turning the crank 180° (½ turn), the #2 and #3 crankpins

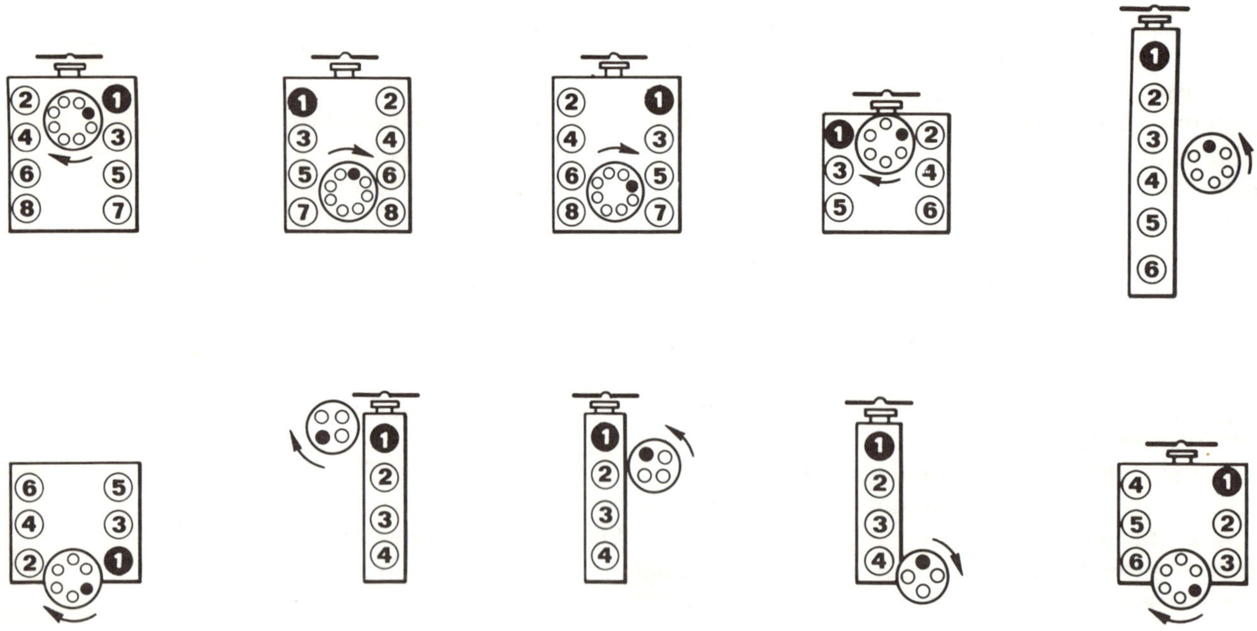

FIG 1-31 Typical engine cylinder numbering patterns and firing orders. (Champion Spark Plug Company)

are at TDC (Fig. 1-33). With a camshaft ground for a 1–3–4–2 firing order, #3 cylinder would have both valves closed and be completing its compression stroke. Rotating the crankshaft another ½ turn would bring it back to the position it was in at the beginning of this discussion, except that the number 1 cylinder would be at the end of its exhaust stroke and the number 4 cylinder at the end of its compression stroke. By the time two complete revolutions (counterclockwise as viewed from the driver's seat) are made, the engine has fired all four cylinders and completed all four strokes in each.

The only difference between an engine with a 1–3–4–2 firing order and one with a 1–2–4–3 firing order is in the camshaft.

COMPANION CYLINDERS

Companion cylinders are those cylinders of an engine whose pistons are in the same position within the four-stroke cycle at all times. As the crankshaft rotates through two revolutions, all the pistons reach TDC and BDC two times. The difference between them at these two times is the valve position regulated by the camshaft. A piston is at TDC in the compression stroke as well as the exhaust stroke. To find the companion cylinders of any engine, simply place the first half of the firing order over the second half. An engine with a 1–8–4–3–6–5–7–2 firing order would have these pairs of companion cylinders:

1–8–4–3
6–5–7–2

Cylinders 1 and 6 are companions; so are 8 and 5, 4 and 7, and 3 and 2.

FIG 1-32 Crankshaft with #1 and #4 in TDC (Top Dead Center) position.

FIG 1-34 Engine dynamometer.
(Engine Systems Development)

FIG 1-33 Crankshaft with #2 and #3 in TDC (Top Dead Center) position.

ENGINE PERFORMANCE

The modern engine is the result of nearly 100 years of development and testing. Its performance and life expectancy are much greater now than at any time in the past because of improvements and changes. These highly perfected machines are designed in a wide variety of sizes and power output capabilities to meet the needs of the intended vehicle.

The actual performance of an engine is first rated by the manufacturer in terms of horsepower and torque. These figures are the result of operating an engine to test its maximum power output on a device called a dynamometer (Fig. 1-34). This machine measures the turning effort and speed an engine makes as it is operated through a wide rpm range.

SERVICE OPERATIONS

On occasion, an engine may be altered, by a repair procedure, to increase its performance capacity. Increasing the bore (diameter of the cylinder), surfacing the cylinder head (increasing compression ratio), or changing the intake or exhaust system are all common engine alterations.

The most common service operation to restore original performance of an engine is the tune-up. This may include repairing or replacing parts of the ignition, fuel, lubrication, cooling, starting, or charging systems. Tune-up is also defined as repair work to restore performance that does not require the removal of the intake or exhaust manifolds, timing cover, cylinder heads, or oil pan. When an engine cannot be restored to near original performance, internal engine work becomes necessary.

There are several terms used frequently to describe engine service operations: *overhaul, recondition,* and *rebuild.* The first two terms can be used interchangeably; to overhaul or recondition means to restore to specified clearance and performance goals. To rebuild means to restore to original dimensions. Because of normal wear that takes place as an engine is operated, it is impossible to restore an engine to its original dimensions.

Detonation

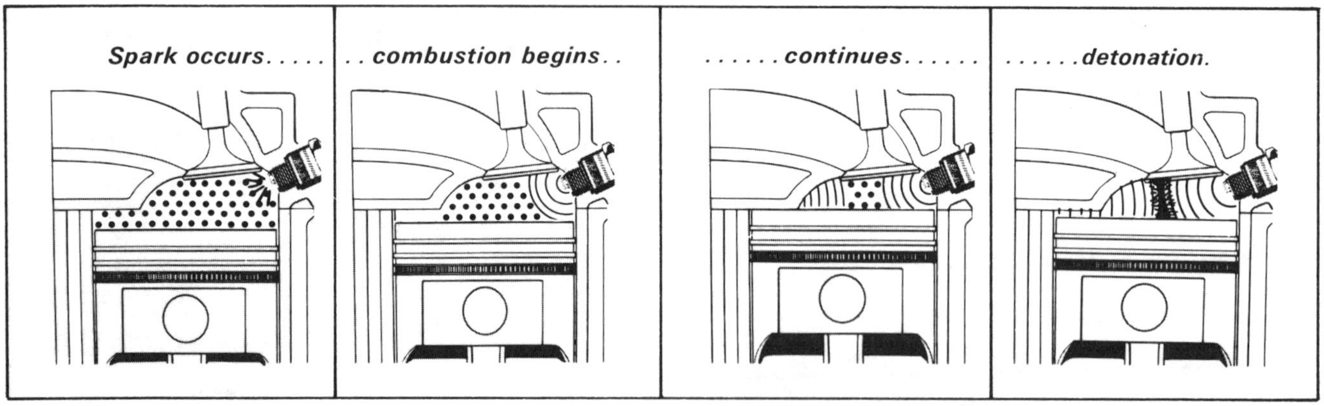

FIG 1-35 Detonation occurs as a second flame front starts following spark ignition. Abnormal heat and stress develop as the flame fronts meet.
(Champion Spark Plug Company)

Engine repair work is usually classified as minor or major. In this text, minor engine repair is the term used for service work performed while the engine is in the vehicle. In-car service can mean as much as a minor overhaul. Major work is considered to be operations that can only be performed with the engine removed. Generally, it takes more time to complete a major engine overhaul than a minor overhaul.

ENGINE WEAR

Regardless of how well an engine is serviced and maintained, it sustains wear each time it is operated. Wherever there is friction, however slight, there will be wear. Wear can be classed as normal or abnormal. If the engine's parts are the proper size, if it is assembled correctly, and if it is operated within reasonable limits, the wear that occurs is considered normal.

Abnormal wear results in premature breakdown of parts and engine failure. The most frequent causes are

1. Improper lubrication
2. Improper cooling
3. Incorrect fuel
4. Improper operation
5. Incorrect clearances between parts

The biggest single cause of abnormal wear in an engine is poor lubrication. This is the direct result of not changing the oil and filter at recommended or sooner-than-recommended intervals. Oil is the substance specifically designed to reduce the friction of the internal engine parts. When its condition is poor, there is more friction and therefore more wear.

The number of miles, days, or hours an engine may be operated between oil and filter changes varies with each manufacturer. Follow manufacturer recommendations for the time between changes and the type of oil for a specific engine.

Oil used in an engine becomes contaminated with undesirable substances and cannot be cleaned. Also, no filtering system yet devised is capable of removing the normal buildup of contamination.

Preignition

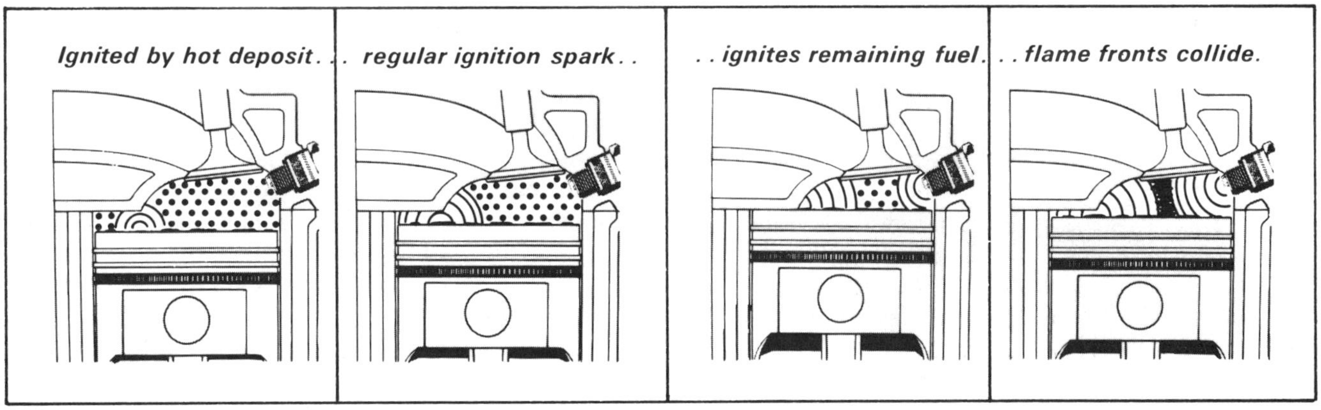

FIG 1-36 Pre-ignition is caused by a hot spot other than an electrical spark that ignites the mixture prior to the correct time.
(Champion Spark Plug Company)

Improper lubrication can mean a lack of oil. This is usually a maintenance problem. The oil level should be checked regularly. Oil consumption is oil that is burned (internal loss) or leaked (external loss) from the engine. Excessive loss of oil will cause engine failure and abnormal wear. Oil consumption diagnosis is covered in Unit 3, and engine lubrication will be discussed in depth in Unit 10, Crankshaft and Bearing Service.

One fact of which many people are unaware is that roughly 75 percent of all engine wear takes place before the engine temperature reaches its normal operating level (*at least* 160° F. or 57° C.). If the cooling system is functioning properly, it will do two main jobs: (1) warm up the engine as quickly as possible; (2) keep the engine at normal operating temperature. Almost everyone understands the second job of the cooling system, but many overlook the first. The faster the engine can reach operating temperature, the less wear it will experience. This makes cooling one of the most important systems of an engine (see Unit 14).

Using the correct fuel is another important requirement to reduce abnormal wear. Every engine manufacturer designates a fuel with a certain octane rating for each of its engines. When fuel with too low an octane level is used, detonation (pinging) results which can damage an engine.

Incorrect engine operation is usually the result of the driver's actions. Automobiles were never designed to tow a heavy trailer or to uproot a tree. Engine abuse is a result of poor maintenance and improper driving habits. Some of the most common operation problems are: (a) lugging, (b) overspeeding, and (c) excessive idling.

Lugging is defined as the point at which an engine will not pick up speed with the throttle wide open. Lugging is harmful because the combustion chamber is fully charged with air and fuel to bring the maximum gas expansion possible, but, because the engine is usually at low speed, there is insufficient lubrication to cushion the load between moving parts. Lugging is usually accompanied by detonation and preignition (Figs. 1-35 and 1-36). Both types of abnormal combustion occur as the piston is on its way to TDC making the engine lose more power. If this continues, internal engine damage occurs.

Overspeeding is another driver error in operation. The internal parts of an engine are designed for a certain operating range. When that range (rpm) is exceeded, the parts become out of balance and tend to fly apart. The parts most often abused during overspeeding are the pistons and connecting rods.

Excessive idling is an operation problem especially for taxicabs and emergency vehicles. During long idling periods, the combustion is not complete, which tends to contaminate the oil in the crankcase. The oil pump does not provide full pressure to the bearing at idle speeds. Some engine parts rely on the splash of oil created by a fast-moving crankshaft for their lubrication, but this does not occur at idle speed.

Incorrect clearances between parts cause abnormal engine wear and eventual engine failure. These clearances are the result of wear or incorrect engine assembly. The amount of incorrect clearance and the length of time the engine has been run with this condition will determine how much damage has been sustained and what repairs are necessary. Generally, the longer the engine has been run and the further the clearances are from specifications, the greater the damage. A step-by-step analysis should be made during disassembly of the engine to determine the cause of incorrect clearances.

Name ______________________ Date ______________

DIRECTIONS

Circle the best answer to each question.

1. In the four-stroke engine, the intake stroke immediately follows the
   I. compression stroke.
   II. power stroke.
   A. I only B. II only C. Both I and II D. Neither I nor II
2. A stroke is
   I. the piston movement in the cylinder from the top to the bottom.
   II. 180 degrees of crankshaft rotation.
   A. I only B. II only C. Both I and II D. Neither I nor II
3. The air-fuel mixture flows into the cylinder on the intake stroke because
   I. engine heat forces the air to move.
   II. pressure outside the cylinder is greater.
   A. I only B. II only C. Both I and II D. Neither I nor II
4. During the compression stroke
   I. both valves are closed.
   II. the air-fuel mixture is heated.
   A. I only B. II only C. Both I and II D. Neither I nor II
5. The power stroke
   I. requires 360 degrees of crankshaft rotation.
   II. occurs immediately following the intake stroke.
   A. I only B. II only C. Both I and II D. Neither I nor II
6. A four-cylinder engine will have one power stroke for every
   I. 180 degrees of crankshaft rotation.
   II. ½ turn of the crankshaft.
   A. I only B. II only C. Both I and II D. Neither I nor II
7. The valve timing of an engine is
   I. stated in degrees of crankshaft rotation.
   II. modified to meet the requirements of the engine application.
   A. I only B. II only C. Both I and II D. Neither I nor II
8. Piston displacement is
   I. a volume measurement of the piston size.
   II. a volume measurement of the entire cylinder.
   A. I only B. II only C. Both I and II D. Neither I nor II
9. Pi (π) is
   I. the ratio of the circumference of a circle to its diameter.
   II. equal to 3.14.
   A. I only B. II only C. Both I and II D. Neither I nor II

82 ⟌ 738
9
738

10. The compression ratio of a cylinder whose volume is 738 cubic centimetres at BDC and 82 cc at TDC is

A. 7:1 B. 8:1 C. 9:1 D. 11:1

11. Among the factors limiting the compression ratio in a gasoline engine is (are)

I. quality of gasoline available.

II. piston materials.

A. I only B. II only C. Both I and II D. Neither I nor II

12. In an engine with a 1–3–4–2 firing order, when #3 cylinder is in the power stroke, #1 cylinder is in the

I. exhaust stroke.

II. intake stroke.

A. I only B. II only C. Both I and II D. Neither I nor II

13. An eight-cylinder engine has a power stroke every 90 degrees of crankshaft movement.

I. 90

II. 180

A. I only B. II only C. Both I and II D. Neither I nor II

14. An eight-cylinder engine with a 1–8–4–3–6–5–7–2 firing order has

I. four sets of companion cylinders.

II. cylinder #8 and cylinder #5 as companion cylinders.

A. I only B. II only C. Both I and II D. Neither I nor II

15. An in-line engine with a 1–2–4–3 firing order

I. completes all four strokes in four crankshaft revolutions.

II. has a crankshaft that rotates clockwise when viewed from the front.

A. I only B. II only C. Both I and II D. Neither I nor II

16. Abnormal engine wear is often the result of

I, excessive filter changes.

II. improper lubrication.

A. I only B. II only C. Both I and II D. Neither I nor II

17. The greatest amount of normal engine wear takes place

I. during the winter.

II. before an engine reaches operating temperature.

A. I only B. II only C. Both I and II D. Neither I nor II

18. The term lugging means an engine is being operated

I. when it is too cold.

II. with fuel that is too low in octane rating.

A. I only B. II only C. Both I and II D. Neither I nor II

19. Lugging an engine can

I. create abnormal combustion.

II. cause overspeeding.

A. I only B. II only C. Both I and II D. Neither I nor II

20. Excessive idling of an engine can be harmful because

I. combustion is not as complete and tends to contaminate the oil.

II. there is a lack of splash oil to some bearing surfaces.

A. I only B. II only C. Both I and II D. Neither I nor II

DATE COMPLETED ______________ INSTRUCTOR'S SIGNATURE ______________________

unit 2

Engine Repair Tools, Fasteners, and Safety Information

Scantron + Pencil #II for Monday Quiz
Valentines Day

This unit presents information about hand and power tools and fasteners commonly used in engine repair. A discussion of lifting devices and safety procedures to be observed while performing engine repair operations is also included.

OBJECTIVE 2-A (written test)

After completing the preparation assignment, you will correctly answer a minimum of 14 out of 20 multiple choice questions in a written test, without using reference material, on the following topics:

1. Hand tools
2. Electric power tools
3. Air power tools
4. Lifting devices

Preparation:

Study pages 29 through 42.

HAND TOOLS

Quite often different tool manufacturers will call the same tool by different names, or the tools may be known by different names in various sections of the country. There are many hand tools that are common to a wide variety of mechanical occupations.

Good mechanics take care of their tools because they know valuable time and effort are saved by doing so. They know that tools must be kept clean and returned to their proper places (Fig. 2-1). Tools with cutting edges need to be kept sharp; delicate measuring instruments need special storage and careful use to maintain their accuracy.

The tools discussed in this unit are those most commonly used by the mechanics. Measuring tools will be covered in Unit 6.

WRENCHES

A wrench is the most basic hand tool used to remove and replace bolts and nuts. The best wrenches are those made from vanadium steel in a drop forging process.

The size of any wrench used on a bolt head or nut is determined by the size of the opening between the jaws of the wrench. This opening is made to fit the flat sides of a nut or bolt head with a clearance of 0.003–0.007 inches (3 to 7 thousandths of an inch). This clearance allows the wrench to slide on and off the nut or bolt with a minimum of play (Fig. 2-2). If the wrench is too large, the points of the nut or bolt head will be rounded off as it is tightened or loosened.

Four designs of wrenches are common to automotive service work: (1) open-end, (2) box-end, (3) combination, and (4) flare-nut (tubing) wrench. Each design has its proper use.

The open end is a solid, nonadjustable wrench with different size openings at each end. They are usually sold in sets ranging from 5/6 inches to 1 inch for domestic cars. Metric wrench sets range from 6 mm to 21 mm. The jaws are commonly set at a 15° angle from the handle to provide access to close working places. A variation of the open-end wrench is the tappet wrench, which is very thin and has extra long handles (Fig. 2-3). These wrenches are used to adjust tappets on engines of the L-head (flathead) where the amount of space is limited.

Box-end wrenches are totally enclosed at the ends. Their opening has six or twelve "points" to fit most six-sided (hexagon) nuts and bolts. Also, the ends have two different size openings (Fig. 2-4). The advantage of a box-end over an open-end is that it is less likely to slip when heavy force is required to turn a nut or bolt. The box-end does, however, need to be lifted off the nut or bolt each turn when the turning space is limited.

FIG 2-1 Rollaway tool storage.

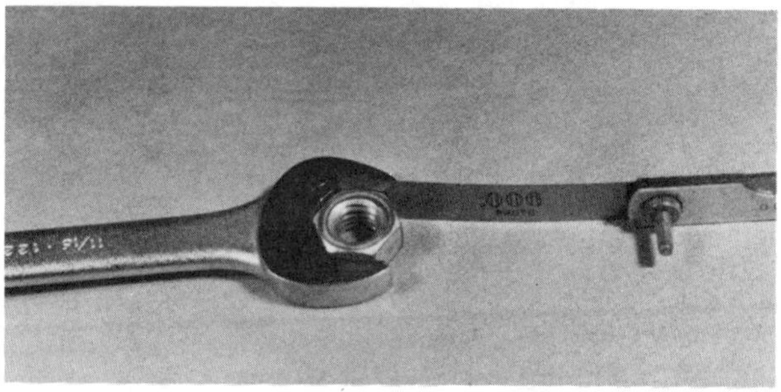

FIG 2-2 The clearance between a wrench and nut is small.

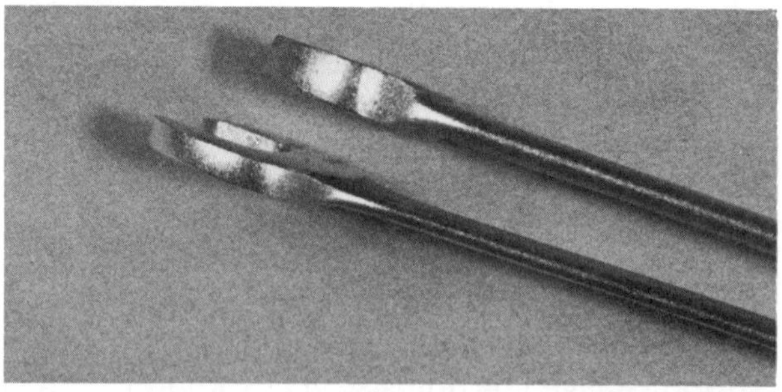

FIG 2-3 The thickness of a regular wrench head (upper) compared to a tappet wrench head (lower).

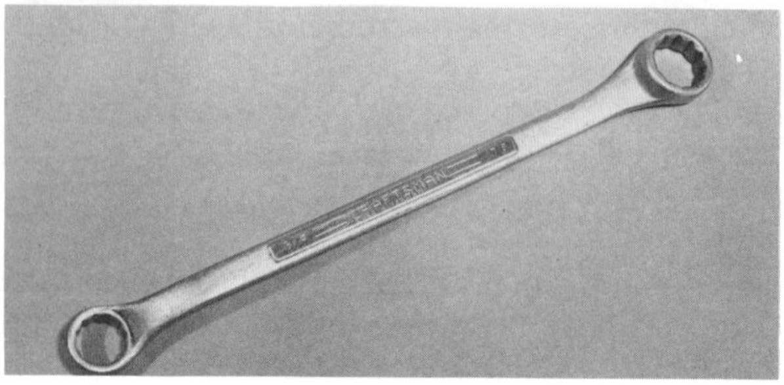

FIG 2-4 Typical forged box wrench.

The combination wrench (Fig. 2-5) makes use of the best features of both the open-end and the box-end. The box-end can be used to break loose a tough nut or bolt and for firm tightening, while the open-end can be used for faster turning. These wrenches commonly are made with the same size of opening at both ends.

The flare-nut (tubing) wrench is a variation of both the box-end and the open-end. As Figure 2-6 shows, the ends have openings where this wrench can be slipped over a piece of tubing and then moved to cover the fitting to be loosened or tightened. This style of wrench is an absolute must for the engine mechanic working with fuel tubing and fittings.

Socket Wrenches and Their Handles—Socket sets contain an assortment of different sizes made to fit different driving handles. The types of handles in use today are: (1) the ratchet, (2) breaker bar, and (3) speed handle (Fig. 2-7). There are also extensions of various lengths. The nut openings of sockets have six or twelve points for hexagon (six-sided) nuts and bolts. A few sockets have eight points for square nuts or pipe plugs. The sizes of the nut ends of the socket are made to fit the flat sides of the nut or bolt head. The end of the sockets opposite the nut opening has a square hole into which the handle fits. A spring-loaded ball on the handle snaps into a groove in the socket to prevent it from falling off. The size of the socket set is designated by the size of the square on the drive-end of the handle (Fig. 2-8).

The auto mechanic will have a complete set of sockets and handles with 3/8 inch and 1/2 inch drive-ends. The nut-size openings will range from 3/8 inch to 1 1/4 inch

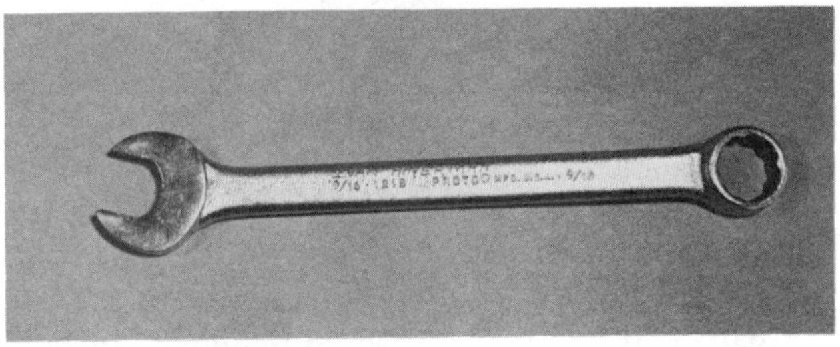

FIG 2-5 Combination (box and open-end) wrench with same size openings.

FIG 2-6 Flare-nut (tubing) wrench.

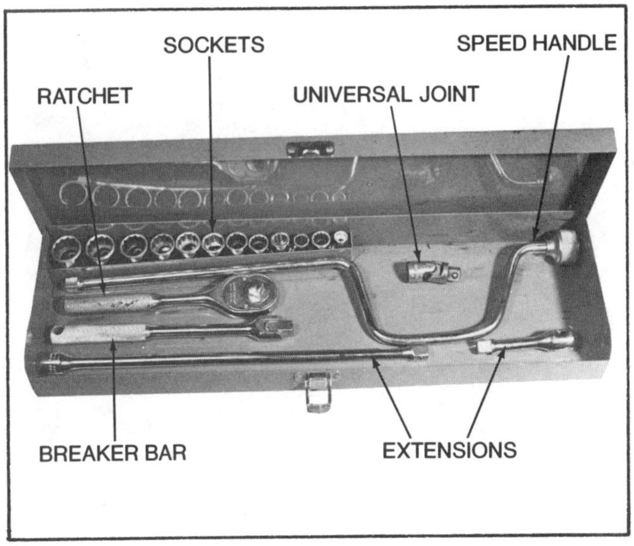

FIG 2-7 Small 3/8 drive socket set.

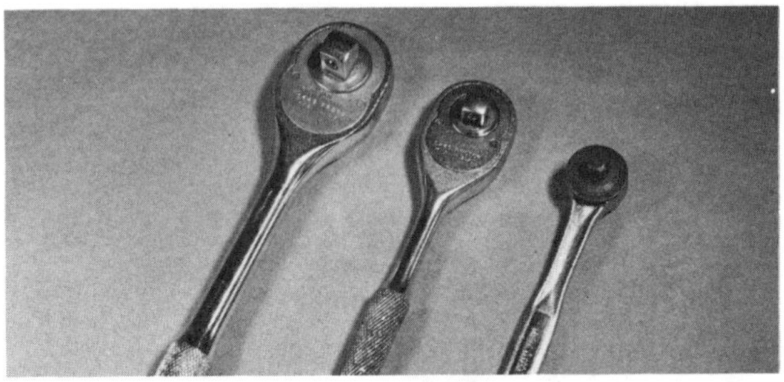

FIG 2-8 Ratchet handles: 1/2 drive (left), 3/8 drive (center), 1/4 drive (right).

for domestic uses. Metric sizes range from 6 mm to 19 mm for 3/8 inch drive handles and 10 mm to 25 mm for 1/2 inch drive handles. A variation of the regular socket is the deep socket, which is three to five inches long. This socket is especially useful for removing and replacing nuts that are a long way down on the bolt (Fig. 2-9). The spark plug socket is considered a deep socket design.

A very useful attachment usually sold with socket sets is called a universal joint. This piece is used for working in places where a straight socket wrench will not work. The universal joint allows the handle to work at an angle with the socket.

There are a few basic rules to be observed when using wrenches:

1. Always use a wrench that fits the nut or bolt properly.
2. Determine the direction a nut or bolt should be turned before trying to loosen it. They usually turn counterclockwise for removal. This may seem obvious, but even journeyman mechanics have damaged parts and wrenches by not turning in the correct direction.
3. Whenever possible, *pull* on a wrench rather than push on it. This rule applies to both tightening and loosening of fasteners. The force developed in the arms and legs of a person's body is much better controlled in pulling than in pushing (Fig. 2-10).
4. Where there is a choice, use sockets rather than hand wrenches.
5. On large nuts and bolts, use the 1/2 inch drive sockets and handles, if possible.

SCREWDRIVERS

The screwdriver is one of the most frequently used hand tools. It is also one of the most frequently abused hand tools. It is designed for one function only: to drive and remove screws. Screwdrivers should not be used as pry bars, chisels, scrapers, or punches.

The engine mechanic makes most frequent use of two designs of screwdrivers: the standard or blade and the Phillips screwdriver. Standard screwdrivers are sold according to length of the blade (Fig. 2-11). The most common sizes range from 2 1/2 to 12 inches. Screwdriver handles are made from wood, plastic, or metal, but plastic is best suited to engine work.

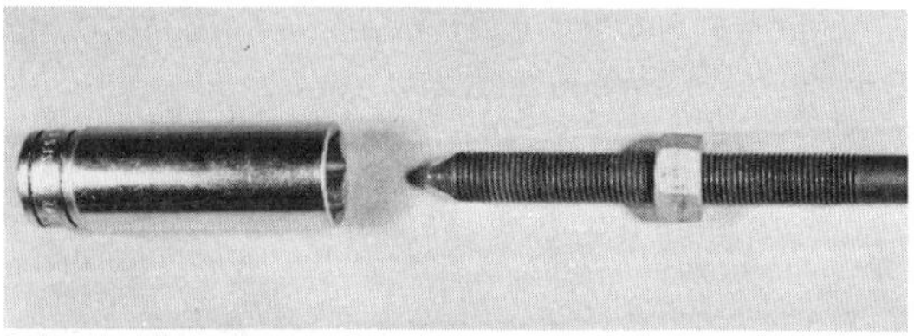

FIG 2-9 A deep socket.

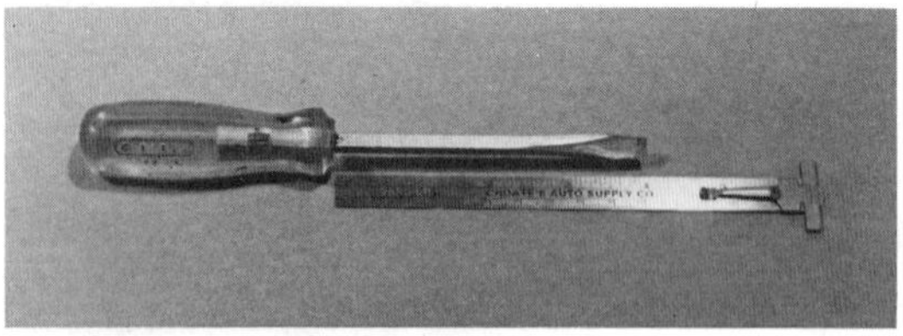

FIG 2-11 Standard (regular) screwdriver designated by the length and width of the blade.

FIG 2-10 Proper position for tightening and loosening fasteners. Pull rather than push on the handle.

Phillips head screws have a four-way slot into which the screwdriver fits. There are three common sizes of Phillips screwdrivers in use today: No. 1, No. 2, and No. 3. All three are pictured in Figure 2-12.

PLIERS

Pliers, made in many styles and sizes, are used to hold or grip small pieces that a hand cannot hold easily. Pliers should never be used as a substitute for a wrench. Most styles of pliers are sold in sizes according to their overall length. The types found most frequently are shown in Figure 2-13.

Slip joint pliers are used for any number of general holding or pulling jobs. Their jaws usually are flat at the ends and curved toward the slip joint. These pliers have two adjustment openings.

Water pump (channel lock) pliers are similar to slip joint pliers except that the jaw range is large and the handles are longer. This provides a greater range of holding size and force.

Vise grip pliers can be used for holding objects, regardless of their shape. This style of pliers has an advantage over others in that if it is clamped on to something, it will hold. This leaves the hands free to do other work.

Needlenose (sometimes called long-nose) pliers are nonadjustable. They are used to hold small objects when removing or replacing them from confined areas. Some styles have a cutting edge between the nose and the pivot. The nose end of this type of pliers is easily damaged, so excessive force should not be applied.

Diagonal cutting pliers, often called "dykes," are used for cutting wire and cotter pins. The word *diagonal* in their name comes from the angle the cutting edges make with the handle. This angle is usually fifteen degrees.

PUNCHES

A punch used for automotive work is made from tool steel with a variety of shapes at the tip (Fig. 2-14). The center punch is most often used to mark or identify parts which must be assembled in a certain order, such as connecting rods. It is also used to indent the surface of metal to act as a locating spot for a twist drill.

The starting punch is used to loosen a pin which is holding two or more pieces together. Once the pin has begun to move out of the hole, a pin punch of the proper size can be used to completely drive it out.

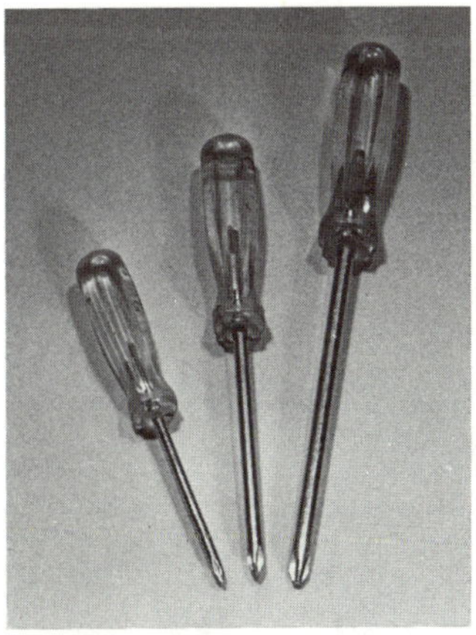

FIG 2-12 Phillips (cross-head) screwdrivers: No. 1 (left); No. 2 (center); No. 3 (right).

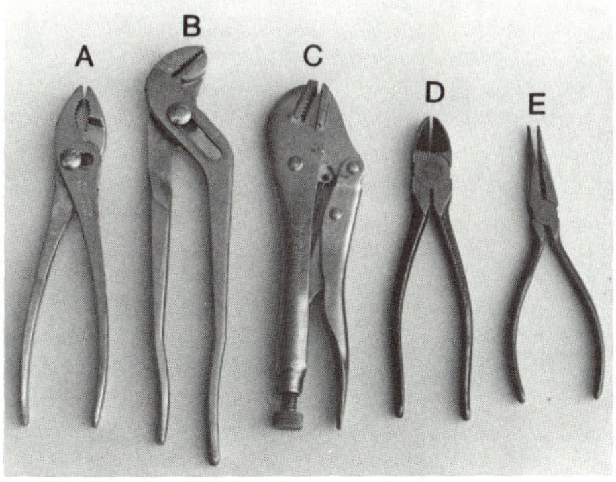

FIG 2-13 Common pliers.

A SLIP JOINT
B WATER PUMP (CHANNEL LOCK)
C VISE GRIP
D DIAGONAL CUTTER
E NEEDLE NOSE

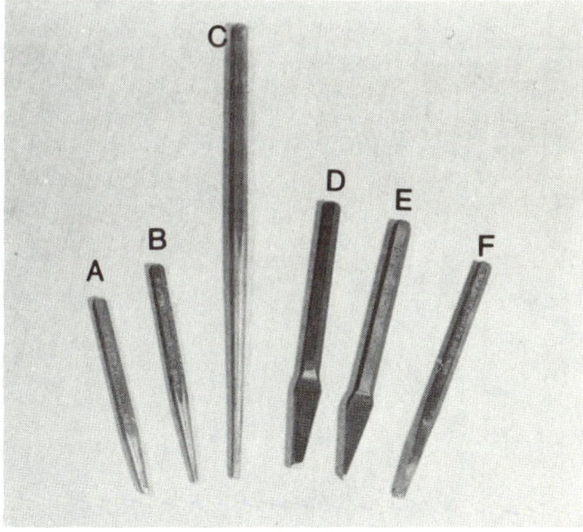

FIG 2-14 Common punches and chisels.

A CENTER PUNCH
B STARTING PUNCH
C ALIGNING PUNCH
D HALF ROUND CHISEL
E CAPE CHISEL
F DIAMOND POINT

The aligning punch is used to line up holes in mating parts. It is usually about twelve inches long with a long gradual taper.

Occasionally, a gasket punch is necessary for making a gasket from scratch. A gasket punch is designed to cut round holes of bolt or stud size in commercially prepared gasket paper (Fig. 2-15). The gasket shape is first outlined, and the large hole(s) are cut with scissors. Gasket punches are then used to cut the bolt holes. This is done by placing the gasket paper between the punch and a piece of wood, then striking the punch with a hammer.

HAMMERS

Two types of hammers are used in the automotive engine repair business: the machinist's hammer (commonly called a ball peen) and the mallet. The ball peen is a hard-faced hammer designed for general use where striking hard blows is necessary. The flat end of the head, called the face, is used for most of the engine repair hammering jobs. The other end of the head, called the peen, is used for peening (mushrooming) soft rivets or pins. Ball peen hammers are sold in different head weights with wooden or fiberglass handles. Figure 2-16 shows hammers and mallets in use today.

The mallet is a soft-faced tool for striking an object which is not to be marred. The materials used in making mallets are rubber vinyl, leather, plastic, and even soft metals such as lead, copper, bronze, or brass. The heads of these mallets can often be replaced, since they break apart after prolonged use.

FILES

Files are metal cutting tools used for a variety of jobs during an engine overhaul. Files are designated by their length and shape. In selecting a file for a job, the

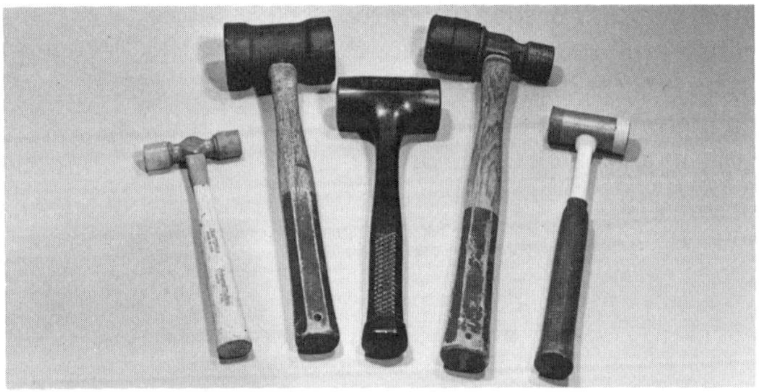

FIG 2-16 Typical soft-head (face) hammers.

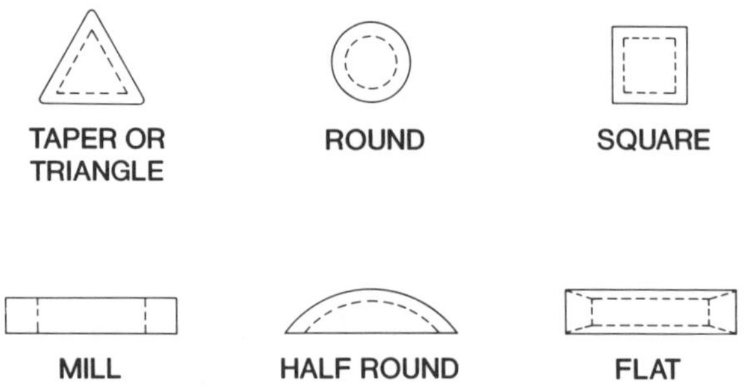

FIG 2-17 Cross-sectional views of commonly used files. (Buick Motor Division, GMC)

FIG 2-15 Gasket (hollow) punch set.

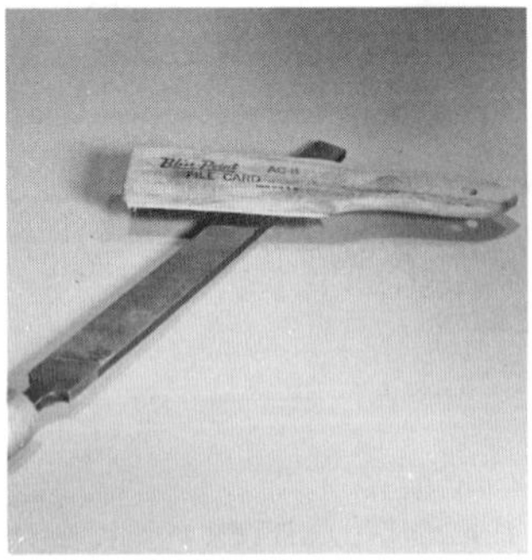

FIG 2-18 A file with a file card.

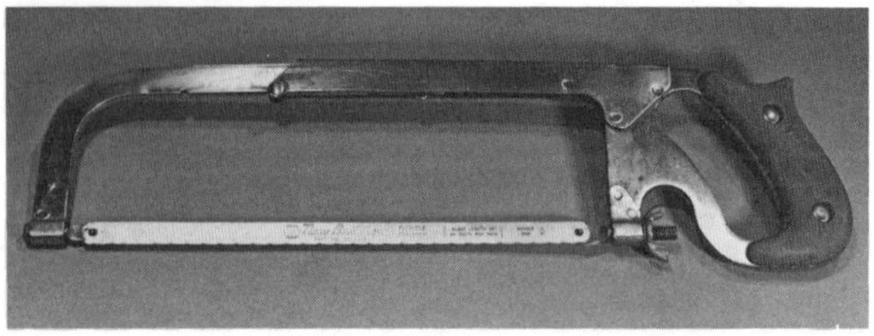

FIG 2-19 Adjustable frame hacksaw designed to hold blades from 8-12 inches.

finished shape of the part to be filed must be considered. Various cross-sectional shapes are shown in Figure 2-17.

Files are graded in two different ways. One way of grading is the number of direction cuts, either single cut or double cut. A second means of classifying a file is by the spacing of the teeth. For automotive use, a smooth or second cut is most often used.

All files should have a tight-fitting handle attached. Every file should be accompanied by a file card, used to clean the chips from the teeth (Fig. 2-18).

HACKSAWS

Another metal cutting tool used by the automotive technician is the hacksaw. A typical hacksaw is adjustable to receive blades of 8, 10, and 12 inches (Fig. 2-19). The blade is held between the end pins and tensioned by the wing nut near the handle.

The piece to be cut should be held as rigid as possible while the cutting is taking place. A bench vise works well. While cutting, the mechanic's body should be in a balanced position. For best cutting performance, the blade of the saw should cut on the forward stroke and be lifted slightly on the return stroke.

Figure 2-20 gives guidelines for blade selection of various shapes of materials.

TAPS AND DIES

Taps and dies are hand tools used to cut threads in metals and other softer materials. Taps are designed to cut threads internally (inside a hole); dies cut threads externally (on the outside of a piece of round stock). For automotive use, two styles of taps are usually selected, the taper and bottoming taps (Fig. 2-21).

The taper tap, whose lower portion is smaller in diameter, is most often used as a starting tap. If it can be used to thread an open hole and pass all the way through it, no other tap will be necessary.

A. Mild Materials In Large Sections	B. Harder Materials In Large Sections	C. Unusual Work Shapes	D. Pipes, Tubing, Conduit

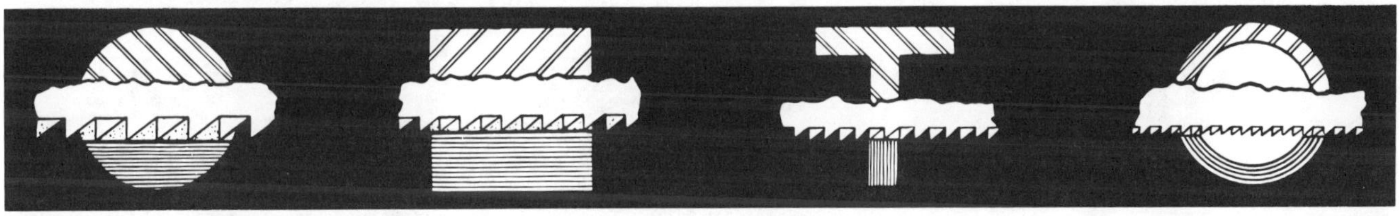

Choose coarse tooth blade to provide plenty of chip clearance, faster cutting.	Choose blade with finer teeth than in A to distribute cutting load over more teeth while still maintaining good chip clearing action.	Choose blade to always keep two or more teeth in contact with narrowest section. Coarse tooth blades stradle work, strip out teeth.	Choose blade with finest teeth per inch to keep two or more teeth in contact with wall. Keep inside of work free of chip accumulation.
Hand Blades— 14 Teeth Per Inch	**Hand Blades—** 18 Teeth Per Inch	**Hand Blades—** 24 Teeth Per Inch	**Hand Blades—** 32 Teeth Per Inch
Power Blades— 4 to 6 Teeth Per Inch	**Power Blades—** 6 to 10 Teeth Per Inch	**Power Blades—** 10 to 14 Teeth Per Inch	**Power Blades—** 14 Teeth Per Inch

FIG 2-20 Blade selection rules. (L.S. Starrett)

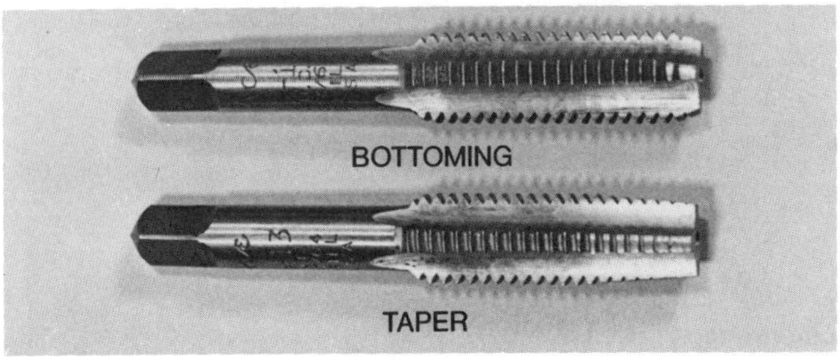

FIG 2-21 Common tap styles for automotive use.

Bottoming taps are used for threading the bottom of a blind hole. Very often these taps are used by the automotive engine mechanic to chase (clean) the threads in engine parts. This is done so that when the engine is assembled, the bolt threads will not bind with the hole threads and give incorrect torque indications.

Pipe taps are used to cut threads for pipe fittings. Engine coolant fittings and oil gallery plugs often have pipe-type threads. The pipe tap diameter is tapered from end to end so that all threads of the tap are cutting at once (Fig. 2-22). When a pipe fitting is screwed into place, the threads will tighten. The more the fitting is turned, the tighter it becomes.

Dies are used by mechanics to chase the threads on bolts or studs prior to reassembly. Some die designs are solid; others are adjustable. A die stock is the handle used to hold dies while in use. The most suitable outside shape for dies used is the hexagon design, since a socket wrench can be used to turn them in tight places.

All taps and dies are marked with their thread size. The set shown in Figure 2-23 covers sizes 6–32 (called six thirty-two) to ½–20 NF. (called one-half twenty, National Fine). Automotive service work rarely demands taps and dies larger than those shown.

Inside the lid of this tap and die set is a tap drill chart, which tells what size hole should be drilled before the threading operation is started. Most tap drill charts allow a 75 percent thread cut. A supplementary tap drill chart is shown on page 37.

DRILLS

Drills (twist drills) used in automotive work are made in three size designations, fraction, number, and letter drills, as identified in the chart on page 37.

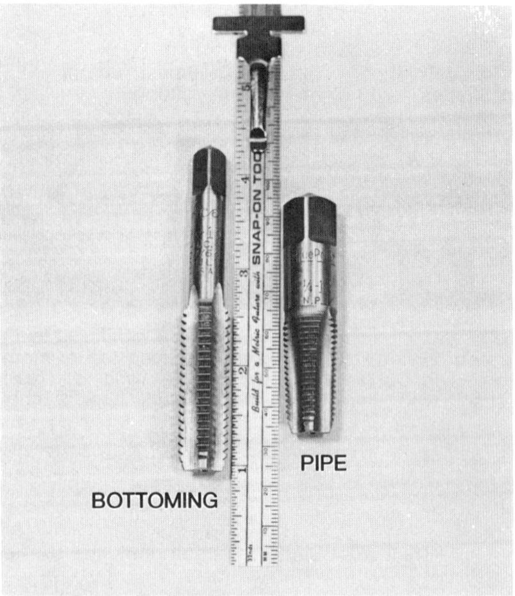

FIG 2-22 A bottoming tap (left) compared to a pipe tap (right).

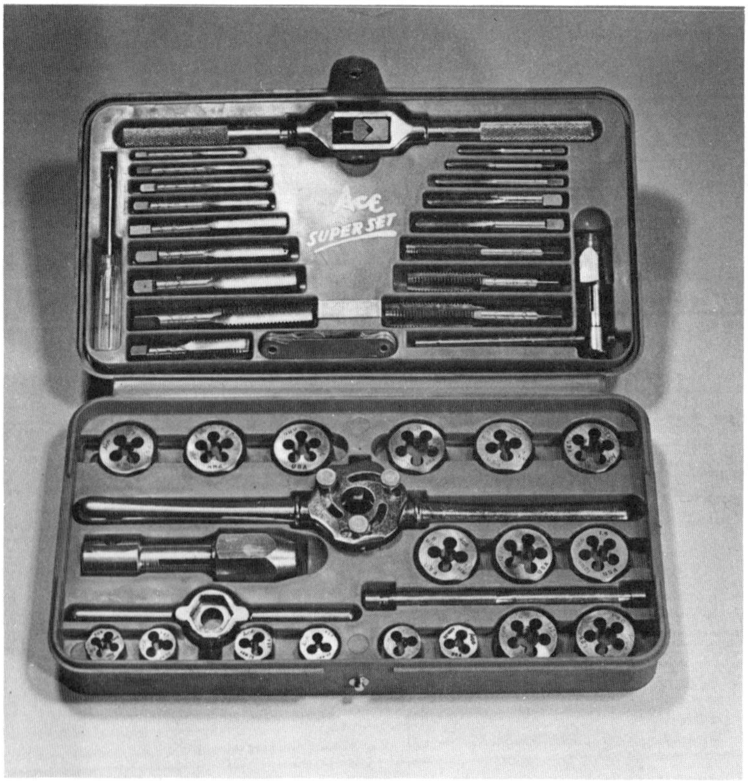

FIG 2-23 Typical tap and die set for automotive repair uses.

Fraction drills usually range from one-sixty-fourth (1/64) to one-half (½) inch. There are 80 (1–80) number drill sizes from 0.0135 to 0.228 inches. Letter drills (A–Z) range from 0.234 to 0.413 inches.

A drill index is the best method of storing drills. It offers a decimal equivalent chart and a closed container to prevent loss (Fig. 2-24).

POWER TOOLS

The power tools discussed here are those most commonly used by the auto mechanic. Shown in Figure 2-25 are some of the hand-held tools used in the auto mechanic operations described in this book. These tools are frequently classified by the type of power used to make them work. There are electrically powered tools and air-powered (pneumatic) tools. Most popular tool manufacturers offer tools that do the same job with either air or electricity as the power source.

Electricity is probably the most widely used power source; however, under general shop conditions, it can be the most dangerous. Electrically powered tools should never be used in areas where explosive gases (gasoline or diesel fuel vapors) are present. Every electric tool should be grounded while in use to protect against an electric shock. Each tool should be equipped with a three-wire conductor cord and a three-blade ground plug end. The plug end should match the receptacle into which it will be placed (Fig. 2-26).

DECIMAL EQUIVALENTS AND TAP DRILL SIZES

DRILL SIZE	DECIMAL	TAP SIZE
1/64	.0156	
1/32	.0312	
60	.0400	
59	.0410	
58	.0420	
57	.0430	
56	.0465	
3/64	.0469	0-80
55	.0520	
54	.0550	1-56
53	.0595	1-64, 72
1/16	.0625	
52	.0635	
51	.0670	
50	.0700	2-56, 64
49	.0730	
48	.0760	
5/64	.0781	
47	.0785	3-48
46	.0810	
45	.0820	3-56,4-32
44	.0860	4-36
43	.0890	4-40
42	.0935	4-48
3/32	.0937	
41	.0960	
40	.0980	
39	.0995	
38	.1015	5-40
37	.1040	5-44
36	.1065	6-32
7/64	.1093	
35	.1100	
34	.1110	6-36
33	.1130	6-40
32	.1160	
31	.1200	
1/8	.1250	
30	.1285	
29	.1360	8-32, 36
28	.1405	8-40
9/64	.1406	
27	.1440	
26	.1470	
25	.1495	10-24
24	.1520	
23	.1540	
5/32	.1562	
22	.1570	10-30
21	.1590	10-32
20	.1610	
19	.1660	
18	.1695	
11/64	.1719	

DRILL SIZE	DECIMAL	TAP SIZE
17	.1730	
16	.1770	12-24
15	.1800	
14	.1820	12-28
13	.1850	12-32
3/16	.1875	
12	.1890	
11	.1910	
10	.1935	
9	.1960	
8	.1990	
7	.2010	1/4-20
13/64	.2031	
6	.2040	
5	.2055	
4	.2090	
3	.2130	1/4-28
7/32	.2187	
2	.2210	
1	.2280	
A	.2340	
15/64	.2344	
B	.2380	
C	.2420	
D	.2460	
E, 1/4	.2500	
F	.2570	5/16-18
G	.2610	
17/64	.2656	
H	.2660	
I	.2720	5/16-24
J	.2770	
K	.2810	
9/32	.2812	
L	.2900	
M	.2950	
19/64	.2968	
N	.3020	
5/16	.3125	3/8-16
O	.3160	
P	.3230	
21/64	.3281	

PIPE THREAD SIZES

THREAD	DRILL	THREAD	DRILL
1/8-27	R	1 1/2-11 1/2	1 47/64
1/4-18	7/16	2-11 1/2	2 7/32
3/8-18	37/64	2 1/2-8	2 5/8
1/2-14	23/32	3-8	3 1/4
3/4-14	59/64	3 1/2-8	3 3/4
1-11 1/2	1 5/32	4-8	4 1/4
1 1/4-11 1/2	1 1/2		

DRILL SIZE	DECIMAL	TAP SIZE
Q	.3320	3/8-24
R	.3390	
11/32	.3437	
S	.3480	
T	.3580	
23/64	.3594	
U	.3680	7/16-14
3/8	.3750	
V	.3770	
W	.3860	
25/64	.3906	7/16-20
X	.3970	
Y	.4040	
13/32	.4062	
Z	.4130	
27/64	.4219	1/2-13
7/16	.4375	
29/64	.4531	1/2-20
15/32	.4687	
31/64	.4844	9/16-12
1/2	.5000	
33/64	.5156	9/16-18
17/32	.5312	5/8-11
35/64	.5469	
9/16	.5625	
37/64	.5781	5/8-18
19/32	.5937	11/16-11
39/64	.6094	
5/8	.6250	11/16-16
41/64	.6406	
21/32	.6562	3/4-10
43/64	.6719	
11/16	.6875	3/4-16
45/64	.7031	
23/32	.7187	
47/64	.7344	
3/4	.7500	
49/64	.7656	7/8-9
25/32	.7812	
51/64	.7969	
13/16	.8125	7/8-14
53/64	.8281	
27/32	.8437	
55/64	.8594	
7/8	.8750	1-8
57/64	.8906	
29/32	.9062	
59/64	.9219	
15/16	.9375	1-12,14
61/64	.9531	
31/32	.9687	
63/64	.9844	
1	1.000	

150-59

FIG 2-24 Drill index.

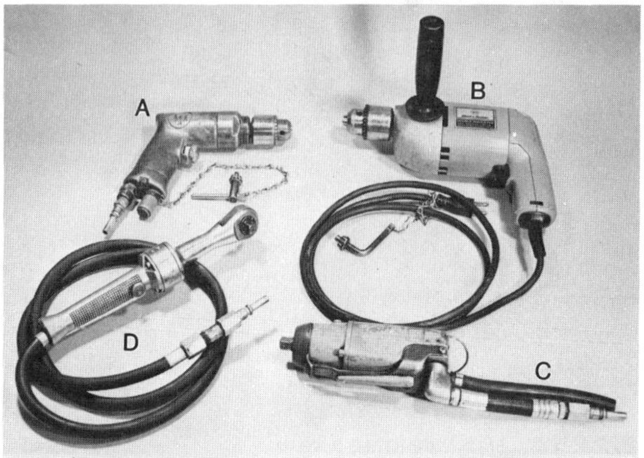

FIG 2-25 Hand-held power tools.

Air-powered tools are especially useful because they are smaller, lighter, and run cooler than comparable electric tools. They require a steady flow of clean, lubricated air. In the absence of a lubricator in the line, most air tool manufacturers recommend a few drops of lube oil be placed into the air opening each day (Fig. 2-27). Never use motor oil, for it tends to gum up the internal parts of the tools.

The recommended air pressure for effective air tool operation is generally 90 psi (620.53 kPa) minimum.

DRILL MOTORS

The electric drill motor is designed to drill holes, but it is used by the auto mechanic for other jobs, such as honing, deglazing, and reaming. All drill motors have a three-jaw chuck which tightly holds the part being serviced. The chuck is tightened and loosened with a matching chuck key. The size designation of a drill motor is often determined by the maximum capacity of the chuck. Most automotive shop drill motors have chuck sizes ⅜ inch and ½ inch (shown in Figure 2-28).

Many drill motors available today have trigger switches which allow them to operate at different speeds. These are known as variable speed drill motors. The further the trigger is depressed, the faster the chuck will turn. For many years this could only be done with air-operated tools. This feature is very useful in establishing the cross-hatch pattern when honing cylinder walls. Another feature available is reversible rotation. The direction of rotation is controlled by a switch (Fig. 2-29). The reversible feature is especially useful when priming the oil pump to prelube an assembled engine, since different engines have different pumping rotations.

Air-powered drill motors are available in the same chuck sizes, and all have the variable speed trigger control (Fig. 2-30). Two of the best features of these drills are that they never overheat even in continuous use and will never cause an electrical shock to the operator.

IMPACT WRENCHES

As its name implies, the electric impact wrench to loosen and tighten most types of fasteners is powered by electricity. It is trigger-operated and applies full impact

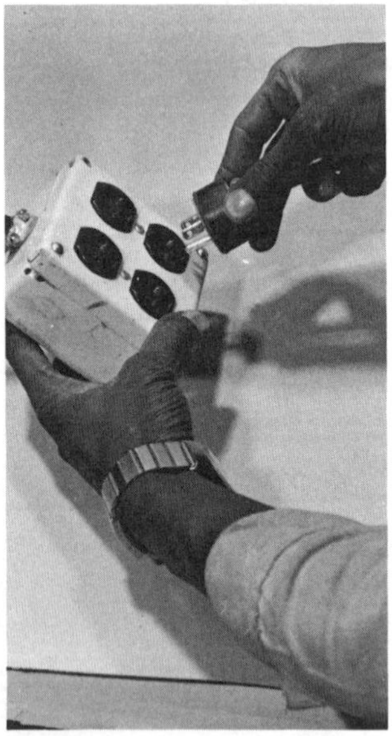

FIG 2-26 Use only three wire cords and receptacles.

FIG 2-27 Lubricating air drill with special oil.

FIG 2-28 Electric drill motor with ½ in. capacity chuck.

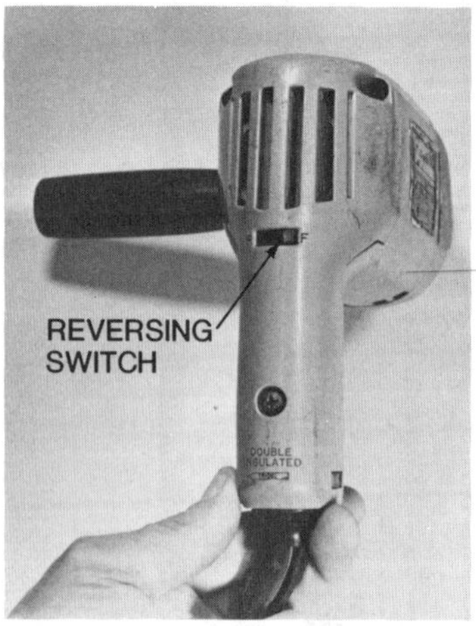

FIG 2-29 Reversing switch position.

power (approximately 130 ft-lbs or, 176.26 N•m,) as the switch is squeezed. The electric version of this wrench is normally available with ½ inch drive only.

Air impact wrenches are designated by the size of the drive end (Fig. 2-31). Most wrenches have some means of regulating the air pressure before it reaches the driving mechanism inside. This determines the speed of the drive end and regulates the amount of torque applied to a nut or bolt. Maximum torque on most ½ inch drive wrenches is 200 ft-lbs (271.16 N•m).

Impact wrenches are used most during engine disassembly; torque wrenches are used only in the assembly operations. Sockets and other attachments used with impact wrenches should be designed for engine disassembly. The ordinary socket will wear excessively or break when used with an impact wrench.

AIR RATCHET

Another air-powered tool which is very popular is the air ratchet (Fig. 2-32). This tool provides a small amount of torque (approximately 20 ft-lbs) (27.12 N•m) for speeding the removal and replacement of nuts and bolts. The operator uses this tool as a ratchet to break loose a nut and then depresses the switch to spin it off.

DRILL PRESS

The drill press is basically a metal cutting machine (Fig. 2-33). Most presses have a ½ inch chuck and an adjustable table for different sizes and angles of work. Small pieces to be drilled should be held in a vise while large pieces need to be clamped to the table. The speed of the chuck is determined by the placement of the belt on the pulleys. Steady pressure on the feed handle will provide the best results. Eye protection should always be worn when operating this machine and all other power equipment.

GRINDER

The bench grinder is an electric motor with wheels mounted at both ends of the turning shaft. Frequently, a grinding wheel is mounted on one end and a wire wheel on the other. The grinding wheel is used to sharpen tools, while the wire wheel is

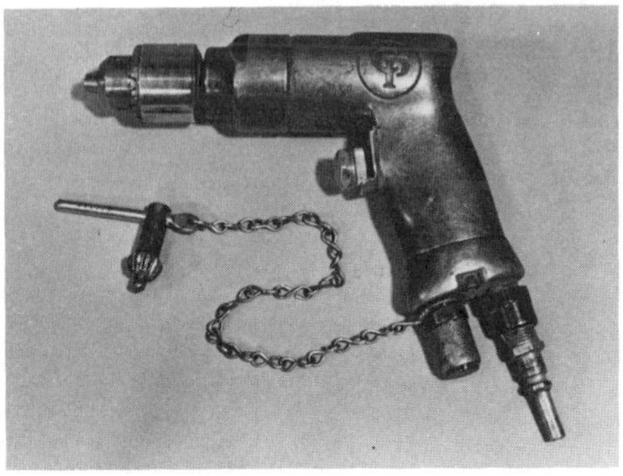

FIG 2-30 Air drill motor with variable speed control.

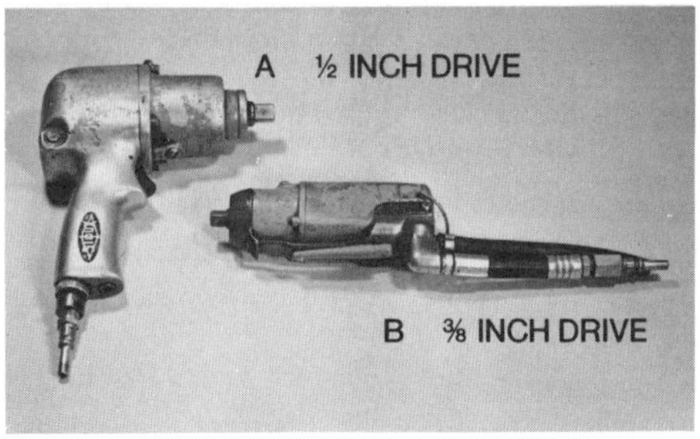

FIG 2-31 Air impact wrenches.

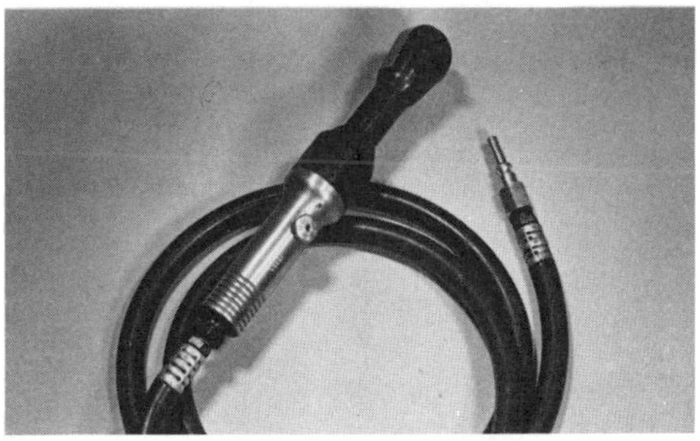

FIG 2-32 Air ratchet with hose.

FIG 2-33 Drill press.

used to remove foreign matter from metal parts. The tool rests should be adjusted so they just clear the wheels. The operator should wear eye protection at all times and should never use excessive pressure to speed the grinding or cleaning processes.

Many engine repairs require power machines specially designed to recondition individual parts of the engine. Their use and safety precautions will be presented as the specific repairs are discussed. General safety procedures are given here.

SAFETY PROCEDURES FOR POWER TOOLS

1. GROUND ALL TOOLS UNLESS DOUBLE INSULATED. If a tool is equipped with a three-prong plug, it should be plugged into a three-hole electrical receptacle. An adapter should not be used to accommodate a two-prong receptacle. Never remove the third prong.
2. KEEP GUARDS IN PLACE and in working order.
3. KEEP WORK AREA CLEAN. Cluttered areas and benches invite accidents.
4. AVOID DANGEROUS ENVIRONMENT. Don't use power tools in damp or wet locations. Keep work areas well lit.
5. USE RIGHT TOOL. Don't force a small tool or attachment to do the job of a heavy duty tool.
6. WEAR PROPER APPAREL. Do not wear loose clothing or jewelry that may get caught in moving parts.
7. USE SAFETY GLASSES with power tools. Also, a face shield or dust mask should be used if the cutting operation is dusty (Fig. 2-34).
8. DON'T ABUSE CORD. Never carry a tool by the cord or yank to disconnect it from a receptacle. Keep the cord from heat, oil, and sharp edges.
9. DON'T OVERREACH. Keep the proper footing and balance at all times.
10. DISCONNECT TOOLS when not in use: before servicing; when changing accessories such as blades, bits, cutters, etc.
11. REMOVE ADJUSTING KEYS AND WRENCHES. Form the habit of checking to see that keys and adjusting wrenches are removed from the tool before turning it on.
12. AVOID ACCIDENTAL STARTING. Don't carry a plugged-in tool with your finger on the switch.

FIG 2-34 Face shield being worn for eye protection.

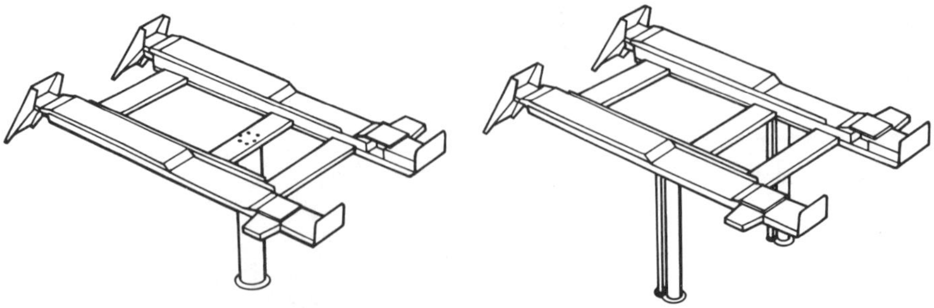

FIG 2-35 Single-post and twin-post lifts.

LIFTING DEVICES

Lifting devices can be classified in two general groups: those which lift the vehicle, and those which lift only the heavy, major parts of a vehicle, namely the engine. Both types are normally used by the engine mechanic. With the exception of the chain hoist, lifting devices commonly found in the automotive shop use hydraulic power. This means the lifting force is developed by using oil under pressure from a compressed air source or hydraulic pump.

VEHICLE LIFTS

Two types of stationary vehicle lifts are normally found in major repair shops: the single-post lift and the twin-post lift (Fig. 2-35). The single-post lift is designed to contact the vehicle's frame, allowing the wheels to hang free. This type of lift is usually preferred by engine mechanics because it allows complete freedom under the entire engine area.

The common twin-post lift is designed to raise a vehicle by its front and rear suspension systems. Other twin-post hoists pick up each side of the car at the frame, allowing the center of the chassis to be free. These hoists are especially useful to the transmission technician.

Most lifts are equipped with some type of safety device to prevent the vehicle from lowering unexpectedly. The weighted lock pin shown in Figure 2-36 is an example. As the lift reaches its maximum height, the lock pin falls into position to prevent the vehicle from lowering without warning. When the technician is ready, the pin is moved up and the lift can be lowered. Hoist stands should be used as shown in Figure 2-37 when the lift does not have a built-in safety feature to prevent the unexpected lowering of the vehicle. Most manufacturers have specific places on the chassis where the vehicle is to be lifted (Fig. 2-38). This is done to prevent damage to the suspension parts. When in doubt, check a shop or repair manual for the lifting points. Be certain all operating procedures are understood before using any vehicle lifts.

FIG 2-36 Lock pin on a single post hoist.

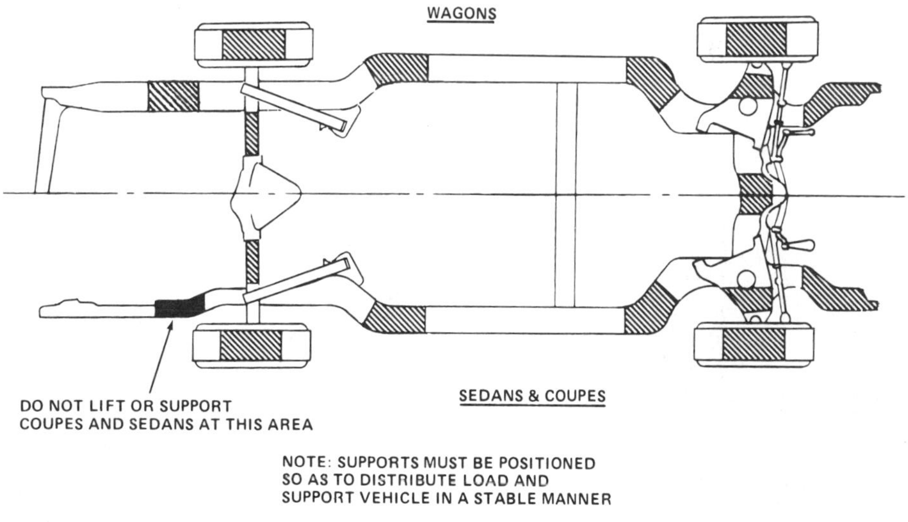

FIG 2-38 Lifting points. (Buick Motor Division,GMC)

FIG 2-37 Hoist safety stand in position under frame of hoist.

The engine lift used by most shops today is a hydraulic-powered portable device as shown in Figure 2-39. A hand-operated hydraulic jack is pumped to raise the engine, and a release valve is loosened to lower the engine.

The chain hoist is another lifting device used frequently for engines. Figure 2-40 shows a chain hoist attached to an "A" frame. Chain hoists usually are labeled with their lifting capacity. Most complete engine assemblies weigh less than 1,000 pounds. Therefore, a chain hoist of at least ½-ton capacity should be used.

FLOOR JACK

This device, sometimes called a service jack, is the most frequently used piece of lifting equipment in any shop (Fig. 2-41). From lifting one wheel of a vehicle to supporting a transmission during engine removal, its uses are many and varied. The controls are usually located on the handle. The knob at the top of the handle controls the release valve in the pump. When the knob is turned fully clockwise, the valve is closed. Pumping the handle will now raise the saddle. Turning the knob counterclockwise will release the pressure and allow the saddle to drop. Most floor jacks for general shop use have a 1½- to 2-ton lifting capacity.

Always use jack (safety) stands to support the weight of a vehicle (Fig. 2-42).

The following safety procedures should always be observed when operating any lifting device.

SAFETY PROCEDURES FOR LIFTING DEVICES

1. All operating instructions and precautions should be understood before using any lifting equipment.
2. Use lifting devices with adequate capacity. Look for rating labels.
3. Be certain to center the weight being lifted. This applies to every type of lifting device.
4. Any evidence of air or oil leakage is a danger sign. Report these problems to your instructor or supervisor.
5. Never work under a vehicle or engine supported by a lift or hoist alone. Use safety (jack) stands under a vehicle. Lower an engine to the floor if work on it must be completed.
6. When in doubt, lower the lift or hoist and seek your instructor's help.

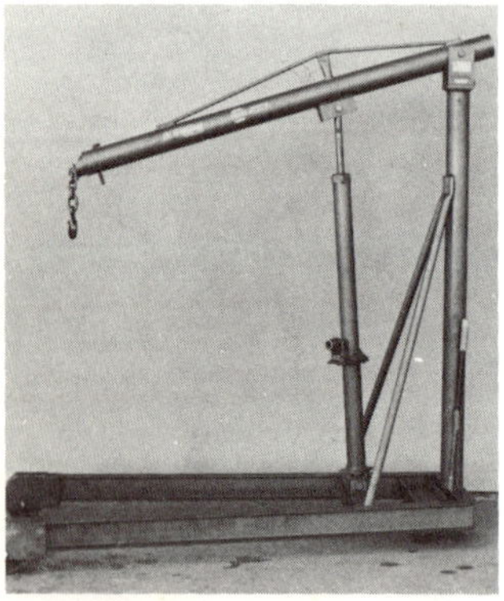

FIG 2-39 Hydraulic engine lift.

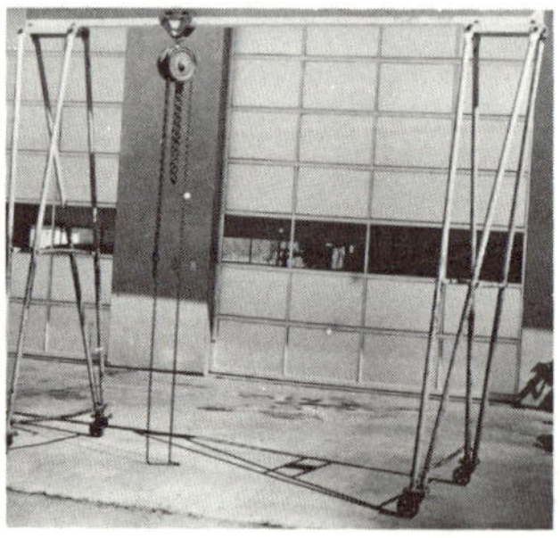

FIG 2-40 "A" frame with chain hoist.

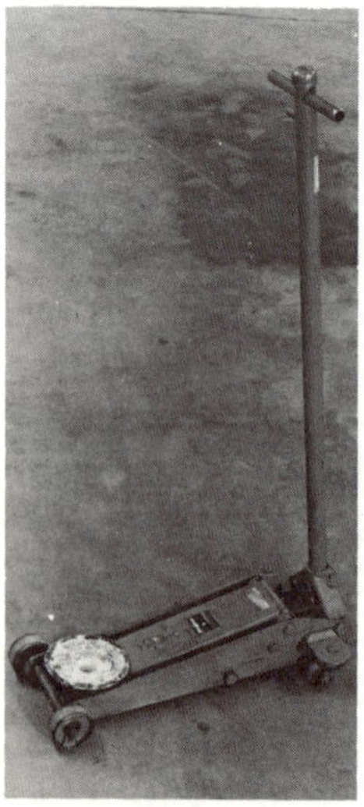

FIG 2-41 Typical floor (service) jack.

FIG 2-42 Jack (safety) stands supporting the outer end of the front suspension.

Name ______________________ Date ______________

DIRECTIONS

Circle the best answer to each question.

1. Wrenches of the highest quality are
   I. made from vanadium steel.
   II. manufactured by the drop forging method.
   A. I only B. II only C. Both I and II D. Neither I nor II
2. The advantage of a box-end wrench over an open-end wrench is that
   I. the box end is lighter in weight.
   II. the open end is less likely to slip.
   A. I only B. II only C. Both I and II D. Neither I nor II
3. Another name for the flare-nut wrench is
   I. box-open combination.
   II. piping.
   A. I only B. II only C. Both I and II D. Neither I nor II
4. A socket with 8 points is used on/for
   I. hexagon nuts and bolt heads.
   II. square-headed fasteners.
   A. I only B. II only C. Both I and II D. Neither I nor II
5. Metric sockets have drive openings in the following sizes:
   I. 8 and 13 mm.
   II. ⅜ and ½ inch.
   A. I only B. II only C. Both I and II D. Neither I nor II
6. Deep sockets are used for
   I. nuts and bolts not reachable by hand.
   II. nuts that are threaded more than 1 inch on a bolt or stud.
   A. I only B. II only C. Both I and II D. Neither I nor II
7. Socket wrenches and their handles should be used in the following manner:
   I. Pull toward oneself when tightening or loosening fasteners.
   II. Where there is a choice, use a socket rather than a hand wrench.
   A. I only B. II only C. Both I and II D. Neither I nor II
8. Pliers should be used when
   I. there is no wrench of correct size immediately available.
   II. the jaws will not damage what is being held.
   A. I only B. II only C. Both I and II D. Neither I nor II
9. A center punch can be used to
   I. make a locating spot prior to drilling.
   II. mark connecting rods and main bearing caps prior to their removal.
   A. I only B. II only C. Both I and II D. Neither I nor II

10. Every file used in the automotive shop
    I. should have a handle.
    II. should be accompanied by a file card.
    A. I only B. II only C. Both I and II D. Neither I nor II
11. The selection of a hacksaw blade is governed by
    I. the hardness of the material.
    II. the size of the material.
    A. I only B. II only C. Both I and II D. Neither I nor II
12. Taps are used for cutting threads on
    I. nuts and bolts.
    II. pipe fittings.
    A. I only B. II only C. Both I and II D. Neither I nor II
13. Electric power tools offer the following advantage over air power tools:
    I. They generally are lighter in weight.
    II. They operate at a cooler temperature.
    A. I only B. II only C. Both I and II D. Neither I nor II
14. Air power drill motors require
    I. daily lubrication with motor oil.
    II. 90 psi air pressure unless otherwise specified.
    A. I only B. II only C. Both I and II D. Neither I nor II
15. Impact wrenches
    I. should be used for disassembly purposes only.
    II. require special sockets.
    A. I only B. II only C. Both I and II D. Neither I nor II
16. Eye protection should be used with the
    I. drill press.
    II. bench grinder.
    A. I only B. II only C. Both I and II D. Neither I nor II
17. Unexpected drops in raised vehicle hoists can be prevented by
    I. locking pins.
    II. hoist stands.
    A. I only B. II only C. Both I and II D. Neither I nor II
18. Hydraulic lifting devices
    I. need not be labeled as to their capacity.
    II. usually leak minor amounts of oil, which causes no problem.
    A. I only B. II only C. Both I and II D. Neither I nor II
19. A floor jack
    I. saddle can be raised when the valve handle is turned clockwise to its stop.
    II. can support a transmission during engine removal and replacement.
    A. I only B. II only C. Both I and II D. Neither I nor II
20. Working on an engine being held by a hoist alone
    I. should never be done.
    II. can be done if the engine is close to the floor.
    A. I only B. II only C. Both I and II D. Neither I nor II

DATE COMPLETED ____________ INSTRUCTOR'S SIGNATURE ____________

After completing the preparation assignment, you will correctly answer a minimum of 7 out of 10 multiple choice questions in a written test, without reference material, on the following topics:

1. Engine fasteners
2. Safety information

OBJECTIVE 2-B (written test)

Preparation:

Study pages 45 through 50.

FASTENERS

Any device used to hold two or more parts or pieces together can be classified as a fastener. The fasteners discussed here are those which the engine technician will most often encounter.

SCREWS AND BOLTS

Although it is classified as a machine screw or cap screw, nearly everyone refers to the fastener shown in Figure 2-43 as a bolt. If this fastener is tightened into a threaded hole, it is properly called a cap screw; if it has a nut threaded onto it, it is properly called a bolt. In this text, such a fastener with a six-sided (hexagon) head will be called a bolt, with or without a nut.

A stud bolt, commonly called a stud, has no head and is threaded on both ends (Fig. 2-44). It usually has coarse threads on an end which is threaded into the largest piece to be joined. The opposite end has fine threads and holds a nut.

Screws will be classified as any fastener turned by a screwdriver or by hand. Machine screws and sheet metal screws very often have similar head shapes (Fig. 2-45). Below the heads of these screws, however, differences can be seen. Machine

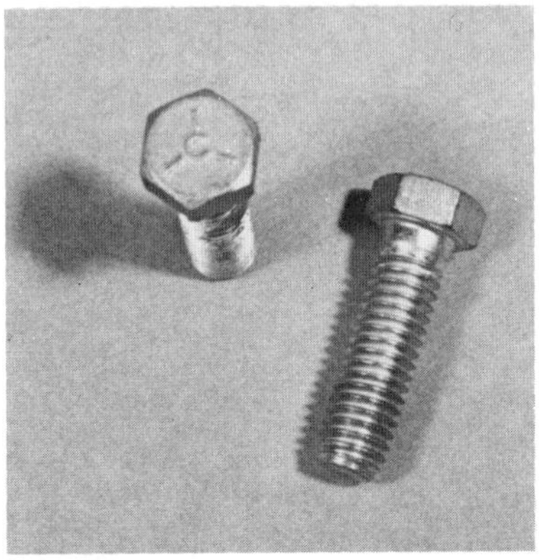

FIG 2-43 Hexagon (six-sided) head bolt.

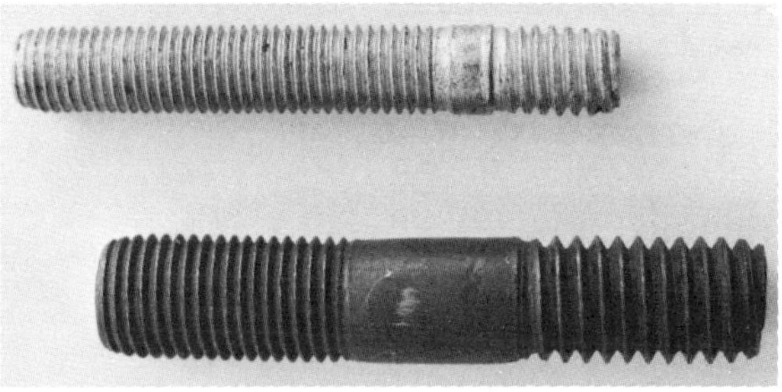

FIG 2-44 Stud bolts.

FIG 2-45 Common head shapes of screws and bolts.

screw threads are usually the same diameter from top to bottom. Sheet metal screws nearly always have tapered threads starting with the small diameter and working up to the largest diameter just below the head. In general shop practice, the machine screw is threaded into a tapped hole (a hole with threads inside) of the correct size. A sheet metal screw is turned into a hole in soft or thin metal, forming its own threads as it is turned. This is called self-tapping.

Bolts, machine screws, and sheet metal screws are sized according to (1) the diameter of the area below the head; (2) the number of threads per inch or per millimetre; and (3) the length measured from just below the head to the tip. A ½–13 NC × 2 inch bolt is an example (Fig. 2-46).

The diameter sizes most often refer to the diameter of the metal before the threads were cut. A ½-inch bolt should measure approximately ½ inch across the area below the head.

Threads, the second designation, are the grooves and points of the bolt or screw below the head. These are called external threads while grooves and points inside a hole or nut are called internal threads. In the example, threads were cut on the metal below the head, which is approximately ½ inch in diameter. Thirteen threads per inch, the National Course series size, was used. To properly measure the number of threads per inch, a thread pitch gauge should be used. This tool (Fig. 2-47) contains a variety of thread-matching size gauges. To use this gauge, select one blade at a time, placing it next to the threads of a fastener. All the teeth of the correct blade will fall into place when it is properly matched (Fig. 2-48).

The third designation, the length, is the total distance measured from below the head surface to the opposite end.

Bolts and screws are further classified by their strength. This depends upon the material used in their manufacture and any heat treatment process used to improve their strength.

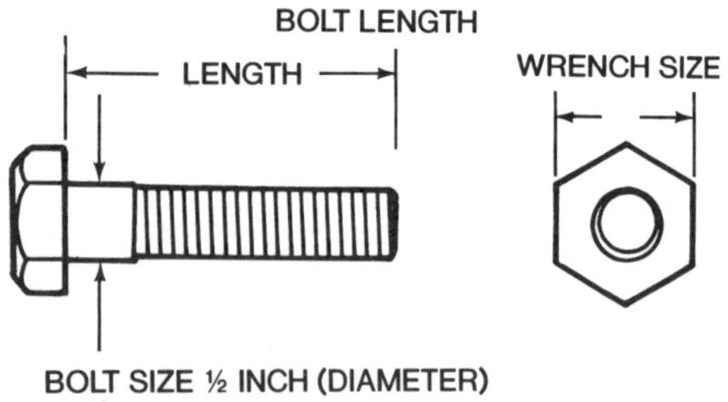

FIG 2-46 Bolt size designations.

FIG 2-47 Thread pitch gauge.

FIG 2-48 A thread pitch gauge matching the threads of the bolt.

Many bolts have markings on their heads. This is a code system for designating tensile (pull) strength. Table 2-1 shows examples of the most common markings on automotive bolt heads. Disregarding any letters or names, the more marks that appear, the greater the strength of the bolt. The greater the strength, the more torque it is capable of withstanding. A bolt with no marks is of the weakest class (1 or 2) and should not be used where high torque is necessary.

The chart (Table 2-1) on general torque specifications limits shows recommended *maximum* torque limits. This chart should be used only when specifications are not available from the vehicle manufacturer. Further, these limits do not apply to assemblies where gaskets, especially of cork, are used. The limits shown are based on solvent-cleaned bolts, bolt holes, and nuts. If sealants (RTV Silicone) or locking chemicals (Loctite) are used on the threads, the torque value should be reduced by 10 percent; if new plated capscrews are used, reduce torque by 20 percent.

Table 2-1 STANDARD TORQUE SPECIFICATIONS AND BOLT HEAD MARKINGS

SAE Grade Number	**1 or 2**		**5**		**6 or 7**		**8**	
Bolt Head Markings Manufacturer's marks may vary. Three-line markings on heads shown below, for example, indicate SAE Grade 5. 	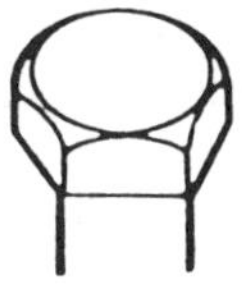				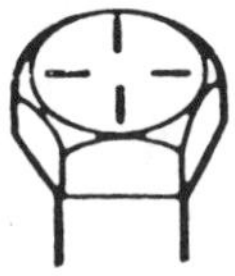		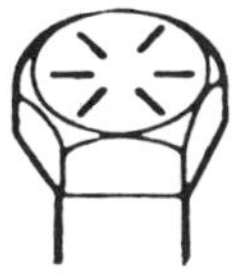	
Usage	**Used Frequently**		**Used Frequently**		**Used at Times**		**Cyl. Heads Main Brg. Caps**	
Bolt diameter and minimum tensile strength psi (Kg/cm)	To ½ - 69,000 (4850.7) To ¾ - 64,000 (4499.2) To 1 - 55,000 (3866.5)		To ¾ - 120,000 (8436.0) To 1 - 115,000 (8084.5)		To ⅝ - 140,000 (9842.0) To ¾ - 133,000 (9349.9)		150,000 (10545.0)	
Quality of Material	**Indeterminate**		**Minimum Commercial**		**Medium Commercial**		**Best Commercial**	
Bolt Body Size (Inches) - (Thread)	Torque		Torque		Torque		Torque	
	Ft-Lb	kg-m	Ft-Lb	kg-m	Ft-Lb	kg-m	Ft-Lb	kg-m
1/4-20	5	0.6915	8	1.1064	10	1.3830	12	1.6596
-28	6	0.8298	10	1.3830			14	1.9362
5/16-18	11	1.5213	17	2.3511	19	2.6277	24	3.3192
-24	13	1.7979	19	2.6277			27	3.7341
3/8-16	18	2.4894	31	4.2873	34	4.7022	44	6.0852
-24	20	2.7660	35	4.8405			49	6.7767
7/16-14	28	3.8132	49	6.7767	55	7.6065	70	9.6810
-20	30	4.1490	55	7.6065			78	10.7874
1/2-13	39	5.3937	75	10.3725	85	11.7555	105	14.5215
-20	41	5.6703	85	11.7555			120	16.5960
9/16-12	51	7.0533	110	15.2130	120	16.5960	155	21.4365
-18	55	7.6065	120	16.5960			170	23.5110
5/8-11	83	11.4789	150	20.7450	167	23.0961	210	29.0430
-18	95	13.1385	170	23.5110			240	33.1920

NUTS AND WASHERS

A nut is used to hold a bolt or screw that passes through a hole. These fasteners are designated by their head design, hexagon (six-sided) or square (four-sided). In some cases, they are also named for their shapes above or below the flat sides of the hexagon or square (Fig. 2-49).

Nuts used within the engine, such as those holding the cap on a connecting rod, are self-locking. This means they are not likely to loosen from vibration, but will require greater effort to loosen and tighten with a wrench.

Washers are companions to fasteners which improve their holding power. A flat washer is used to distribute tension force and to protect the base material. Lock washers, regardless of shape, are used to prevent loosening of a nut, bolt, or screw once it is secured. Washers are designated by their general shape and the size (diameter) of the bolts over which they fit (Fig. 2-50).

PINS

Pin fasteners have a variety of shapes and sizes (Fig. 2-51). The most common type is the cotter pin. Like other pin fasteners, it is used to secure two or more pieces together. The cotter pin is normally used where it can be easily removed and replaced. Once a cotter pin is removed, it should be discarded and a new one of like size installed.

The roll pin relies on its expanding spring pressure to hold itself in place; therefore, its diameter is slightly larger than the hole into which it is driven.

Dowel pins are frequently used to align two or more parts and their gaskets. These are not usually removed during engine repair.

Sizes of pins are designated by their diameter and length. A 1/8 inch × 2 cotter pin is 1/8 inch across and 2 inches in length (Fig. 2-52).

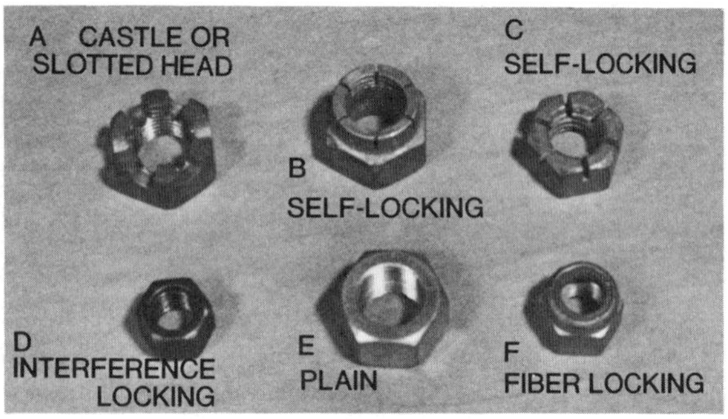

FIG 2-49 Common shapes of nuts.

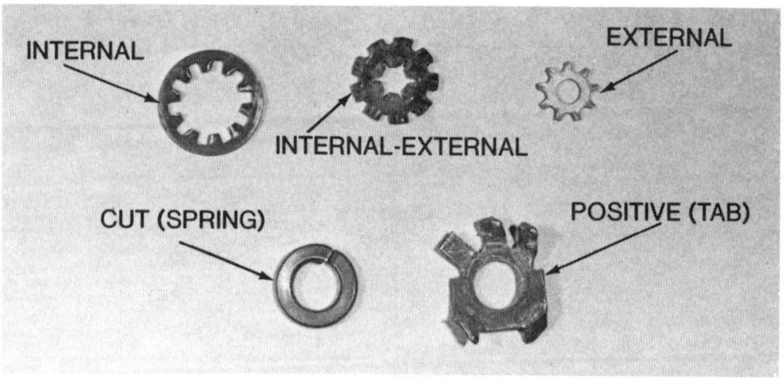

FIG 2-50 Typical lock washers.

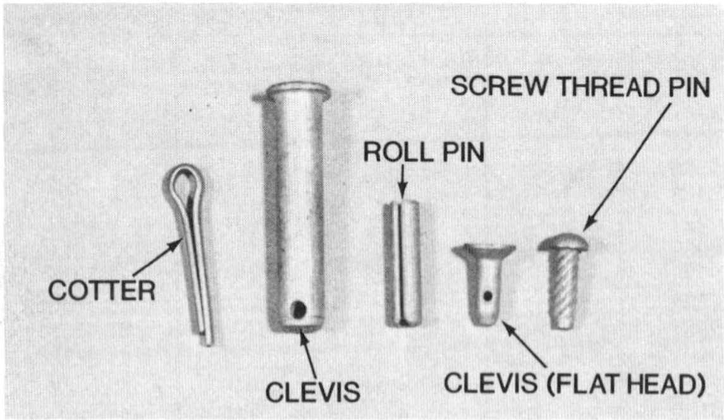

FIG 2-51 Common automotive pin type fasteners.

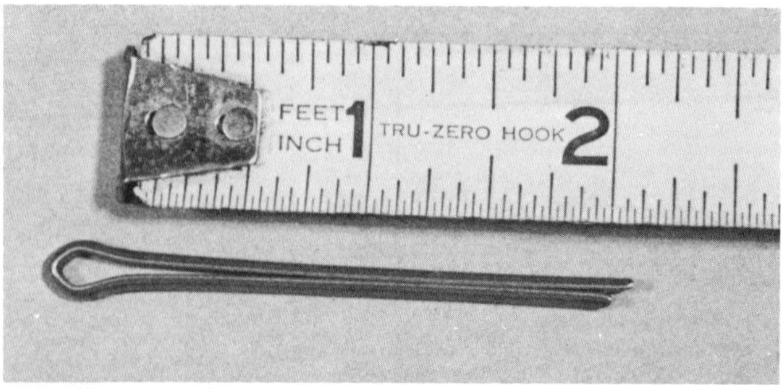

FIG 2-52 Cotter pin length measurement.

KEYS

Keys are designed to keep pulleys and gears which are attached to a shaft from turning. Two types used in automotive engines are the square key and the woodruff key (Fig. 2-53). Roughly half of the key sets in the shaft, while the other half holds the gear in position.

SAFETY INFORMATION

An injured athlete cannot perform at top capacity; nor can the automotive mechanic who is injured be fully productive. In many shop situations, the mechanic is paid a percentage of the labor cost to complete a repair job. Any injury, however slight, prevents the mechanic from earning as much if he were not injured. The most tragic injury accidents result in a permanent disability which prevents the person from doing repair work again.

Accidents can be avoided by a combination of a proper attitude and the use of proper work procedures. The auto mechanic should clearly understand the proper use of all tools and equipment and potential hazards before any work is performed. The student has the responsibility to seek the instructor's help when any procedure is unclear. The safety signs posted in the shop should be periodically reviewed and always adhered to.

A person learning the automotive service trade must continually strive to be a good housekeeper. Good housekeeping in the shop means, maintaining the organization of all tools and equipment for the most efficient and safest work. Tools and equipment should always be returned to their proper place if they are not immediately needed. Liquid spills, especially oil or grease, should be wiped up immediately (Fig. 2-54). Loose parts or other trash should be disposed of properly. Good housekeeping in the shop usually means that work will be accomplished more efficiently and in a safe manner.

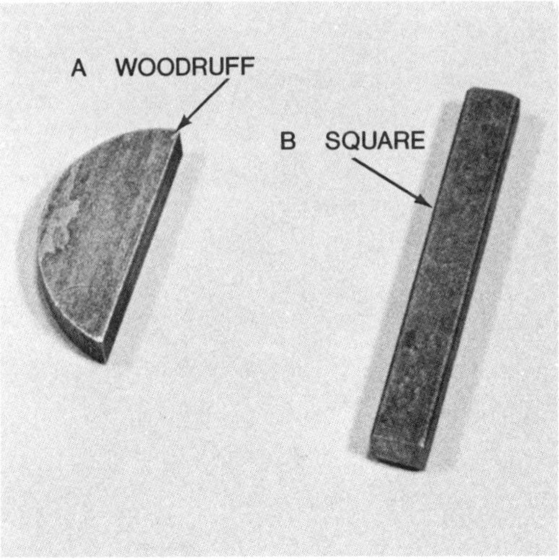

FIG 2-53 Keys.

FIG 2-54 Good shop practice requires wiping up any spills immediately.

EYE PROTECTION

The most common eye hazards in the automotive shop are flying matter such as dirt or metal chips and splashing liquids such as solvents or steam cleaning liquid.

Flying matter is most often the result of incorrect use of a blow gun. Blow guns are necessary to blow away foreign material from clean parts just prior to their assembly. Be certain all workers are out of a danger area of all flying particles. When working under a car, mechanics are usually looking upward while loosening or tightening fasteners, and soil particles will fall toward the eyes. In these situations, eye protection should always be worn.

During the cleaning operations required to remove dirt or soils from a part, chemicals can be splashed into the eyes. All cleaning operations require either heat, pressure, and chemicals. The pressure phase can mean high pressure water is being used, pressure applied with a solvent brush or other means. This often allows splashing into the eyes to occur.

All machines in the shop are labeled with the eye protection equipment necessary. Follow the directions.

Eye protection commonly comes in two forms. The face shield is used to guard against splashing liquids, and the goggles are used to guard against flying particles.

FIRST AID FOR EYES

Follow these simple rules when eyes have foreign particles or solution in them:

1. Don't rub—rubbing usually worsens the problem.
2. Notify your instructor.
3. Flush with clean water.
4. Seek treatment from a doctor.

Name ________________________________ Date ________________

DIRECTIONS

Select the best answer to each question and circle the letter at the left.

I. The term hexagon indicates
  I. a six-sided figure.
  II. the head shape of most engine bolts.
    A. I only
    B. II only
    C. Both I and II
    D. Neither I nor II

2. When the size of a bolt is designated ½ inch–13 N.C.,
  I. ½ is the distance across the points of the head.
  II. ½ is the distance across the flat sides of the head.
    A. I only
    B. II only
    C. Both I and II
    D. Neither I nor II

3. The line markings on a bolt head
  I. indicate its relative strength.
  II. indicate the manufacturer's production process.
    A. I only
    B. II only
    C. Both I and II
    D. Neither I nor II

4. Lock washers are used to prevent a fastener from loosening, while flat washers are used
  I. to distribute the tension force evenly.
  II. to protect the base material.
    A. I only
    B. II only
    C. Both I and II
    D. Neither I nor II

5. Washers are designated by
  I. their general shape.
  II. their outside diameter.
    A. I only
    B. II only
    C. Both I and II
    D. Neither I nor II

6. A cotter pin
  I. is often referred to as a roll pin.
  II. expands to help hold itself in place.
    A. I only
    B. II only

C. Both I and II
D. Neither I nor II

7. Avoiding accidents in the shop involves
   I. having a proper attitude
   II. using recommended work procedures.
   A. I only
   B. II only
   C. Both I and II
   D. Neither I nor II

8. A person who is not sure about the operation of a tool or machine should:
   I. ask the help of another student.
   II. ask the help of the instructor.
   A. I only
   B. II only
   C. Both I and II
   D. Neither I nor II

9. Good housekeeping in the shop means
   I. cleaning up only at designated times.
   II. removing any safety hazards as soon as they are noticed.
   A. I only
   B. II only
   C. Both I and II
   D. Neither I nor II

10. The responsibility for maintaining/wearing eye protection in the shop rests with the
   I. instructor.
   II. student.
   A. I only
   B. II only
   C. Both I and II
   D. Neither I nor II

DATE COMPLETED ____________________ INSTRUCTOR'S SIGNATURE ____________________

unit 3

Quiz tuesday

Reference Materials

Competent automotive service work can be performed only when the proper identification information and specifications are available. This unit provides instruction on the uses of reference materials frequently found in auto shops. Diagnosis of engine problems prior to reconditioning is also included.

OBJECTIVE 3-A (written test)

After completing the preparation assignment, you will correctly answer a minimum of 7 out of 10 multiple choice questions in a written test, without reference material, on the following topics:

1. Identification procedures
2. Specifications
3. Work authorization

Preparation:

Study pages 53 through 73.

IDENTIFICATION PROCEDURES

Just as a builder needs plans and specifications for materials before construction can begin, a mechanic must be able to identify the automobile and its components before diagnosis and correction of any operational problems on both import (manufactured outside the U.S.) and domestic (U.S.-built) engines.

TYPES OF MARKINGS

In the explanations that follow, three types of identification markings are mentioned. One method, referred to as embossing, uses raised numbers and/or letters such as in Figure 3-1, a typical Vehicle Identification Number or VIN.

Another marking method, shown in Figure 3-2, is called stamping. These identification marks are placed on engines with steel stamps at the factory. Since the marks are applied after the piece is finished, they are indented (below the surface).

A third type of marking is called casting or forging marks. Figure 3-3 shows the casting number of a cylinder head. These markings are part of the mold in which the piece is cast or forged into shape. Casting numbers are raised above the surrounding surface and have the same texture. Practically all major engine parts have a marking of this type somewhere on the surface. Sometimes these are referred to as part numbers.

VEHICLE IDENTIFICATION

The identification process usually begins with the automobile itself. Correct information as to the make, model, year, and whether it is domestic or imported is the first requirement. Information on the make of the car is usually found on the outside body or inside instrument area near the driver's seat.

The single best source for identification information is the vehicle's registration slip or title card. Figure 3-4 shows a typical registration card, with the make, year, model, and date the automobile was first sold. If registration information is not available, look at the driver's side door jamb or the inside door area on a 1968-and-later car to find year and month manufactured (Fig. 3-5).

Either on one of the two locations mentioned or on the metal label or tag attached to the instrument panel (dash) close to the windshield on the driver's side, the official Vehicle Identification Number (commonly abbreviated VIN) should be

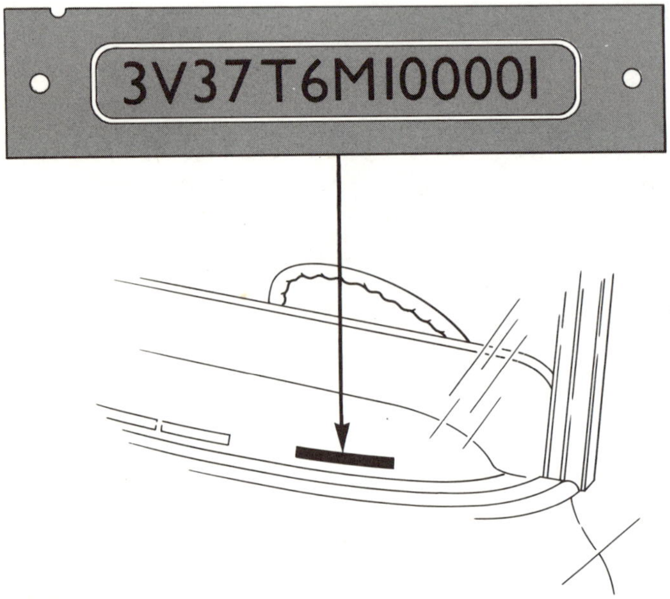

FIG 3-1 Embossed identification number.
(Oldsmobile Division, GMC)

FIG 3-2 Stamped figures on a cylinder block.

FIG 3-3 Cylinder head casting number.

found. Another source of identification (a poor one at best) is the marking on a tail lamp or parking lamp lens. The marking in Figure 3-6 happens to match the model year of the vehicle. But the number 76 simply indicates the first year that this lens design was approved for use on a vehicle as tested by standards of the Society of Automotive Engineers (SAE). The lens had to pass a series of light passage and reflection tests prior to its approval. The car maker is not required to change these numbers each year, nor is it illegal to retain the same lens shape over a period of years without change. In other words, the lens will bear the same number as the year it was approved for use, not the year it was installed. On import vehicles, the lens labeling numbers and letters are even less reliable. The lens number or year is frequently one year behind (older) than the actual year model. Since import manufacturers usually make fewer design changes, the same lens shape and labeling are installed for several years.

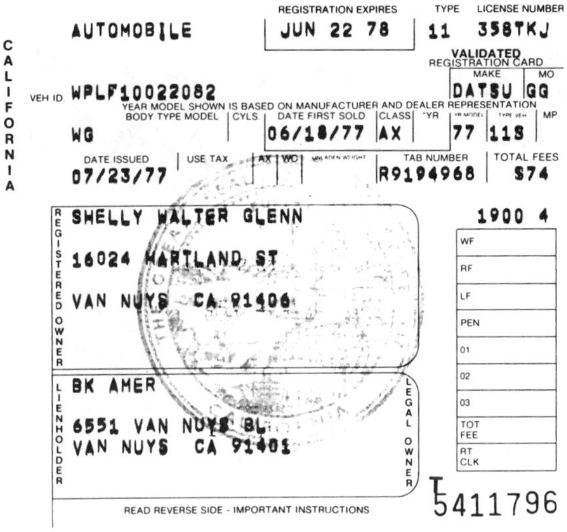
CALIFORNIA
AUTOMOBILE
REGISTRATION EXPIRES JUN 22 78
TYPE 11
LICENSE NUMBER 358TKJ
VALIDATED REGISTRATION CARD
VEH ID WPLF10022082
MAKE DATSU
MO GG
YEAR MODEL SHOWN IS BASED ON MANUFACTURER AND DEALER REPRESENTATION
BODY TYPE MODEL WG
CYLS
DATE FIRST SOLD 06/18/77
CLASS AX
YR 77
MP 11S
DATE ISSUED 07/23/77
USE TAX
AX WO
TAB NUMBER R9194968
TOTAL FEES $74
REGISTERED OWNER
SHELLY WALTER GLENN
16024 HARTLAND ST
VAN NUYS CA 91406
1900 4
WF
RF
LF
PEN
01
02
03
TOT FEE
RT CLK
LIENHOLDER
BK AMER
6551 VAN NUYS BL
VAN NUYS CA 91401
LEGAL OWNER
READ REVERSE SIDE - IMPORTANT INSTRUCTIONS
5411796

FIG 3-4 Typical registration form.

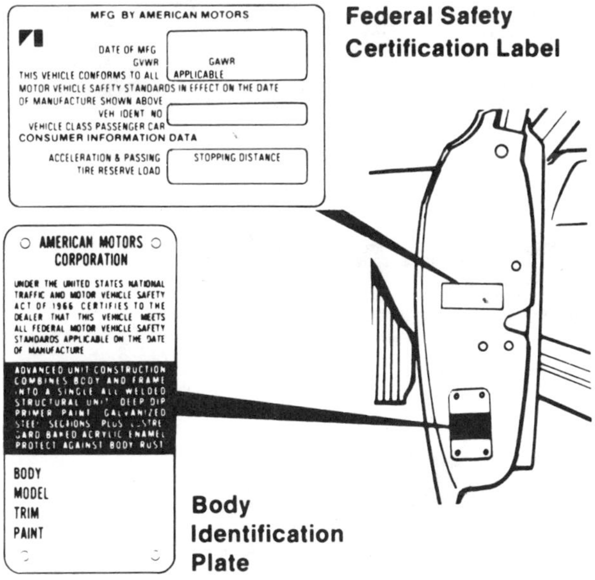

FIG 3-5 Safety sticker and body identification plate (left door).
(American Motors Corporation)

FIG 3-6 Typical lens markings.

Another source of identification information is the front end pictures (Fig. 3-7) found in some repair manuals. The technician should compare the front of the vehicle to be identified with one of the pictures in the manual. When the year of the vehicle is accurately determined, the rest of the information can be gathered from under the hood or door jamb.

On vehicles manufactured by General Motors, a body number and trim plate is attached to the firewall, the metal shield separating the engine and passenger compartments (Fig. 3-8). Notice that the year numbers follow the first letters of identification.

Chrysler Corporation and Ford Motor Company vehicles can often be identified

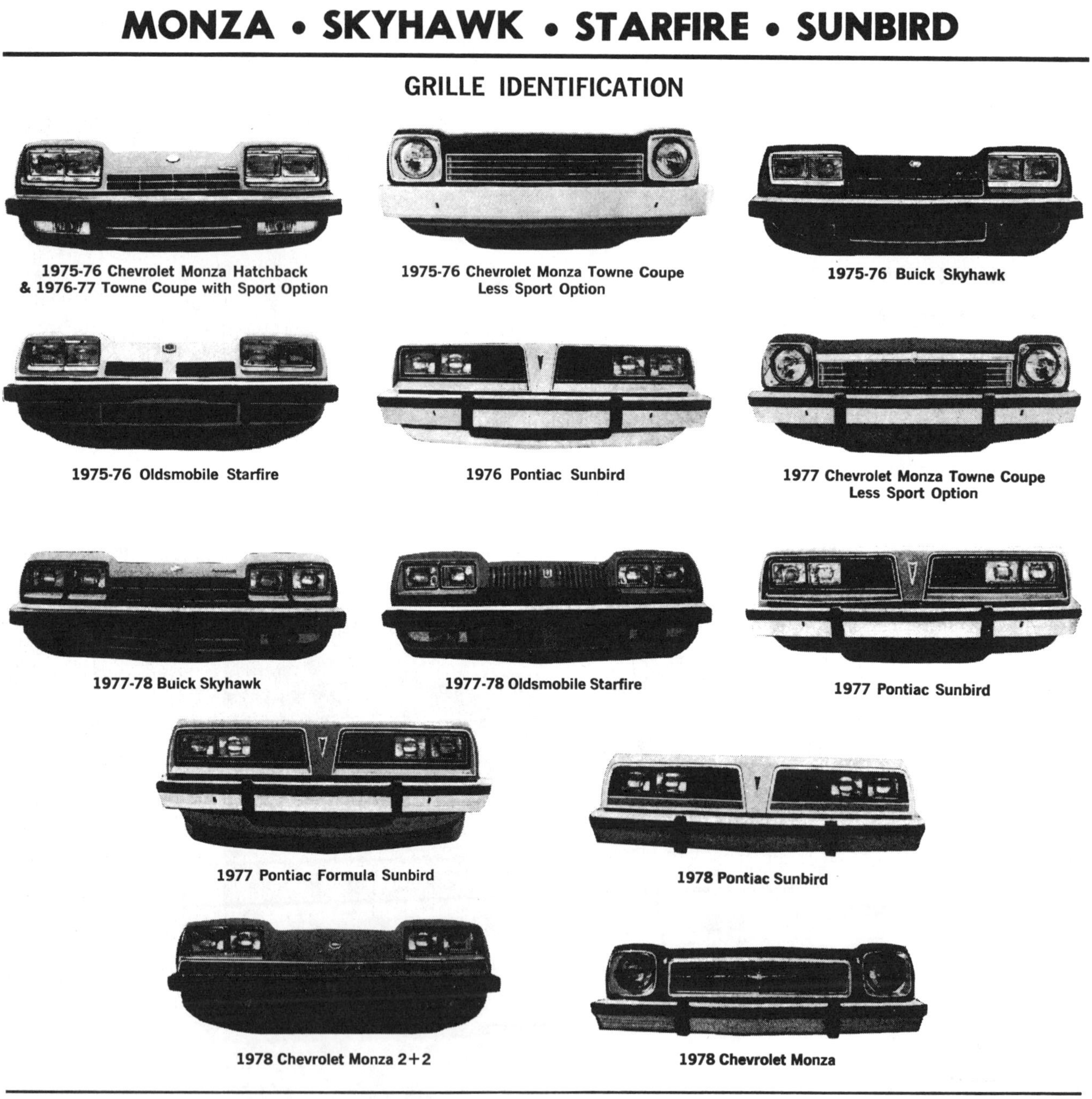

FIG 3-7 Front end identification pictures.

by the plate attached to the driver's side door or door jamb. A typical identification plate is shown in Figure 3-9.

The first section of the manufacturer's repair manual is devoted to vehicle identification (Fig. 3-10). Very specific details to identify body, engine, and other component parts are given. Some manufacturers also give identification information in the car owner's manual.

Note: In identifying engine or other parts, the right and left side designations are determined as viewed from the driver's seat. Engine crankshaft rotation is usually stated as clockwise when viewed facing the front of the vehicle. *All references to crank rotation in this text will be as viewed facing the vehicle front.*

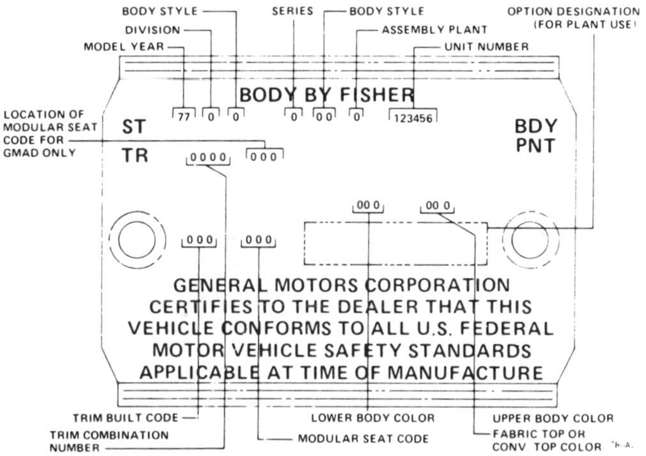

FIG 3-8 Body number plate.

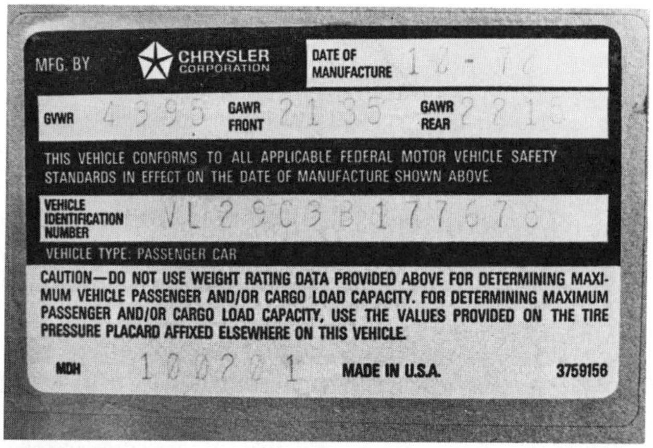

FIG 3-9 Door jamb identification plate.

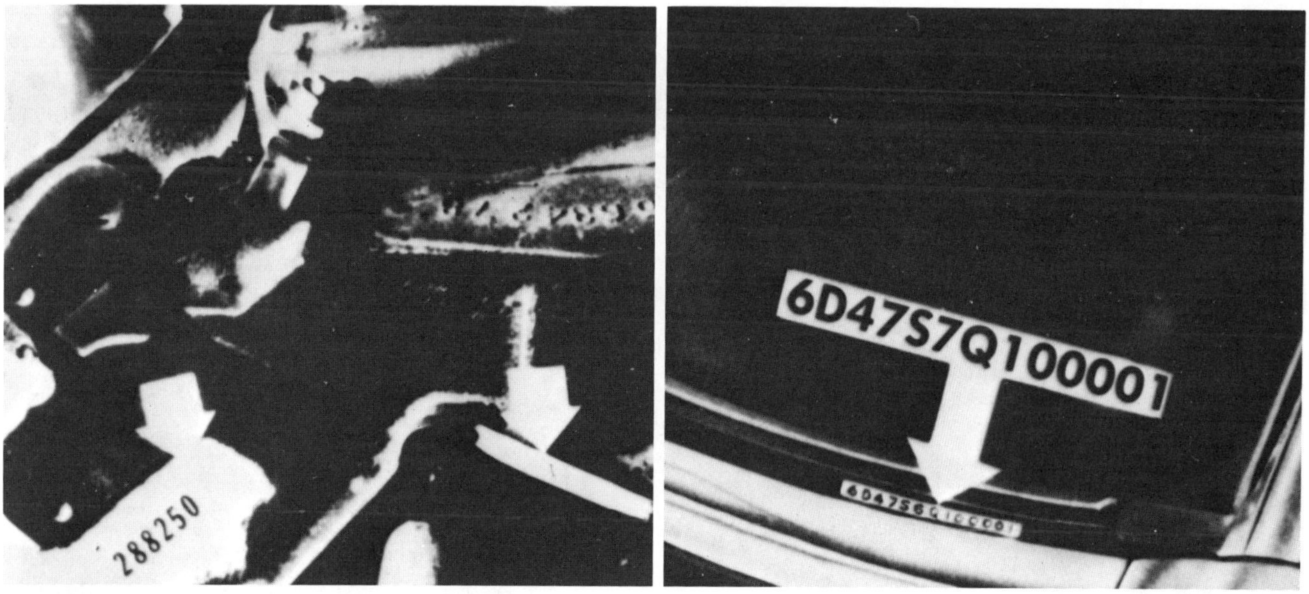

FIG 3-10 Identification information found in manufacturer's shop manuals.
(Cadillac Motor Division, GMC)

ENGINE IDENTIFICATION

The exact identification of an engine is most often related to its cubic inch or cubic centimetre displacement. Who the manufacturer is and whether it is a V-8 or a six- or a four-cylinder is usually obvious. Engine identification is simplified somewhat by a little practice and, more importantly, knowing the year of manufacture of the engine and/or vehicle. Generally speaking, the older the vehicle, the greater the chance for exchange of parts or entire engines.

By 1975, most domestic manufacturers used the fifth digit or letter within the Vehicle Identification Number (VIN) as an engine code designation.

The following information relates specific manufacturers' methods of engine identification used since the late 1950s.

American Motors V-8 engines are most often identified by a metal tag attached to the right side valve cover as shown in Figure 3-11. The code numbers and letters stamped on the tag must be interpreted through the use of a repair or shop manual. This system works well as long as the valve covers are not changed from side to side or replaced for some reason. The six-cylinder engines of American Motors have a one-letter designation. A copy of their code, usually found in their manuals, is necessary to determine engine size.

Chrysler Corporation stamps the cubic inch displacement on a milled pad near the front of the cylinder block. Figure 3-12 shows a typical example.

Ford Motor Company engines can be identified by the stamped metal tag attached to the intake manifold or cylinder head (Fig. 3-13). The upper numbers nearest the hole indicate the cubic inch displacement, while the lower numbers and letters nearest the hole indicate the date of manufacture. The fifth letter of the Ford VIN is a code for the engine's CID. Refer to a repair or shop manual, which is accurate as long as there has not been an engine change.

General Motors engines are labeled for identification with letters stamped on a machined pad somewhere near the front of the cylinder block. These letters pertain to designations within one year's production, so are only accurate if you know the year of manufacture and when an engine change has not been made.

FIG 3-11 Identification tag on American Motors V-8 engines. (American Motors Corporation)

FIG 3-13 Ford engine tag. (Ford Motor Company)

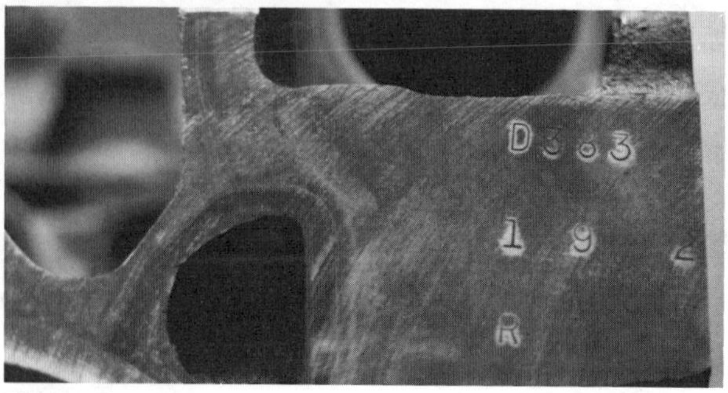

FIG 3-12 Typical Chrysler Corporation engine identification. (Chrysler Corporation)

When an engine mechanic is asked to rebuild an unknown engine, the only available means of identification may be the casting number of the block (Fig. 3-14). As the engine is further disassembled, other casting or forging numbers may be visible on the head or crankshaft.

In recent years, the engines installed in vehicles first sold in California sometimes have slightly different internal parts, emission controls, or tune-up specifications. To insure the proper completion of any type of engine repair on such a vehicle, the engine must be identified. The easiest, though possibly not the most reliable, source of information is the bottom line of the under-hood tune-up sticker (Fig. 3-15). Vehicles first sold in California should include that state's name in the bottom line.

SPECIFICATIONS

The word *specification* can have many meanings for the engine technician. Specification can mean

1. The size of a part, such as the diameter of a cylinder (bore) or a main bearing journal;
2. The difference in dimensions (sizes) of two mating parts, which is usually referred to as the clearance;
3. The performance of an engine, such as 350 horsepower at 4600 rpm;
4. The method of designating or labeling, such as the cylinder numbering pattern, or a procedure by which something may be repaired;
5. The strength of materials, the materials used, or even the finish texture of a part.

In repairing an engine, the first four definitions of *specification* will be used.

Specifications are found in one of three common publications:

1. Vehicle manufacturers' repair manuals
2. General repair manuals, published independently
3. Parts and equipment manufacturers' shop manuals

FIG 3-14 Engine block casting number.

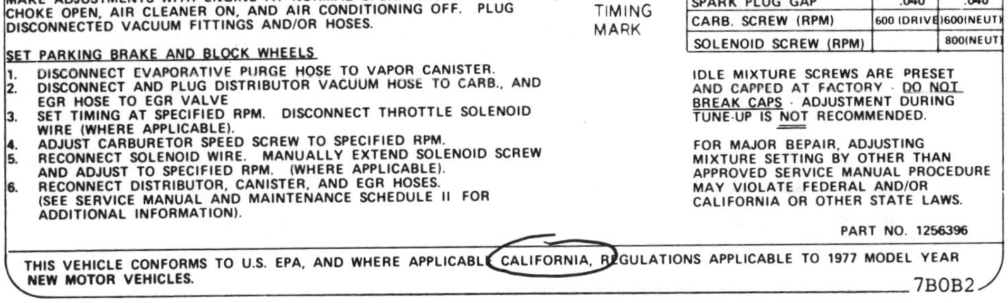

JC
231 CU. IN.
740E2LU

VEHICLE EMISSION CONTROL INFORMATION
GM
GENERAL MOTORS CORPORATION

TIMING MARK

CATALYST
EGR · EFE
LOW ALTITUDE CERTIFICATION

	TRANSMISSION	
	AUTO.	MAN.
TIMING (°B.T.D.C. @RPM)	12° @600	12° @800
SPARK PLUG GAP	.040	.040
CARB. SCREW (RPM)	600 (DRIVE)	600(NEUT)
SOLENOID SCREW (RPM)		800(NEUT)

MAKE ADJUSTMENTS WITH ENGINE AT NORMAL OPERATING TEMPERATURE, CHOKE OPEN, AIR CLEANER ON, AND AIR CONDITIONING OFF. PLUG DISCONNECTED VACUUM FITTINGS AND/OR HOSES.

SET PARKING BRAKE AND BLOCK WHEELS

1. DISCONNECT EVAPORATIVE PURGE HOSE TO VAPOR CANISTER.
2. DISCONNECT AND PLUG DISTRIBUTOR VACUUM HOSE TO CARB., AND EGR HOSE TO EGR VALVE
3. SET TIMING AT SPECIFIED RPM. DISCONNECT THROTTLE SOLENOID WIRE (WHERE APPLICABLE).
4. ADJUST CARBURETOR SPEED SCREW TO SPECIFIED RPM.
5. RECONNECT SOLENOID WIRE. MANUALLY EXTEND SOLENOID SCREW AND ADJUST TO SPECIFIED RPM. (WHERE APPLICABLE).
6. RECONNECT DISTRIBUTOR, CANISTER, AND EGR HOSES. (SEE SERVICE MANUAL AND MAINTENANCE SCHEDULE II FOR ADDITIONAL INFORMATION).

IDLE MIXTURE SCREWS ARE PRESET AND CAPPED AT FACTORY · DO NOT BREAK CAPS · ADJUSTMENT DURING TUNE-UP IS NOT RECOMMENDED.

FOR MAJOR BEPAIR, ADJUSTING MIXTURE SETTING BY OTHER THAN APPROVED SERVICE MANUAL PROCEDURE MAY VIOLATE FEDERAL AND/OR CALIFORNIA OR OTHER STATE LAWS.

PART NO. 1256396

THIS VEHICLE CONFORMS TO U.S. EPA, AND WHERE APPLICABLE CALIFORNIA, REGULATIONS APPLICABLE TO 1977 MODEL YEAR NEW MOTOR VEHICLES.

7B0B2

FIG 3-15 Under-hood tune-up label. (Buick Motor Division, GMC)

VEHICLE MANUFACTURERS' MANUALS

Vehicle manufacturers' manuals are published for use by dealership service departments. Very often a single volume covers more than one make or model of automobile (Fig. 3-16). These manuals cover all necessary information on the automobile, including specifications and repair procedures (Fig. 3-17). Information on operation principles is often presented at the beginning of each section. An understanding of the function of the parts or systems can be especially helpful during the diagnosis of malfunctions. This type of manual gives the greatest amount of information, although the purchase of a manual for each vehicle that a shop is asked to repair is not always economically practical.

GENERAL REPAIR MANUALS

General repair manuals are condensed versions of the manufacturers' manuals (Fig. 3-18), combining makes and often six or seven model years in one publication. The index shown in Figure 3-19 is typical of this type of repair manual, which has less information than manufacturers' manuals, but is less costly. The information necessary for the average engine repair job can usually be found in the general repair manual.

PARTS AND EQUIPMENT MANUFACTURERS' MANUALS

As a service to the people who use their products, many parts and equipment manufacturers assemble and publish specifications. The volume shown in Figure 3-20 is typical of those furnished by engine bearing manufacturers. This type of manual usually will include identification information as well as specification of engines and the dimensions of manufacturers' parts. It often covers all engine manufacturers and as many year models as the product services.

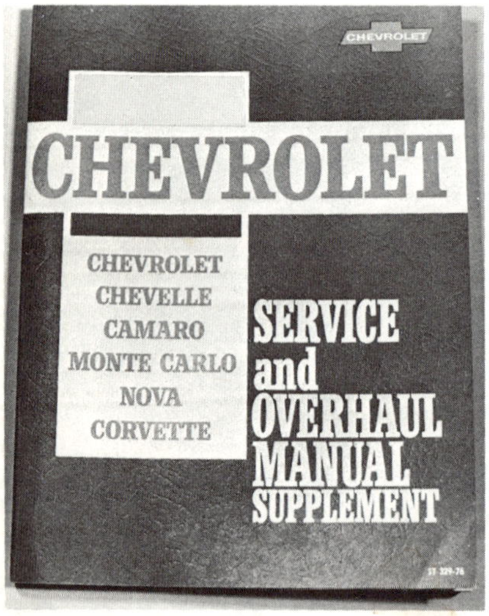

FIG 3-16 Manufacturer's repair manual.

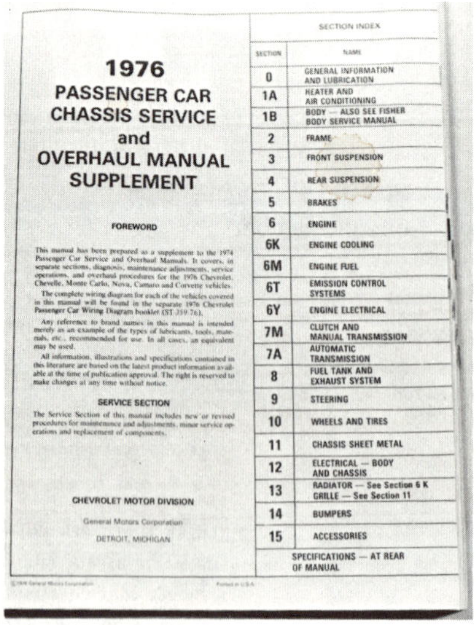

1976
PASSENGER CAR
CHASSIS SERVICE
and
OVERHAUL MANUAL
SUPPLEMENT

FOREWORD

This manual has been prepared as a supplement to the 1974 Passenger Car Service and Overhaul Manuals. It covers, in separate sections, diagnosis, maintenance adjustments, service operations, and overhaul procedures for the 1976 Chevrolet, Chevelle, Monte Carlo, Nova, Camaro and Corvette vehicles.

The complete wiring diagram for each of the vehicles covered in this manual will be found in the separate 1976 Chevrolet Passenger Car Wiring Diagram booklet (ST-359-76).

Any reference to brand names in this manual is intended merely as an example of the types of lubricants, tools, materials, etc., recommended for use. In all cases, an equivalent may be used.

All information, illustrations and specifications contained in this literature are based on the latest product information available at the time of publication approval. The right is reserved to make changes at any time without notice.

SERVICE SECTION

The Service Section of this manual includes new or revised procedures for maintenance and adjustments, minor service operations and replacement of components.

CHEVROLET MOTOR DIVISION
General Motors Corporation
DETROIT, MICHIGAN

SECTION INDEX

SECTION	NAME
0	GENERAL INFORMATION AND LUBRICATION
1A	HEATER AND AIR CONDITIONING
1B	BODY — ALSO SEE FISHER BODY SERVICE MANUAL
2	FRAME
3	FRONT SUSPENSION
4	REAR SUSPENSION
5	BRAKES
6	ENGINE
6K	ENGINE COOLING
6M	ENGINE FUEL
6T	EMISSION CONTROL SYSTEMS
6Y	ENGINE ELECTRICAL
7M	CLUTCH AND MANUAL TRANSMISSION
7A	AUTOMATIC TRANSMISSION
8	FUEL TANK AND EXHAUST SYSTEM
9	STEERING
10	WHEELS AND TIRES
11	CHASSIS SHEET METAL
12	ELECTRICAL — BODY AND CHASSIS
13	RADIATOR — See Section 6 K GRILLE — See Section 11
14	BUMPERS
15	ACCESSORIES
SPECIFICATIONS — AT REAR OF MANUAL	

FIG 3-17 Index page of a service manual with edge markings.

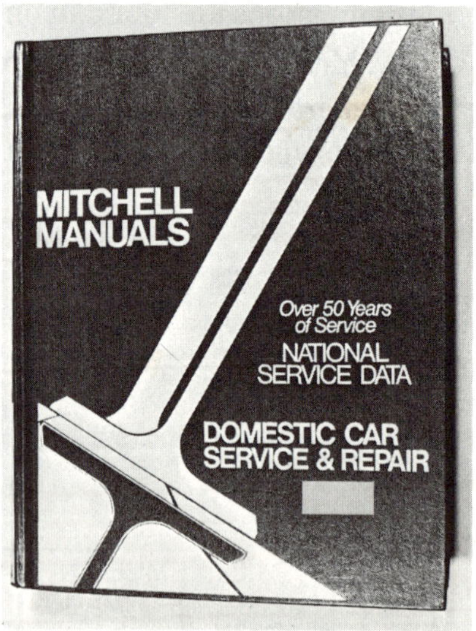

FIG 3-18 General repair manual.

TEARDOWN AND ASSEMBLY SHEET

The teardown and assembly sheet is a document used by the student during service work. It contains:

1. Ownership and identification information
2. Specifications required for engine service
3. Measurement fill-in space
4. Checkoff points for work progress by students and evaluation by the instructor

When the decision has been made to recondition (overhaul) the engine, the teardown sheet should be accurately filled out. The ownership and identification information and the specifications should be completed and understood before any work is started.

The student will have to contact the owner to discuss problems with the engine. The owner can be reached through information from the teardown sheet. In order to reduce the problem of soiled and damaged shop or repair manuals, the specifications should be filled in on the sheet before the engine work is started.

Check points are provided for use by the student and instructor as the engine work progresses. This often prevents repeating an operation and may prevent damage to an engine part. Spaces are also provided for measurement indications when the inspection phase of the work is in progress. The measurements may indicate what parts need machine work or replacement. The teardown sheet should remain available with each engine.

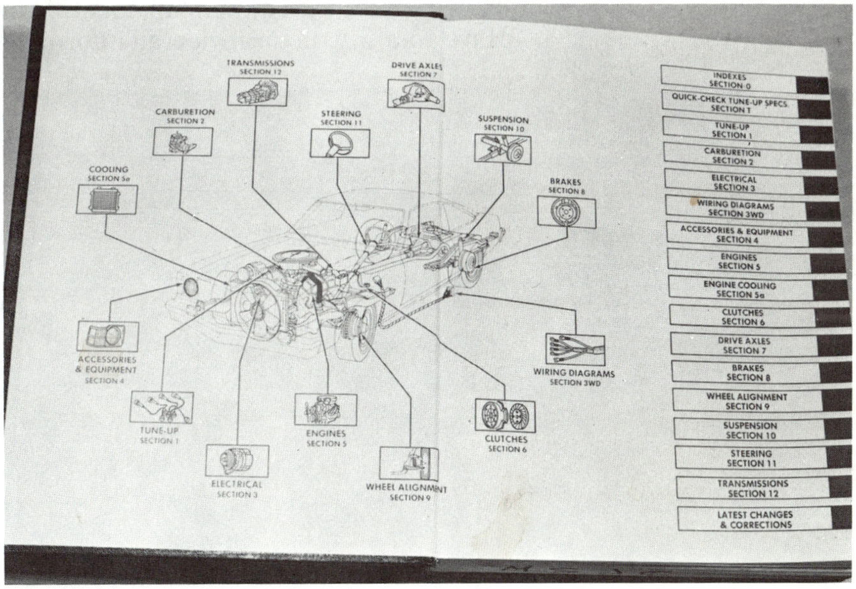

FIG 3-19 Index of a general repair manual with vehicle areas labeled.

FIG 3-20 Bearing catalog and shop manual.

OBJECTIVE 3-A1 (manipulative)

Working as a team with one other student and with Job Report Sheet 3-1A, you will locate and fill in all ownership identification and specification information in the engine teardown and assembly report sheet for an instructor-assigned vehicle and engine.

Time Recommended:

2 hours

Preparation:

1. Review pages 53 through 62.
2. Study pages 63 through 72.
3. Study manufacturer's or general repair manuals for the vehicle or engine being serviced.
4. Attend scheduled lecture on engine specifications.

Tools and Materials Needed:

1. Vehicle manufacturers' shop manual
2. General repair manual
3. Bearing manufacturers' shop manual

Procedure:

Fill in all spaces with ownership, identification, and specification information. Have your instructor review all information.

ENGINE TEARDOWN AND ASSEMBLY

Fill in all specifications (marked ■) before starting engine work. Keep this sheet with the engine until all work is completed. Then turn in to your instructor.

Make of engine ______________ Year ______________ Engine number ______________

No. of cylinders __________ Bore __________ Stroke __________ Cu. in. or cc __________

Compression pressure ______________ psi (kPa) Compression ratio ______________

Valve arrangement ______________ Cylinder arrangement ______________

Firing order ______________ Oil capacity ______________ Oil pressure ______________

Cylinder numbering pattern

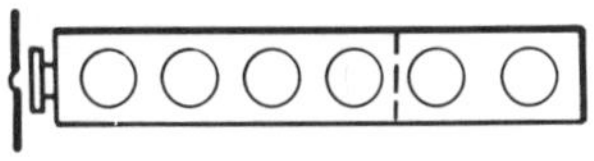

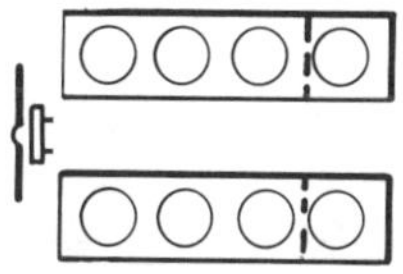

■ Journal Specifications

Main bearing journal diameter ______________________________

Connecting rod bearing journal diameter ______________________________

Camshaft journal diameter ______________________________

■ Engine Tightening Specifications (in foot-pounds unless specified)

Main bearing caps ______________ Connecting rod cap ______________ Oil pump ______________

Water pump ______________ Timing (front) cover ______________ Water outlet ______________

Spark plugs ______________ Cylinder head ______________ Intake manifold ______________

Exhaust manifold ______________ Rocker arm shaft/stud ______________ Rocker arm cover ______________

Vibration damper ______________ Flywheel ______________ Clutch pressure plate ______________

■ Clearance Specifications

Pistons ______________ Rod bearings ______________ Main bearings ______________

Piston ring end gap: Compression ______________ Oil ______________

Piston ring side clearance: Compression ______________________ Oil ______________________

Main thrust bearing number ______________________ Crankshaft end play ______________________

Connecting rod end play or side clearance ______________________

Oil pump side clearance ______________________ End clearance ______________________

■ Valve Specifications

Valve lash: Intake ______________________ Exhaust ______________________

Valve seat width: Intake ______________________ Exhaust ______________________

Valve seat angle: Intake ______________________ Exhaust ______________________

Valve face angle: Intake ______________________ Exhaust ______________________

Valve interference angle ______________ Valve margin ______________ (1/32 in. or 0.794 mm min.)

Spring tension: Inner ____________ @ ____________ Outer ____________ @ ____________

Installed height: Inner spring ______________________ Outer spring ______________________

Inner valve spring free length ______________ Outer valve spring free length ______________

Intake valve stem diameter ______________ Intake valve guide clearance ______________

Exhaust valve stem diameter ______________ Exhaust valve guide clearance ______________

Valve lift: Intake ______________ Exhaust ______________ Rocker arm ratio ______________

Type of rocker arm ______________________

DATE______________________ INSTRUCTOR'S SIGNATURE______________________

■ Compression pressure ______________________ (Spec.)

Comp. test—dry 1 ______ 2 ______ 3 ______ 4 ______ 5 ______ 6 ______ 7 ______ 8 ______

Engine only removal ______ *OR* engine + transmission removal ______

■ Engine Disassembly (Save everything removed)

Disconnect, remove, and store battery ______________________

Drain engine oil and water ______________ Remove carb. and fuel lines ______________

Remove valve covers ______________________ Remove oil pan ______________________

Remove timing gear or chain cover ______________________

Remove dist. cap and wires: Yes ________ No ________

Remove alt. fan belt ________ Remove fuel pump ________ Remove starter motor ________

Remove manifold: Intake ________ Exhaust ________ and valley cover ________

Remove rocker arm assemblies ________ What type? (Shaft) ________ (Ball and stud) ________

Remove pushrods (keep order) ________ What type? (Solid) ________ (Tubular) ________

Remove cylinder heads: Right ________ Left ________ Interchangeable ________

Remove water pump ________ Remove timing gear or timing sprockets and chain ________

Inspect all parts as they are removed from the engine. Save everything!

Cylinder ridge out: Yes ________ No ________

Connecting rod caps marked before removal ________ How ________

DATE________ INSTRUCTOR'S SIGNATURE ________

Remove rod and piston assemblies ________ Use guide sleeves on rod bolts ________

Present cylinder bore ________ Bore: Yes ________ No ________ How much ____

(TDC—Just below ring ridge; BDC—Bottom of ring travel)

#1A	TDC ________	#2A	TDC ________	#3A	TDC ________	#4A	TDC ________
	BDC ________		BDC ________		BDC ________		BDC ________
#5A	TDC ________	#6A	TDC ________	#7A	TDC ________	#8A	TDC ________
	BDC ________		BDC ________		BDC ________		BDC ________

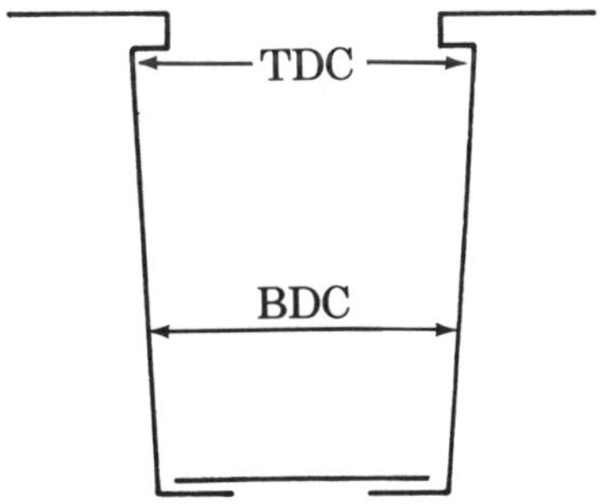

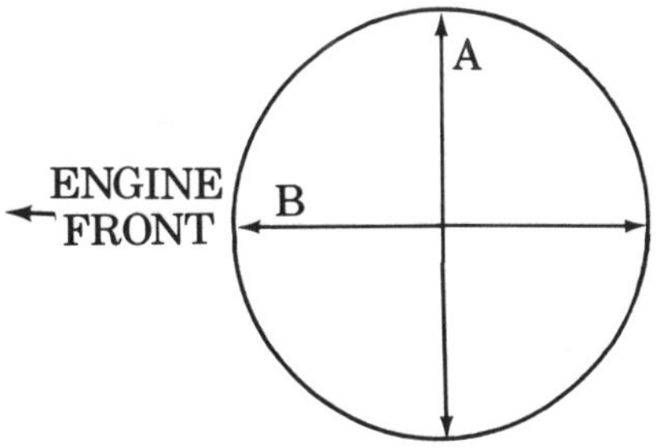

Amount of taper (difference between TDC and BDC)

#1 ________	#2 ________	#3 ________	#4________
#5 ________	#6 ________	#7 ________	#8________

#1B TDC ________ #2B TDC ________ #3B TDC ________ #4B TDC ________

BDC ________ BDC ________ BDC ________ BDC ________

#5B TDC ________ #6B TDC ________ #7B TDC ________ #8B TDC ________

BDC ________ BDC ________ BDC ________ BDC ________

Amount of out-of-round (difference between A and B in the same cylinder)

#1 ________ #2 ________ #3 ________ #4 ________

#5 ________ #6 ________ #7 ________ #8 ________

Amount of wear (difference between bore specification & "A" TDC measurements)

#1 ________ #2 ________ #3 ________ #4 ________

#5 ________ #6 ________ #7 ________ #8 ________

Crankshaft main bearing journal size (diameter)

#1 A ________ #2 A ________ #3 A ________

B ________ B ________ B ________

C ________ C ________ C ________

D ________ D ________ D ________

#4 A ________ #5 A ________ #6 A ________

B ________ B ________ B ________

C ________ C ________ C ________

D ________ D ________ D ________

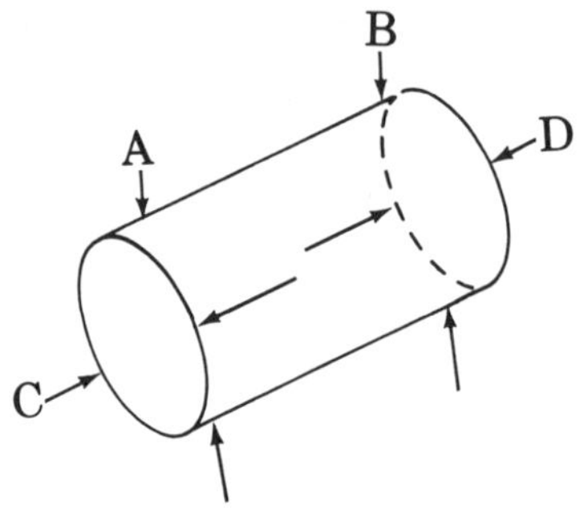

Journals undersize? 0.010 in. 0.020 in. 0.030 in. or STD. (Circle one)

Journals tapered: Yes ________ No ________ How many ________ (Max. 0.010 in. or .025 mm)

Journals out-of-round: Yes ________ No ________ How many ________ (Max. 0.0012 in. or 0.031 mm)

■ Original connecting rod journal size ____________________ (Spec.)

Undersize: Yes ________ No ________ How much ________________

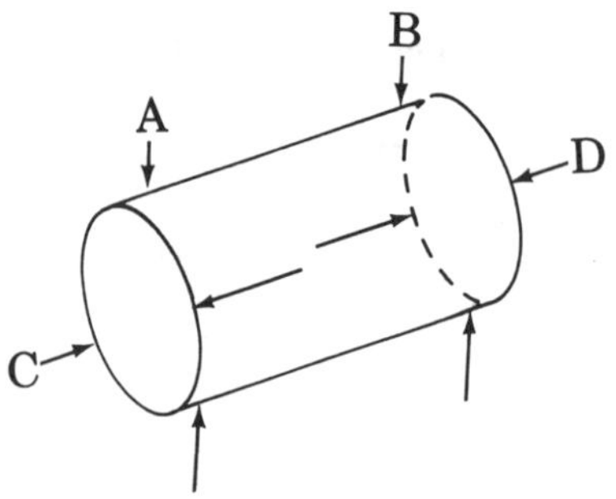

Connecting rod journal measurement

#1 A ____________ #2 A ____________ #3 A ____________ #4 A ____________

B ____________ B ____________ B ____________ B ____________

C ____________ C ____________ C ____________ C ____________

D ____________ D ____________ D ____________ D ____________

#5 A ____________ #6 A ____________ #7 A ____________ #8 A ____________

B ____________ B ____________ B ____________ B ____________

C ____________ C ____________ C ____________ C ____________

D ____________ D ____________ D ____________ D ____________

Journals tapered: Yes ________ No ________ How many ________ (Max. 0.001 in. or 0.025 mm)

Journals out-of-round: Yes ________ No ________ How many ________ (Max. 0.0012 in. or 0.031 mm)

DATE________________________ INSTRUCTOR'S SIGNATURE ________________________

Remove camshaft ____

■ Original camshaft journal measurement

#1 ________ #2 ________ #3 ________ #4 ________ #5 ________ (Spec.)

Present camshaft journal measurement

#1 ________ #2 ________ #3 ________ #4 ________ #5 ________

#1 ________ #2 ________ #3 ________ #4 ________ #5 ________

#6 ________ #7 ________ #8 ________

#9 ________ #10 ________ #11 ________ #12 ________ #13 ________

#14 ________ #15 ________ #16 ________

Camshaft lobe wear max. diff. ________________________ (Max. diff. 0.010)

Largest lobe ________________ Smallest lobe ________________ Diff. ________

DATE________________________ INSTRUCTOR'S SIGNATURE ________________________

■ Original connecting rod bore measurement ________________________________ (Spec.)

Connecting rod bore measurements (rod without precision insert, cap torqued)

#1 A __________ #2 A __________

B __________ B __________

#3 A __________ #4 A __________

B __________ B __________

#5 A __________ #6 A __________

B __________ B __________

#7 A __________ #8 A __________

B __________ B __________

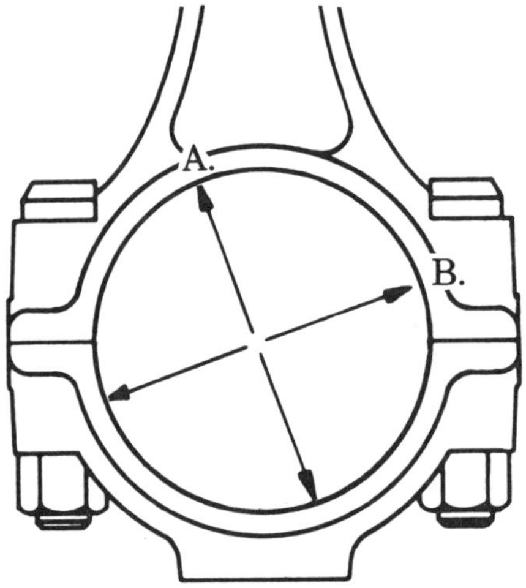

Rod bores out-of-round: Yes ______ No ______ How many ______ (Max. 0.0010 in. or 0.025 mm)

■ Original block main bearing bore measurements ________________________________ (Spec.)

Cylinder block main bearings bore measurements: (Main bearing bore without precision insert, cap torqued)

#1 A ________ #2 A ________ #3 A ________ #4 A ________

B ________ B ________ B ________ B ________

#5 A ________ #6 A ________ #7 A ________

B ________ B ________ B ________

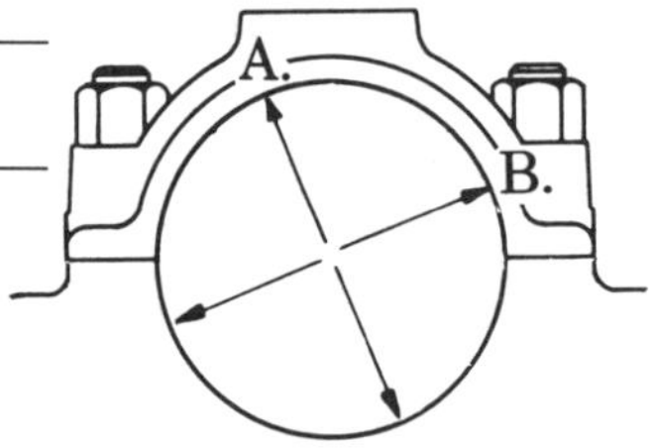

Block bores out-of-round: Yes ______ No ______ How many ______ (Max. 0.0010 in. or 0.025 mm)

DATE ____________________ INSTRUCTOR'S SIGNATURE ____________________

Remove cam bearings, label the number and position ____________________

Remove all oil gallery plugs ____________________

Remove all core (freeze) plugs (save for measurement) ____________________

Type of plug ____________________ Size ____________________

DATE ____________________ INSTRUCTOR'S SIGNATURE ____________________

Hot tank block at this time if needed (ferrous metals only).

Bore cylinders or deglaze cylinder walls, wash with soap and water 3 times, and oil to prevent rusting.

Clean oil galleries with rifle brush ____________________

All threads in the block chased ____________________

DATE ____________________ INSTRUCTOR'S SIGNATURE ____________________

New camshaft bearings installed: Yes __________ No __________ Camshaft fit checked ______

Camshaft bearing oil holes checked for alignment: Yes __________ No __________

New camshaft expansion plug installed: Yes __________ No __________

Install new core (freeze) plugs ____________________

Install valve lifters, if mushroom-type ____________________

How is camshaft retained in engine ____________________

DATE ____________________ INSTRUCTOR'S SIGNATURE ____________________

■ Main bearing torque ____________________ (Spec.)

■ New main bearing oil clearance ____________________ (Spec.)

Main bearing oil clearance this engine

#1 __________ #2 __________ #3 __________ #4 __________

#5 __________ #6 __________ #7 __________

■ Crankshaft end play—new ____________________ (Spec.)

Crankshaft end play—this engine ____________________

Crankshaft to oil seal installed __________ Sealant used __________

DATE ____________________ INSTRUCTOR'S SIGNATURE ____________________

Timing gears or chain sprockets worn: Yes __________ No __________ Replaced __________

How is camshaft timed to crankshaft? ____________________

DATE ____________________ INSTRUCTOR DOUBLE CHECK ____________________

Fuel pump eccentric (optional), crankshaft front oil slinger (optional), oil seal, front cover, crankshaft pulley, and/or damper assembly installed ____________________

Piston measurements (new or used). Circle those requiring replacement.

#1 __________ #2 __________ #3 __________ #4 __________

#5 __________ #6 __________ #7 __________ #8 __________

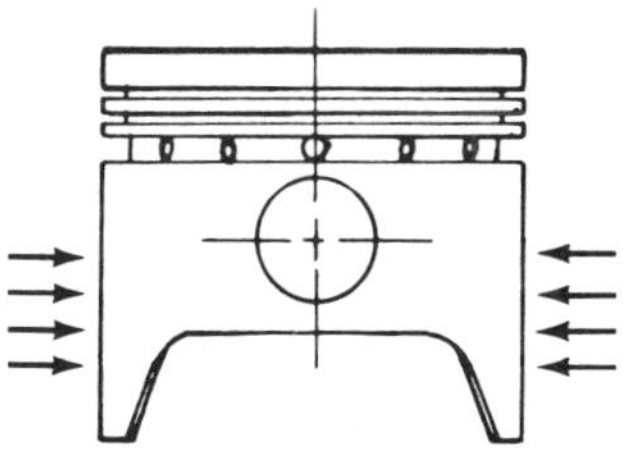

■ Piston to Cylinder Clearance ____________________ (Spec.) Pistons Knurled ____________________

■ Original piston pin diameter ____ (Spec.)

Piston pin fit

#1 ________________ #2 ________________ #3 ________________ #4 ________________

#5 ________________ #6 ________________ #7 ________________ #8 ________________

Piston pins secured in pistons ________________________ How ______________________________

■ Piston ring end gap: Comp. ________________________ Oil ________________________ (Spec.)

■ Piston rings installed on pistons (side clearance): New ________________________________ (Spec.)

Used ________________________ (Spec.)

Ring gap location on piston checked ____

DATE________________________________ INSTRUCTOR'S SIGNATURE ____________________

■ Use guide sleeves to install piston assemblies.

■ New connecting rod bearing clearance ____ (Spec.)

Pistons and connecting rod bearings installed, connecting rod bearing clearance

#1 ______ #2 ______ #3 ______ #4 ______ #5 ______ #6 ______ #7 ______ #8 ________

Connecting rod end play checked ____________ How much ____________ Max. ______________

■ Con. rod bolt or nut torque ____________. Double check by instructor ____________ Piston position in block checked ____

DATE________________________________ INSTRUCTOR'S SIGNATURE ____________________

Oil pump or distributor drive checked ______________ Worn: Yes ____________ No ____________

Oil pump or distributor drive installed ______________________________________ (Ford or Mopar)

Oil pump checked: Yes ________________ Repaired ________________ Replaced ________________

■ Side clearance ________________________ End Clearance ________________________ (Spec.)

Pressure relief valve checked ________________________ What type ________________________

DATE________________________________ INSTRUCTOR'S SIGNATURE ____________________

Oil pump with drive shaft and oil pan and filter installed ______________ plug tight

Engine valves, parts, and guides cleaned, valve guides worn: Yes ____________ No ______________

Use small hole gauge and one-inch micrometer to check.

Valve guides replaced: Yes ____________ No ____________

Valve guides knurled: Yes ____________ No ____________

Valves refaced: Yes ____________ No ____________ What angle ____________

Valve stem ends faced: Yes ____________ Valve seats refaced angle ____________

Interference angle used: Yes ____________ No ____________ How much ____________

Valve margin ____________ (1/32 in. or 0.794 mm) minimum

Valve springs checked: Yes ____________

Valve spring installed height checked: Yes ____________

DATE____________ INSTRUCTOR'S SIGNATURE ____________

■ Diagram sequence used in cylinder head and intake manifold tightening:

Install valve lifters ____________ Install push rods ____________

Cylinder heads and gaskets installed ____________

Valve clearance (lash) preset ____________

Push rod or lifter (valley) cover and intake manifold installed ____________

Fill with oil ____________ quarts. Prerun oil pump (engine pre-lube) ____________

DATE____________ INSTRUCTOR'S SIGNATURE ____________

Water pump installed ____________ Thermostat housing ____________

Flywheel and clutch installed ____________ Clutch disc alignment OK ____________

Torque converter properly set in transmission ____________

Transmission attached to engine ____________

Engine sling properly attached to engine ____________ Instructor double check ____________

Engine installed ______________________________

Distributor points and condenser checked ______________ Replaced ______________

Distributor vacuum and mechanical advance checked ______________________

Distributor timed to engine and distributor in correct position ______________

■ Point gap or dwell ______________ (Spec.) Static time ______________

Distributor wires: OK ____________ Replaced ____________ Type ____________

Oil filter replaced: Yes __________ Oil in engine __________ No of quarts __________

Fuel pump and lines installed __________ Engine starter checked: Yes ________ No ________

Repaired ______________ Replaced ______________

PCV valve serviced ______________________________

Starter, generator, fan, fan belts, power steering pump (optional), exhaust manifold, breather pipe installed ______________ Engine ready to start ______________

Secure fire extinguisher ______________

DATE______________ INSTRUCTOR'S SIGNATURE ______________

Start engine with instructor present ______________________

Oil pressure at idle ______________ at 1,000 rpm ______________

Oil to rocker arms: Yes ______________ ■ Idle speed set at ______________ rpm (Spec.)

■ Valve lash set: Hot __________ Cold __________ How much ______________ (Spec.)

Cylinder head bolt (or nut) torque rechecked: Valve covers replaced (overhead valve engine) ________

Engine oil or water leaks checked ______________________

Set timing with timing light ______________ Timing location ______________

■ Number of degrees __________ BTDC or ATDC in Drive __________ @ ______ rpm (Spec.)

■ Each mark equals ______________________________ degrees (Spec.)

Final carburetor adjustment ______________________________

Remarks:

DATE ______________ INSTRUCTOR'S SIGNATURE ______________

WORK AUTHORIZATION

Regardless of the type of repair shop (independent, dealership, or school shop) a legal document, usually referred to as a work order, must be properly filled out. This form states the type of work to be done, the costs or arrangements made for parts and labor, and serves as a liability release when signed by the vehicle owner.

A work order that is properly completed is shown in Figure 3-21. It supplies all the required information for the owner, the school, the instructor, and the student. A description of the vehicle and the owner's identification are also included. The middle area, left side, gives the job description, while the lower left gives the charges made by the school for materials. Depending on the arrangement for replacement parts or machine work done outside the school shop, the upper right side may or may not be completed. The middle right side area should be signed by the student mechanic and the instructor after approval of the work to be completed is granted by the owner by signing the bottom line.

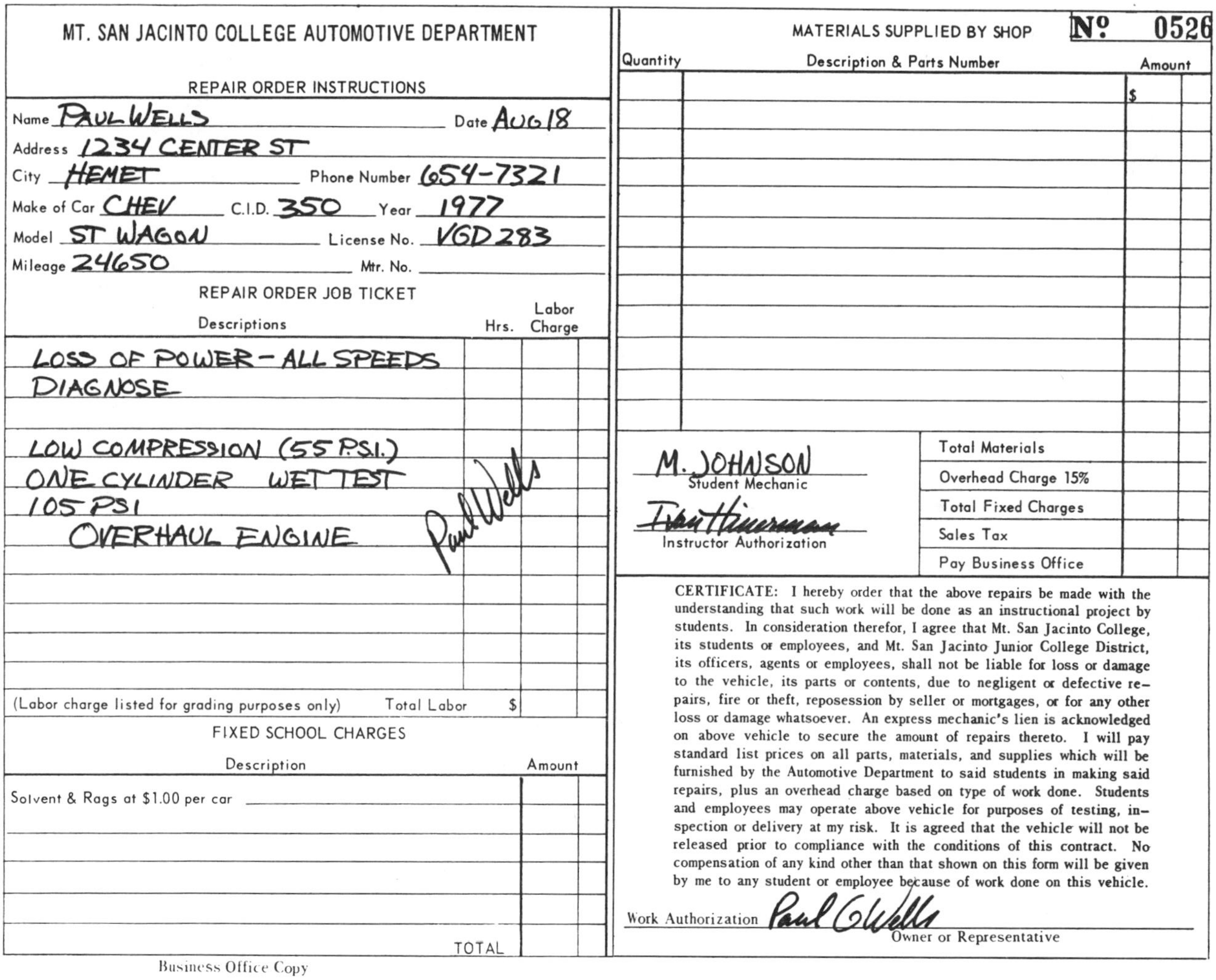

MT. SAN JACINTO COLLEGE AUTOMOTIVE DEPARTMENT

REPAIR ORDER INSTRUCTIONS

Name PAUL WELLS Date AUG 18
Address 1234 CENTER ST
City HEMET Phone Number 654-7321
Make of Car CHEV C.I.D. 350 Year 1977
Model ST WAGON License No. VGD 283
Mileage 24650 Mtr. No.

REPAIR ORDER JOB TICKET

Descriptions	Hrs.	Labor Charge
LOSS OF POWER – ALL SPEEDS		
DIAGNOSE		
LOW COMPRESSION (55 P.S.I.)		
ONE CYLINDER WET TEST		
105 PSI		
OVERHAUL ENGINE		
Paul Wells		
(Labor charge listed for grading purposes only) Total Labor		$

FIXED SCHOOL CHARGES

Description	Amount
Solvent & Rags at $1.00 per car	
TOTAL	

Business Office Copy

MATERIALS SUPPLIED BY SHOP № 0526

Quantity	Description & Parts Number	Amount
		$

M. JOHNSON
Student Mechanic

[signature]
Instructor Authorization

Total Materials	
Overhead Charge 15%	
Total Fixed Charges	
Sales Tax	
Pay Business Office	

CERTIFICATE: I hereby order that the above repairs be made with the understanding that such work will be done as an instructional project by students. In consideration therefor, I agree that Mt. San Jacinto College, its students or employees, and Mt. San Jacinto Junior College District, its officers, agents or employees, shall not be liable for loss or damage to the vehicle, its parts or contents, due to negligent or defective repairs, fire or theft, reposession by seller or mortgages, or for any other loss or damage whatsoever. An express mechanic's lien is acknowledged on above vehicle to secure the amount of repairs thereto. I will pay standard list prices on all parts, materials, and supplies which will be furnished by the Automotive Department to said students in making said repairs, plus an overhead charge based on type of work done. Students and employees may operate above vehicle for purposes of testing, inspection or delivery at my risk. It is agreed that the vehicle will not be released prior to compliance with the conditions of this contract. No compensation of any kind other than that shown on this form will be given by me to any student or employee because of work done on this vehicle.

Work Authorization Paul G Wells
Owner or Representative

FIG 3-21 Work authorization form.

Name ______________________ Date ______________

DIRECTIONS

Select the best answer to each question and circle the letter at the left.

1. Among the common identification markings used in vehicle manufacturing is
   I. stamping.
   II. forging.
   A. I only
   B. II only
   C. Both I and II
   D. Neither I nor II

2. Vehicle Identification Numbers can be found in/on
   I. Registration-Title card.
   II. Instrument panel—driver's side near the windshield.
   A. I only
   B. II only
   C. Both I and II
   D. Neither I nor II

3. Engine parts are always identified as to right and left when
   I. viewed from the front of the vehicle or engine.
   II. viewed from the driver's seat.
   A. I only
   B. II only
   C. Both I and II
   D. Neither I nor II

4. Crankshaft rotation of all front to rear mounted engines made today is
   I. clockwise when viewed from the front.
   II. counterclockwise when viewed from the driver's seat.
   A. I only
   B. II only
   C. Both I and II
   D. Neither I nor II

5. Casting or forging numbers are
   I. required by the Society of Automotive Engineers.
   II. raised above the surrounding surface.
   A. I only
   B. II only
   C. Both I and II
   D. Neither I nor II

6. The safety sticker required by law can be found
   I. on the firewall.
   II. under the hood on the left side.
      A. I only
      B. II only
      C. Both I and II
      D. Neither I nor II
7. Lens markings
   I. indicate the year the lens was made.
   II. are not as reliable as other sources of identification information.
      A. I only
      B. II only
      C. Both I and II
      D. Neither I nor II
8. General repair manuals
   I. always cover several years of automobiles.
   II. are less expensive than purchasing all the vehicle manufacturers' manuals included in one general repair volume.
      A. I only
      B. II only
      C. Both I and II
      D. Neither I nor II
9. The teardown and assembly sheet contains
   I. a parts list for the engine being repaired.
   II. instructions for the use of measurement tools.
      A. I only
      B. II only
      C. Both I and II
      D. Neither I nor II
10. A fully completed work authorization form should provide
   I. a liability release from the vehicle owner.
   II. an estimate as to the cost of repairs.
      A. I only
      B. II only
      C. Both I and II
      D. Neither I nor II

DATE COMPLETED ______________________ INSTRUCTOR'S SIGNATURE ______________________

OBJECTIVE 3-B (written test)

After completing the preparation assignment, you will correctly answer a minimum of 7 out of 10 multiple choice questions in a written test, without using reference material, on the following topics:

1. Diagnosis procedures
2. Engine testing
3. Engine repair information

Preparation:

Study pages 77 through 97.

DIAGNOSIS PROCEDURES

Under normal operating conditions and with proper care, an automotive engine will provide many thousands of miles of reliable service. However, an engine is a mechanical device with many moving parts that are subject to wear. Over a period of time, any engine may lose its ability to perform well; it may even fail completely.

One of the most important tasks an auto mechanic must perform is to diagnose engine problems. Diagnosis, in this text, means determining the nature of a problem through careful investigation. This includes asking questions and using one's sight (visual inspection) and hearing (listening for abnormal noises), along with the proper use of test equipment to check specifications. The very best diagnosis of a problem is made when the greatest amount of available information is obtained. The word *troubleshooting* is often substituted for *diagnosis* of problems.

Most automobile problems can be successfully diagnosed by following a step-by-step procedure. This statement applies especially to engine malfunctions, but it is also true of problems with other parts of the automobile. The general steps listed below are adequate for all situations:

1. Gather all available information.
2. Verify the complaint by visual inspection and road testing.
3. Test the engine with diagnostic equipment.

Diagnosis procedures will vary somewhat depending on the running condition of the vehicle.

GATHERING INFORMATION

To gather all the necessary service information, it is usually best to talk to the owner or driver of the vehicle who made the complaint about its performance. Just as a doctor will quiz a patient about symptoms, so should the automotive mechanic quiz the owner or driver about operating problems. Ask questions like, "When did the car first make unusual noise?" "How many miles per gallon do you get?" Or "How often do you need to add oil?" The more questions asked and answered, the better basis you will have for making an accurate diagnosis.

VERIFYING COMPLAINTS

To verify the complaint means to be certain of the accuracy of the problem description. This part of the diagnosis usually begins with a visual inspection. Whether or not the engine will run, a careful visual inspection should always be made. Record all observed operating irregularities so that the final diagnosis judgment will include all available information and be as accurate as possible.

Oil consumption and abnormal noise problems which would be detected in this phase of diagnosis are fully described later in this unit. These two types of troubles are the ones most frequently encountered by the auto mechanic.

Road testing is another important step in verifying the complaint. Whenever it is possible, the owner or driver should be taken along on a short test drive so that the exact problem can be discussed as it is happening. If the road testing is done alone, take the vehicle through as many different operation situations as possible until the complaint is fully recognized.

ENGINE TESTING

After the presenting complaint has been verified and all the necessary information has been compiled, engine testing can begin.

Whether or not the crankshaft will turn is important. If it does turn but the engine will not run, then electrical or fuel system service may solve the problem. If, however, the crankshaft will not turn at all, it can be assumed that there is severe internal damage and that no further diagnosis is possible without partial disassembly of the engine.

If the engine runs, a number of diagnosis procedures can be used:

1. Compression testing
2. Cylinder leakage testing
3. Cylinder power balance

COMPRESSION TEST

The compression tester is the most widely used diagnostic tool for determining internal engine condition. The amount of air pressure an engine can produce is directly related to its operating efficiency and power output.

Note: Some manufacturers caution against using this test on engines with catalytic converters (many 1975 and later models). Check a repair manual for the car being tested.

Compression Specifications—Engine manufacturers develop compression specifications for each engine. These figures are based on the product of the compression ratio times the atmospheric pressure at sea level, roughly 15 psi. An engine with a 9:1 compression ratio should have a compression pressure of approximately 135 psi ($9 \times 15 = 135$). Combustion chamber designs, cranking speed of the starter, and several other factors modify the actual compression capability of an engine. For these reasons, compression pressures will vary upward and downward from the mathematical norm calculation.

A typical compression specification may be listed as 150 pounds per square inch (psi) plus or minus ($\pm$) 10 percent (%). The plus or minus tolerance is the acceptable range of compression pressure. This means the engine pressure should be somewhere between 135 and 165 psi.

Another specification may state that the compression should be 140 psi with a maximum variation of 20 psi between cylinders. This means the engine should have 140 psi, but, if all cylinders are somewhere between 110 and 130 psi, its performance is considered to be acceptable, though not optimal.

In the absence of manufacturers' specifications, Table 3-1 can be used. This chart is calculated to show a lowest cylinder reading that is 75 percent of the highest reading, a method of interpreting compression readings that is accepted by most manufacturers.

Table 3-1 COMPRESSION TEST CHART FOR COMPARING HIGHEST AND LOWEST PRESSURES

Maximum PSI	Minimum PSI	Maximum PSI	Minimum PSI
134	101	174	131
136	102	176	132
138	104	178	133
140	105	180	135
142	107	182	136
144	108	184	138
146	110	186	140
148	111	188	141
150	113	190	142
152	114	192	144
154	115	194	145
156	117	196	147
158	118	198	148
160	120	200	150
162	121	202	151
164	123	204	153
166	124	206	154
168	126	208	156
170	127	210	157
172	129	212	158

* Highest and lowest recommended compression pressures.

Results and Indications—If the compression builds up to 70–75 percent of the total specification on the first stroke and meets the specification pressure by the fourth stroke, then the cylinder is in good condition.

When there is low compression between adjacent (next to one another) cylinders, the most likely cause is a blown gasket.

Worn or sticking piston rings and worn cylinder walls will be indicated by low compression on the first stroke which builds up gradually on the following strokes.

Valve and/or valve seat trouble is indicated by a low compression reading on the first stroke and failure to build on succeeding strokes.

Carbon buildup, which takes place within the combustion chamber during engine operation, can result in higher-than-normal compression pressures. Carbon buildup reduces the size (volume) of the combustion chamber, which raises the compression pressure. This, in turn, can affect other compression loss defects in the cylinder.

OBJECTIVE 3-B1 (manipulative)

Working as a team with one other student and with the aid of Job Report Sheet 3-B1, you will test the compression of an engine, using a compression tester on an instructor-assigned vehicle engine.

Time Recommended:

Twice the flat rate time allowed in *Motors, Chilton,* or *National Service Data* manuals; approximately 3 minutes per cylinder.

Preparation:

1. Review pages 77 through 78.
2. Study pages 79 through 81.

3. Study manufacturer's or general repair manuals for the engine being serviced.
4. Attend lectures on engine diagnosis.

Tools and Materials Needed:

1. Compression tester and adapter
2. Assorted hand tools
3. Fender covers
4. Compression specifications

Procedure:

Pretest Procedure:

1. Be certain the battery in the vehicle can crank the engine at normal speeds. If additional cranking power is necessary, a jumper battery can be used.
2. Run the engine until it reaches normal operating temperatures; then shut it off. Some aluminum head engines may require spark plug removal with the engine cold; consult a repair manual.
3. Carefully remove the spark plug wires by twisting the rubber insulation boot until it turns freely. Then, pull upward on the boot.
4. Use compressed air to remove any foreign matter around the base of the spark plug. Use suitable eye protection.
5. Remove all spark plugs, keeping them in order as to their respective cylinders.
6. Compare the electrode area of each plug with the spark plug chart in the shop. Consult with your instructor if any abnormal indications are observed.
7. Remove the air cleaner from the carburetor. Block the choke and throttle valves wide open.
8. Turn ignition switch ON.
9. Pull secondary lead from the center of the distributor cap and ground it.
10. Connect a remote starter switch (Fig. 3-22). If the switch being used is not part of the compression tester, the connections shown are still the same.

Compression Testing:

1. Select the adapter with the same size and length of threaded area as the spark plug just removed (Fig. 3-23).
2. Screw adapter into spark plug hole and connect the other end of the hose to the quick connector of the compression tester.
3. Move the control switch to the crank position and crank continuously until four to five full compression strokes are completed (Fig. 3-24).
4. Note the gauge readings at the end of the first and fourth strokes. Record your results on Job Report Sheet 3-B1.
5. Release the pressure from the gauge by moving the tester control switch to the vent position. This releases the pressure from the gauge and the hose.
6. Disconnect the adapter from the gauge and remove the adapter from the spark plug hole.
7. Repeat steps 1 through 6 on all remaining cylinders. Crank each cylinder the same number of revolutions (four or five) and record the readings.
8. Complete Job Report Sheet 3-B1. Review page 85, Results and Indications. Return completed Report Sheet to your instructor.

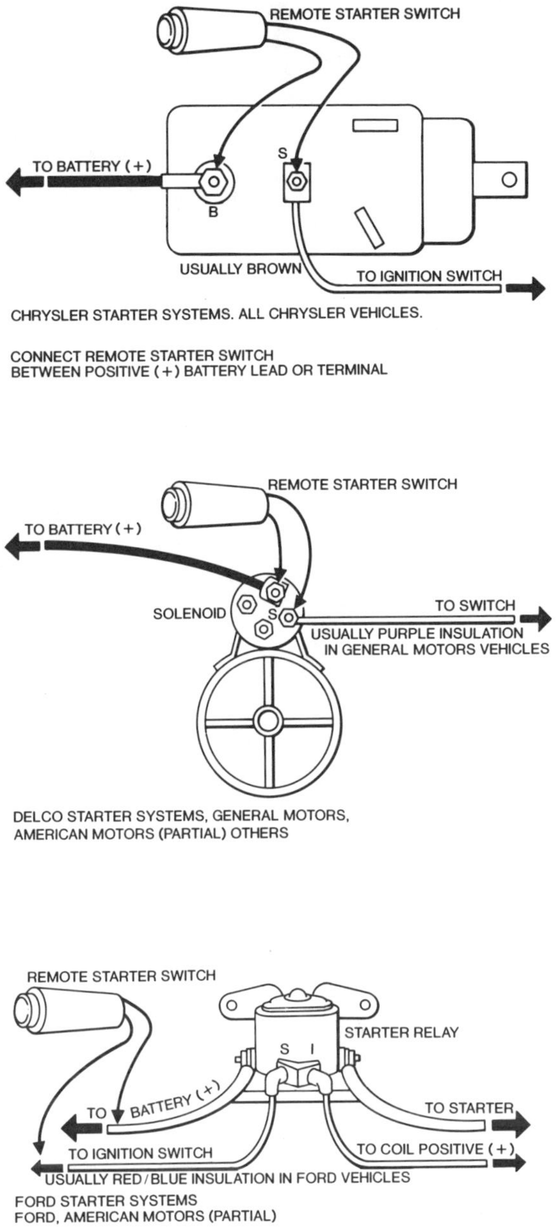

FIG 3-22 Common remote control starter connections.

FIG 3-23 Comparing compression tester adapter size with a spark plug removed from the engine.

FIG 3-24 Read the maximum pressure after 4–5 compression strokes.

COMPRESSION TESTING

Student name ______________________________ Date ______________

Make of Car ____________ Model ________________ Year ______________

Time Started ______ Time finished ________________ Total time ______________

Flat rate time ____

Specifications:

Manufacturer's compression pressure ______________________________

Allowable variation ______________________________________

Label cylinder numbering pattern:

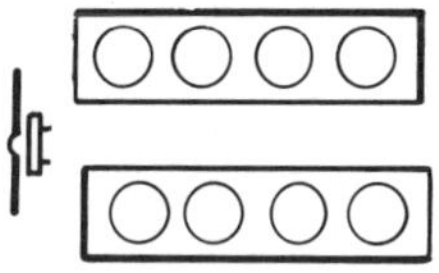

Results (in psi or kPa):

Cylinder No.

1 ________________ (1st) ________________ (4th)

2 ________________ (1st) ________________ (4th)

3 ________________ (1st) ________________ (4th)

4 ________________ (1st) ________________ (4th)

5 ________________ (1st) ________________ (4th)

6 ________________ (1st) ________________ (4th)

7 ________________ (1st) ________________ (4th)

8 ________________ (1st) ________________ (4th)

Checkup Questions:

1. What precautions should be observed when using a remote starter switch?

____

____

____

2. Why is it important to watch and record the first needle jump as well as the fourth?

____

____

3. What is the probable cause of heavy compression loss in two adjacent (next to one another) cylinders?

____

Diagnosis:

Write a brief statement (your opinion) of the condition of the engine.

____

____

DATE COMPLETED ______________ INSTRUCTOR'S SIGNATURE ______________

CYLINDER LEAKAGE TESTING

Every engine sustains a certain amount of wear as it is operated. Some causes of improper performance are directly related to combustion chamber leakage, which is not always detected in a compression test. Even if leakage is detected, sometimes it is difficult to trace the exact leakage source.

Purpose—The cylinder leakage tester provides an accurate means of testing the internal condition of an engine. Cylinder leakage testing will locate intake and exhaust valve leaks, leakage into the cooling system, or any other cause of compression loss. It is the only internal engine condition test recommended by many automotive manufacturers for vehicles with catalytic converters installed.

Results and Indications—Gauge readings should be comparatively even and show less than twenty percent leakage. Engines with high mileage may have increased percentage readings. As long as the readings are consistent and less than thirty percent, major engine service should not be necessary.

If escaping air can be heard through the carburetor, the intake valve is leaking. Air escaping through the exhaust pipe means the exhaust valve is leaking. Bubbles in the radiator coolant indicates a leaking head gasket or a crack in the cylinder head or block. A high percentage of leakage in two adjacent cylinders usually indicates a leaking head gasket. Air escaping through the oil filler opening of the engine usually means worn rings or cylinder walls, stuck or broken rings, or a cracked or burned piston.

OBJECTIVE 3-B2 (manipulative)

Working as a team with one other student and with the aid of Job Report Sheet 3-B2, you will determine the percentage of cylinder leakage, using a cylinder leakage tester on an instructor-assigned vehicle engine.

Time Recommended:

Twice the flat rate time allowed in *Motors, Chilton,* or *National Service Data* manuals; approximately 8 minutes per cylinder.

Preparation:

1. Review pages 79 through 81.
2. Study pages 85 through 88.
3. Study manufacturer's or general repair manuals for the engine being serviced.
4. Attend lectures on engine diagnosis.

Tools and Materials Needed:

1. Cylinder leakage tester and adapter
2. Indicator light or test (continuity) light
3. Assorted hand tools
4. Fender covers
5. Firing order specifications

Procedure:

Pretest Procedure:

1. Run the engine until it reaches normal operating temperature; then shut it off. Some aluminum head engines may require spark plug removal with the engine cold; consult a repair manual.
2. Carefully remove the spark plug wires by first twisting the rubber insulation boot until it turns freely. Then, pull upward on the boot.
3. Use compressed air to remove any foreign matter around the base of the spark plug. Use suitable eye protection (Fig. 3-25).
4. Remove all spark plugs, keeping them in order as to their respective cylinders.
5. Compare the electrode area of each plug with the chart in the shop. Consult with your instructor if any abnormal indications are observed.
6. Remove the air cleaner from the carburetor. Move the fast idle cam to its highest step, which will place the throttle in a slightly open position.
7. Remove the cap from the crankcase oil filler opening.
8. Connect a remote starter switch as shown previously in Figure 3-22.
9. Turn the radiator cap to its first stop to relieve any pressure buildup. Be ready to catch any coolant that overflows. Slowly remove the radiator cap.

Tester Calibration:

1. Turn the CLT (Cylinder Leakage Tester) regulator counterclockwise until it turns freely.
2. Connect the tester hose to the cylinder connection fitting (Fig. 3-26).

FIG 3-25 Blowing foreign matter from spark plug area.

FIG 3-26 Connecting tester hose to proper outlet.

3. Connect the shop air supply (70 psi minimum, 200 psi maximum) (482.63–1378.95 kPa) to the fitting labeled INPUT AIR CONNECTION.
4. Turn the CLT REGULATOR knob clockwise until the leakage gauge reads zero (Fig. 3-27). Momentarily connect, then disconnect the test adapter. The pointer of the gauge should return to ZERO. If not, readjust the CLT REGULATOR knob until ZERO is reached.

Cylinder Leakage Testing:

1. Select the adapter with the same size and length of threaded area as the spark plug just removed from #1 cylinder. (The adapter for cylinder leakage testing has NO check valve in the threaded end.)
2. Install the adapter in the spark plug hole. Place your thumb over the quick disconnect end and slowly "jog" (rotate) the engine until the pressure begins to build up.
3. Continue to jog the engine until the timing marks are aligned. This places the piston of #1 cylinder in the TDC position on the compression stroke (Fig. 3-28).
4. Remove the distributor cap. The brass rotor tip should be pointing toward the spark plug wire leading to #1 cylinder.
5. Remove the secondary lead from the center of the distributor cap and ground it.
6. Attach the indicator light or test (continuity) light to breaker ignition systems. One lead should be connected to the primary terminal of the distributor or negative (−) terminal of the coil; the other lead connects to a good ground (Fig. 3-29). An indicator light cannot be used on breakerless ignition systems.

FIG 3-27 Adjusting CLT regulator.

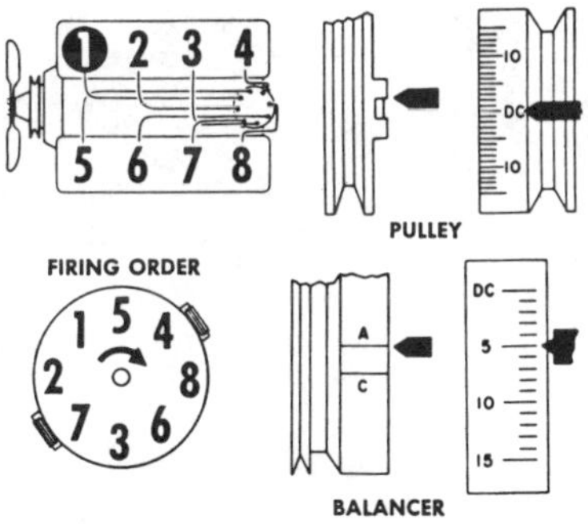

FIG 3-28 Typical timing marks and their locations.
(Delco-Remy Division, GMC)

FIG 3-29 Test light connected to ignition primary circuit.

7. Turn ON the ignition switch. The light attached in the last step should be ON if the timing marks are properly aligned as in Step 3.
8. Connect the tester hose from the machine to adapter hose. Note the percentage of leakage and record the indication on Job Report Sheet 3-B2.
9. While the air is pressuring the cylinder, listen for escaping air through the carburetor, the tailpipe, and the crankcase oil filter opening. Also, check for bubbles in the radiator coolant.
10. Before testing the leakage of the next cylinder in the firing order, mark the crankshaft pulley with chalk (Fig. 3-30).
11. Disconnect the tester hose from the adapter and "jog" the engine until the next chalk mark is on its way to the stationary timing mark. The indicator or test lamp should have gone out as the engine was jogged toward the next mark.
12. Jog the engine further toward the next mark until the light comes ON. The tester is then ready for the next cylinder in the firing order.
13. Remove the adapter from the cylinder just tested and install it in the next cylinder in the engine's firing order.
14. Repeat steps 8, 9, 10, 11, and 12 until all cylinders have been tested.

FIG 3-30 Pulley markings for an eight-cylinder engine leakage test.

JOB REPORT SHEET 3-B2

CYLINDER LEAKAGE

Student name ______________________ Date ______________

Make of car ______________ Model ______________ Year ______________

Time started ______ Time finished ______________ Total time ______________

Flat rate time ______________________________

Specifications:

Equipment manufacturer's leakage limit—20%

Maximum allowable leakage—30%

Label cylinder numbering pattern:

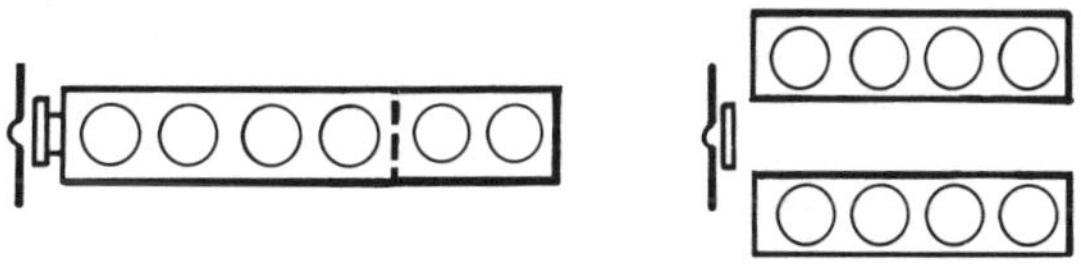

Results:

Cylinder #	Air leakage noted for each cylinder: in exhaust, carburetor, crankcase, radiator?
1. ______________%	______________
2. ______________%	______________
3. ______________%	______________
4. ______________%	______________
5. ______________%	______________
6. ______________%	______________
7. ______________%	______________
8. ______________%	______________

Checkup Questions:

1. How was TDC determined for cylinders other than #1?

____

____

2. Was the engine capable of running? (Yes/No) ____________________

3. Was the engine at normal operating temperature during the test? (Yes/No) ____________

4. What influence would a cold engine have on percentage of leakage?

____

____

____

Diagnosis:

Write a brief statement (your opinion) of the condition of the engine.

____

____

____

____

DATE COMPLETED ____________________ INSTRUCTOR'S SIGNATURE ____________________

OBJECTIVE 3-B3 (manipulative)

Working as a team with one other student and with the aid of Job Report Sheet 3-B3, you will complete a power balance test on an instructor-assigned vehicle engine.

Time Recommended:

Twice the flat rate time allowed in *Motors, Chilton,* or *National Service Data* manuals; approximately 15 minutes.

Preparation:

1. Review pages 85 through 88.
2. Study pages 91 through 92.
3. Study manufacturer's or general repair manuals for the engine being serviced.
4. Attend lectures on engine diagnosis.

Tools and Materials Needed:

1. Tachometer
2. Electronic shorting device or shorting probe
3. Assorted hand tools
4. Fender covers
5. Oscilloscope, if available

Procedure:

Pretest Procedure:

1. Start the engine and bring it to normal operating temperature.
2. Calibrate tachometer, if necessary, then connect to the engine.
3. Connect electronic shorting device.

Power Balance Testing (all cylinders except one):

1. Adjust idle speed to 1000 rpm or power set line (Fig. 3-31).
2. Beginning with cylinder #1, depress the button to short that cylinder. For testing without the electronic device, slip shorting probe under spark plug boot and connect a ground lead (Fig. 3-32).

FIG 3-31 Setting the idle speed at the power set line (1,000 rpm).

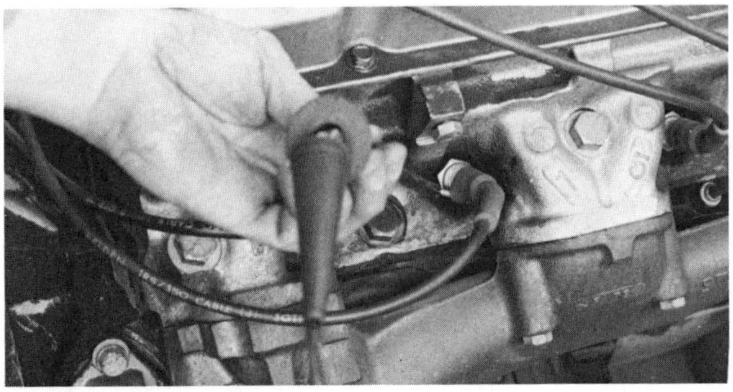

FIG 3-32 Using a shorting probe under the spark plug boot.

3. Observe and record the rpm drop from 1,000 rpm on Job Report Sheet 3-B3.
4. Momentarily press the number 1 button again to release it, or remove grounding probe from spark plug. This will place that spark plug in operation again.
5. Repeat steps 2, 3, and 4 until all cylinders have been shorted.
6. Complete Job Report Sheet 3-B3. Review pages 93 and 94, Results and Indications, before returning completed report sheet to your instructor.

CYLINDER POWER BALANCE

Student Name ______________________________ Date ______________

Make of Car ____________ Model ______________ Year ______________

Time Started ____________ Time Finished ____________ Total Time ____________

Flat Rate Time ____

Specifications:

Recommended fast idle rpm for test ______________________________ (800–1000)

Recommended rpm drop of one cylinder from normal operating engine.

4-cylinder engines	90–100
6-cylinder engines	60–70
8-cylinder engines	40–50

Label cylinder numbering pattern:

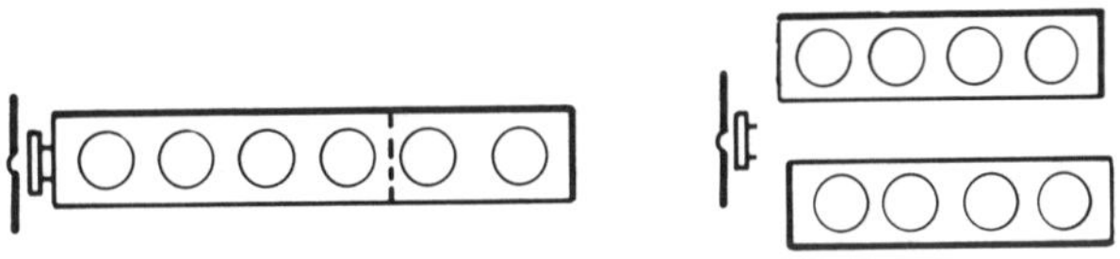

Results:

Rpm Drop in Cylinder #

1. ____________ 2. ____________ 3. ____________ 4. ____________

5. ____________ 6. ____________ 7. ____________ 8. ____________

Checkup Questions:

1. Would a misfiring spark plug give the same indication as a cylinder with a defective valve? ________

   Why or why not? ____

   ____

   ____

2. Was an oscilloscope used during this test? ________________________________

3. Would a defective intake valve cause more rpm drop than an exhaust valve in the same condition? (Yes/No)

4. List other problems noted. ________________________________

   ____

Diagnosis:

Write a brief statement (your opinion) of the condition of the engine.

____

____

____

____

DATE COMPLETED ________________ INSTRUCTOR'S SIGNATURE ________________

OIL CONSUMPTION

It is important that the engine technician make a thorough examination and diagnosis of suspected oil loss before the engine work begins. Once the engine is disassembled, it becomes very difficult to determine causes of high oil consumption.

Excessive oil consumption can occur in one or both of the following ways. When oil leaks or drips from the engine, it is considered *external loss*. When oil is burned within the combustion chamber with the fuel, it is considered *internal loss*.

External Loss—External oil loss (leakage) is usually the result of *one or both* of the following engine malfunctions: (1) Gasket or seal leakage; (2) Excessive crankcase pressure.

As the engine runs, oil is splashed around in its lower interior area. This splashing helps create an oil vapor or mist. The gaskets on many parts of the engine are designed expressly to keep the oil within the engine. When these gaskets are broken or improperly installed, oil loss in liquid or vapor form will occur (Fig. 3-33).

Oil seals around the crankshaft are often sources of leakage. These seals apply pressure against the crankshaft as it turns. Over a period of high mileage and engine heat, these seals lose their ability to keep the oil within the engine.

Every internal combustion engine allows part of its combustion gas to pass by the piston rings and enter the crankcase. This gas, referred to as blow-by, occurs on each compression and power stroke as the engine runs. This condition is considered normal as long as the amount of gas remains at a low level.

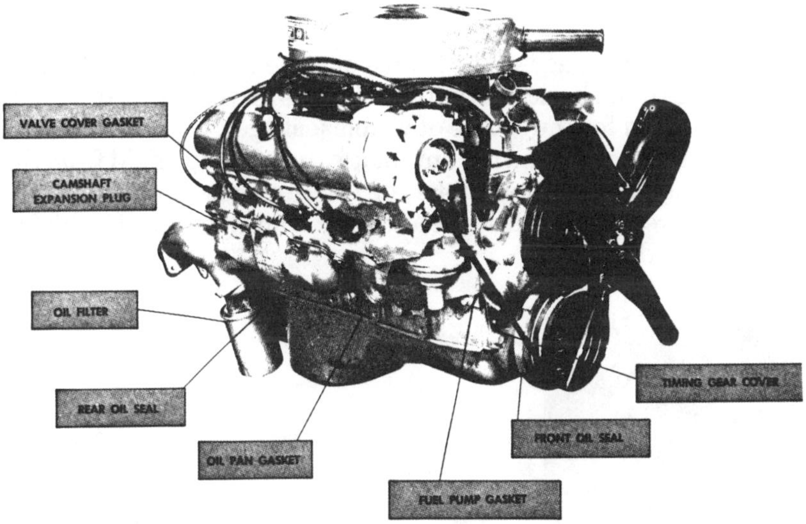

FIG 3-33 Locations of external oil loss.

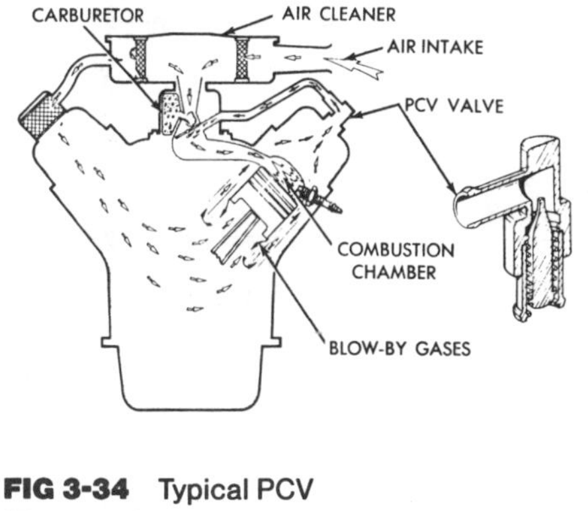

FIG 3-34 Typical PCV (Position Crankcase Ventilation) system. (Chevrolet Motor Division, GMC)

All engines have some means of ventilating the crankcase to keep the crankcase pressure at or below atmospheric pressure. By using intake manifold vacuum, blow-by gases can be recycled into the combustion chamber, as shown in Figure 3-34. Since the mid sixties, all domestic engines have had a PCV (Positive Crankcase Ventilation) system for regulating crankcase pressure.

An engine with slight to average wear usually has little blow-by. An engine with high mileage or continued improper operation will normally have excessive blow-by, which creates excessive crankcase pressure. Excessive crankcase pressure will force oil out of the crankcase where the gaskets are broken or weak.

Making an accurate oil consumption diagnosis requires getting as much information about the engine as possible. Ask the owner/driver about the oil loss conditions and amount of oil used.

External Loss Diagnosis—The first step in excessive oil consumption diagnosis would be visual inspection. This involves careful examination of the entire engine and under areas of the vehicle. Slight amounts of oil loss will usually be covered with dust or dirt (Fig. 3-35). Heavy oil loss shows a washed effect (Fig. 3-36), the result of the oil flow washing away the dirt, giving the area a clean, but wet, appearance.

The next step, if necessary, is to wipe off the suspected area and drive the vehicle through different conditions of speed and load. Then, visually inspect again for the source of oil loss.

Another method of oil leak detection is to park the vehicle over a large piece of paper or other light-colored material immediately after driving the vehicle (Fig. 3-37). The oil will usually drip from the leakage area. Check to be certain oil was not blown toward the back by a wind draft during vehicle movement. After a period of waiting, again inspect the vehicle underside to find the actual opening in the engine where the leakage is occurring.

Hints for Detecting External Leakage—The source of some oil leakage is difficult to detect because air blows the oil backward as the vehicle is driven forward. Therefore, look forward and upward for the source of visible leakage.

Be certain the leakage is engine oil and not automatic transmission fluid from the power steering system, the transmission cooler, or the transmission itself. Automatic transmission fluids are usually red, while standard transmission oil is thicker and has a definite sulphur odor.

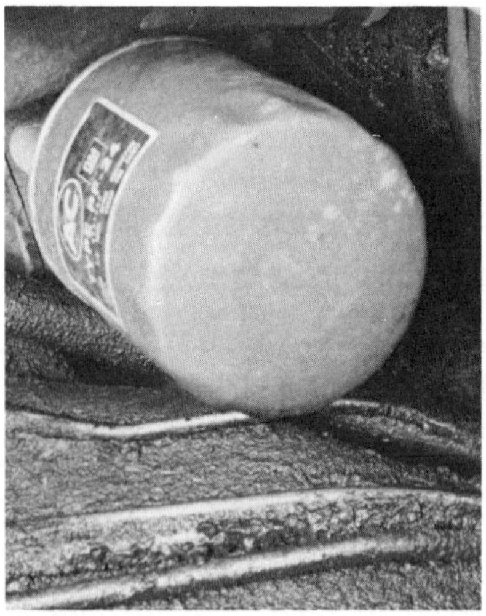

FIG 3-35 Minor oil leakage (dust- or dirt-covered).

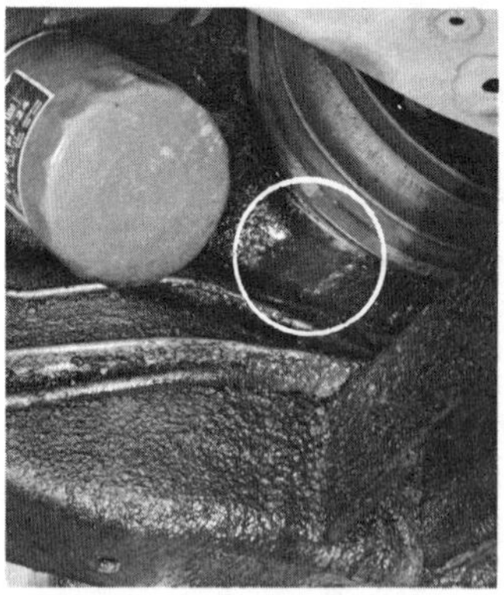

FIG 3-36 Major oil leakage showing a washed effect.

FIG 3-37 Paper beneath the vehicle will assist in locating oil leaks.

Occasionally the owner of a leaking engine will want to have it steam cleaned in order to find the source of leakage. Sometimes this is helpful but more often it cleans away the dirt, allowing the leaking oil to spread much further from the source. This can make detection more difficult.

What amount of oil consumption or loss is considered excessive or abnormal? Automotive service people have different answers to this question, but, generally speaking, no leakage (external loss) is necessary or proper. An engine that burns (internal loss) more than one quart per thousand miles of average driving has excessive loss. Keep in mind the mileage on the engine, its past history of service, and the overall value of the vehicle when making decisions on repair work for oil loss.

Internal Loss—Most people associate high oil consumption with worn piston rings. This is one possibility, but a very careful check should be made for all possibilities. In most cases, internal oil consumption is more difficult to trace than external leakage.

1. When checking for internal oil loss (burning), start by operating the engine.
2. White exhaust smoke is normally water vapor or steam, and true black smoke results from an overly rich fuel mixture.
3. If the exhaust gases are blue to blue-gray in color, oil is being burned in the combustion chamber.
4. If the blue smoke disappears quickly, the oil loss is probably going through the valve guides into the combustion chamber (Fig. 3-38).
5. If roughly the same amount of blue smoke continues to show while the engine is at a fast idle, the problem may be caused by several malfunctions:
   a. Oil loss past the intake manifold gasket
   b. Faulty PCV valve
   c. Plugged oil return holes in the head
   d. Worn or stuck piston rings

The intake manifold on many V-8 engines also serves as a lifter chamber cover or valley cover. If this is the case, a good seal must be maintained by the gasket separating the manifold and cylinder head. A gasket leak would allow oil to be drawn into the intake manifold runners leading to the combustion chambers. An engine with this problem usually has a rough idle.

To test for gasket leak, connect a vacuum gauge to the engine and test for a steady reading of 17–21 inches of mercury (Hg). If the reading is low but steady at approximately 13–16 inches, intake manifold leakage is a possibility. Before condemning the intake manifold gasket, check for late (retarded) timing and low compression or a faulty EGR valve. These problems will give the same low vacuum gauge readings as intake manifold leakage.

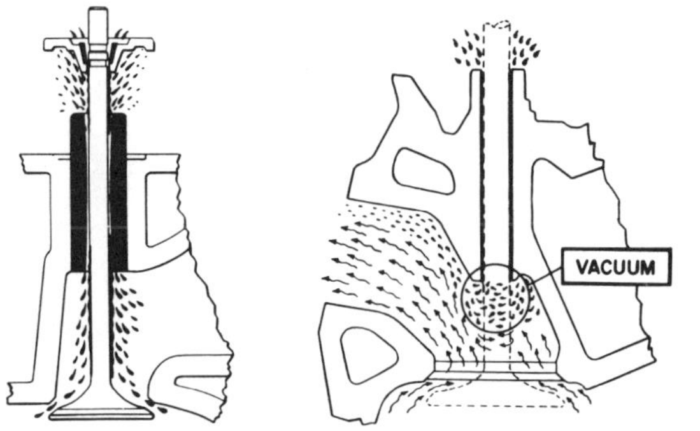

FIG 3-38 Oil loss through worn valve guides.

A faulty PCV valve can cause excessive oil consumption if the pin remains in the maximum flow position. To test for this problem, pinch off the hose between the intake manifold and the crankcase while the engine idles. The rpm will drop as the hose is pinched. After five to ten minutes, the amount of blue smoke should be reduced if the valve is stuck.

Plugged oil return holes in the cylinder head are most common to V-type engines. As the engine runs, the oil level is raised on the valve guides to a point where engine vacuum sucks the oil past the valve stem. The oil then enters the combustion chamber where it is burned and passed out the tailpipe as blue smoke. This condition is easily remedied by removing the valve covers and pushing a rod into the return hole to unplug the opening.

Stuck or worn piston rings are very common causes of high oil consumption. To test for this problem, make a compression test as outlined in this unit. As the test is being conducted on each cylinder, watch the readings of the first and fourth or fifth jump of the needle. If the rings are in good condition, the first jump will be 70–75 percent of the fourth. A reading of less than 50 percent of the fourth jump indicates piston ring problems.

A conclusive test for very poor piston rings can be made by driving the vehicle up a grade of at least one-half mile. While going up the hill some blue smoke may be seen from the tailpipe. When coming down the hill with the throttle closed (deceleration), increasingly more smoke should be seen as the vehicle travels with a fully closed throttle. Also, when the engine is accelerated once again, huge amounts of blue smoke will appear.

Other problems which produce similar oil consumption symptoms are scored cylinder walls, cracked pistons, and broken piston rings. Regardless of what the problem is, the engine will require major service work.

ABNORMAL NOISE ANALYSIS

Every automotive technician is frequently called upon to determine engine problems by listening to their unusual sounds. If the technician fully understands the internal engine functions and analyzes the location and frequency (number of times per revolution) of noises, the listening diagnosis can be very accurate.

Many of the sounds an engine makes in its operation are considered normal. The information that follows concerns only the abnormal noises, the ones that are usually trouble signs.

Two effective tools for the diagnosis of unusual engine noises are the stethoscope and a 3-foot (1-metre) piece of hose of ¼ to ½ inch (8–13 mm) inside diameter.

The stethoscope is used to locate problem noise and determine its frequency (Fig. 3-39). As an engine is run, the probe is held to the suspected problem areas. When the noise is loudest and corresponds to the abnormal sounds, the problem area has been located. A disadvantage in the use of the stethoscope is that it tends to pick up normal noises as well as those that are abnormal.

FIG 3-39 Technician using a stethoscope to locate engine noise.

A hose the size of common shop air hose can be used most effectively for noise detection. One end of the hose is held near an ear while the other end is directed toward the sound (Fig. 3-40). Usually, noises can be found without actual hose-engine contact. This is especially true of the valve train noises which cannot be touched while they are in motion. When the general area of noise is located, contact between the hose and the engine can be made, allowing the abnormal noise to be pinpointed.

A variation of both noise detectors can be made by removing the handle and probe of the stethoscope, leaving a hose. This provides the benefits of both tools.

Table 3-2 can be used to assist in detecting abnormal engine noises.

Note: The engine being analyzed for abnormal noise should be capable of running. When possible, all minor adjustments should have been made previously, including checking for loose bolts or nuts and exhaust gas leakage that create unusual noises.

Table 3-2 ABNORMAL NOISE ANALYSIS

Location	Frequency (No. of Times Per Rev)	Engine Speed	Type of Sound	Probable Cause-Source
Lower engine-oil pan	Crankshaft speed	Float no load	Light knock	Connecting rod bearing
Lower engine-oil pan	Double crankshaft speed	Float no load	Rattle, clatter	Connecting rod bearing-very poor condition
Lower engine-oil pan	Crankshaft speed	Low speed-under load	Heavy, dull knock	Main bearing
Lower engine-rear	Crankshaft speed	All speeds	Clicking or squeak Squeal	Clutch throw-out bearing or torque converter to flywheel bolts
Middle engine	Crankshaft speed	Fast idle-under load	Light knock Loudest when cold	Piston slap
Middle engine	Crankshaft speed	Fast idle-under load	Light knock Loudest when hot	Piston crack
Middle engine-fuel pump area	Camshaft speed	All speeds	Clicking, tapping	Fuel pump rod or arm spring
Oil filler opening	Camshaft speed	All speeds	Popping	Burned piston
Vapor engine at valve cover	Camshaft speed	All speeds	Clicking, tapping	Lifter, pushrod, or rocker arm
Carburetor	Camshaft speed	Fast idle	Popping	Intake valve burned or stuck
Exhaust pipe	Camshaft speed	Fast idle	Popping	Exhaust valve
Intake manifold-upper engine	Steady noise Loudest at idle	Idle	Hissing	Vacuum leakage
Upper engine-ignition parts	Intermittant	Accelerator	Snapping	Secondary ignition circuit
Upper engine	Steady-intermittant	Idle	Squeaks or squeal	Accessories-dry bearings or belts

FIG 3-40 Using piece of air hose (¼ in.) to locate engine noises.

ENGINE REPAIR INFORMATION

When the initial diagnosis process has been completed, and there are indications that the engine needs more than minor repair, several factors should be considered before the work begins. Among the considerations are

1. Cost of parts and labor
2. Time necessary to complete the work
3. Availability of replacement parts
4. Owner authorization

The parts and labor and the length of time needed to complete the job are often difficult to estimate before the engine is actually disassembled. To give the vehicle owner the most accurate cost estimate, the results of the initial diagnosis should be checked against the information in a flat rate manual.

The flat rate manual is a reference publication used to assist in writing estimates for automobile repair. These manuals are produced by vehicle manufacturers for use in their warranty work and by independent publishers (Fig. 3-41).

The independent flat rate manuals contain information identifying automobiles and their engines. A description of a typical repair job includes the time the factory allows for warranty payment to the dealer as well as the time the publisher estimates it will take the average journeyman mechanic to complete that job. These manuals also have typical parts descriptions with their factory numbers and prices at the time of publication. This information assists in preparing an owner cost estimate of parts to be replaced.

Note: When using one of these manuals, always refer to the directions inside the front cover (Fig. 3-42). Flat rate manual publishers list information in slightly different ways. Time estimates given in flat rate manuals represent the average time required by a journeyman mechanic to properly complete the operation described under that job title. In a school situation, the instructor is the best person to consult on the approximate time needed to complete a job.

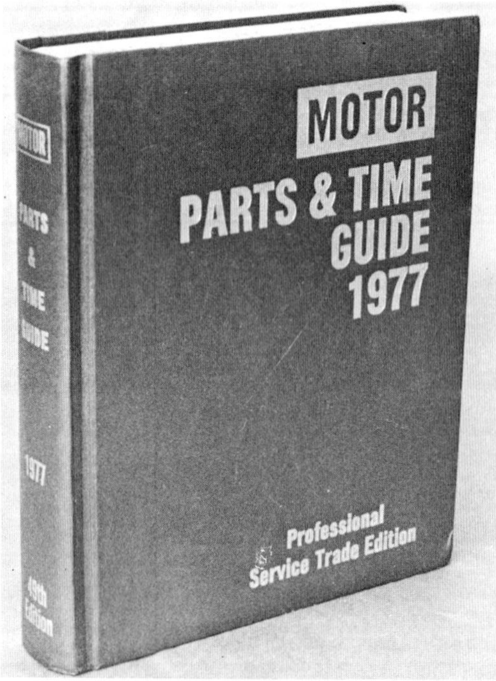

FIG 3-41 Flat rate manual.

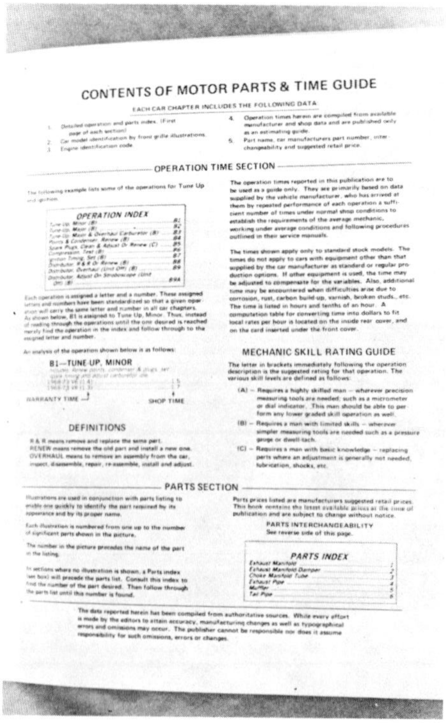

CONTENTS OF MOTOR PARTS & TIME GUIDE

EACH CAR CHAPTER INCLUDES THE FOLLOWING DATA

1. Detailed operation and parts index. (First page of each section)
2. Car model identification by front grille illustrations.
3. Engine identification code.
4. Operation times herein are compiled from available manufacturer and shop data and are published only as an estimating guide.
5. Part name, car manufacturers part number, interchangeability and suggested retail price.

OPERATION TIME SECTION

The following example lists some of the operations for Tune Up and ignition.

OPERATION INDEX

Tune-Up, Minor (B)	B1
Tune-Up, Major (B)	B2
Tune-Up, Major & Overhaul Carburetor (B)	B3
Points & Condenser, Renew (B)	B4
Spark Plugs, Clean & Adjust Or Renew (C)	B5
Compression, Test (B)	B6
Ignition Timing, Set (B)	B7
Distributor, R & R Or Renew (B)	B8
Distributor, Overhaul (Unit Off) (B)	B9
Distributor, Adjust On Stroboscope (Unit Off) (B)	B9A

Each operation is assigned a letter and a number. These assigned letters and numbers have been standardized so that a given operation will carry the same letter and number in all car chapters. As shown below, B1 is assigned to Tune Up, Minor. Thus, instead of reading through the operations until the one desired is reached merely find the operation in the index and follow through to the assigned letter and number.

An analysis of the operation shown below is as follows:

B1—TUNE UP, MINOR

WARRANTY TIME — SHOP TIME

The operation times reported in this publication are to be used as a guide only. They are primarily based on data supplied by the vehicle manufacturer, who has arrived at them by repeated performance of each operation a sufficient number of times under normal shop conditions to establish the requirements of the average mechanic, working under average conditions and following procedures outlined in their service manuals.

The times shown apply only to standard stock models. The times do not apply to cars with equipment other than that supplied by the car manufacturer as standard or regular production options. If other equipment is used, the time may be adjusted to compensate for the variables. Also, additional time may be encountered when difficulties arise due to corrosion, rust, carbon build-up, varnish, broken studs, etc. The time is listed in hours and tenths of an hour. A computation table for converting time into dollars to fit local rates per hour is located on the inside rear cover, and on the card inserted under the front cover.

DEFINITIONS

R & R means remove and replace the same part.
RENEW means remove the old part and install a new one.
OVERHAUL means to remove an assembly from the car, inspect, disassemble, repair, re-assemble, install and adjust.

MECHANIC SKILL RATING GUIDE

The letter in brackets immediately following the operation description is the suggested rating for that operation. The various skill levels are defined as follows:

(A) — Requires a highly skilled man — wherever precision measuring tools are needed; such as a micrometer or dial indicator. This man should be able to perform any lower graded skill operation as well.
(B) — Requires a man with limited skills — wherever simpler measuring tools are needed such as a pressure gauge or dwell-tach.
(C) — Requires a man with basic knowledge — replacing parts where an adjustment is generally not needed, lubrication, shocks, etc.

PARTS SECTION

Illustrations are used in conjunction with parts listing to enable one quickly to identify the part required by its appearance and by its proper name.

Each illustration is numbered from one up to the number of significant parts shown in the picture.

The number in the picture precedes the name of the part in the listing.

In sections where no illustration is shown, a Parts index (see box) will precede the parts list. Consult this index to find the number of the part desired. Then follow through the parts list until this number is found.

Parts prices listed are manufacturers suggested retail prices. This book contains the latest available prices at the time of publication and are subject to change without notice.

PARTS INTERCHANGEABILITY
See reverse side of this page.

PARTS INDEX

Exhaust Manifold	1
Exhaust Manifold Damper	2
Choke Manifold Tube	3
Exhaust Pipe	4
Muffler	5
Tail Pipe	6

The data reported herein has been compiled from authoritative sources. While every effort is made by the editors to attain accuracy, manufacturing changes as well as typographical errors and omissions may occur. The publisher cannot be responsible nor does it assume responsibility for such omissions, errors or changes.

FIG 3-42 Read the instructions inside the front cover for the correct use of the manual.

Name ____________________ Date ____________

DIRECTIONS

Select the best answer to each question and circle the letter at the left.

1. Proper diagnosis should include
   I. visual inspection.
   II. gathering as much information about the problem(s) as possible.
   A. I only
   B. II only
   C. Both I and II
   D. Neither I nor II

2. When an engine has a crankshaft that will not turn,
   I. no further diagnosis is necessary.
   II. engine removal is necessary.
   A. I only
   B. II only
   C. Both I and II
   D. Neither I nor II

3. A compression specification has a ± 10 percent tolerance. Which of the following is/are within this tolerance?
   I. 120 psi
   II. 140 psi
   A. I only
   B. II only
   C. Both I and II
   D. Neither I nor II

4. Increased compression pressures are often caused by
   I. piston expansion as the engine ages.
   II. carbon buildup.
   A. I only
   B. II only
   C. Both I and II
   D. Neither I nor II

5. Compression testing is best performed while
   I. the engine is cool.
   II. all the spark plugs are removed.
   A. I only
   B. II only
   C. Both I and II
   D. Neither I nor II

6. The first compression stroke is important to note because
   I. it indicates relative piston ring condition.
   II. it indicates any leakage in an adjacent cylinder.
      A. I only
      B. II only
      C. Both I and II
      D. Neither I nor II

7. Air escaping through the carburetor during a cylinder leakage test can mean
   I. the intake valve is defective.
   II. the intake valve is open.
      A. I only
      B. II only
      C. Both I and II
      D. Neither I nor II

8. Cylinder leakage testing requires
   I. air pressure equal to the compression pressure.
   II. the piston be in BDC position of the compression stroke.
      A. I only
      B. II only
      C. Both I and II
      D. Neither I nor II

9. Abnormal internal oil consumption is indicated by
   I. blue exhaust smoke.
   II. oil spots on the ground.
      A. I only
      B. II only
      C. Both I and II
      D. Neither I nor II

10. Engine noise frequency is
   I. the number of times per engine revolution it is heard.
   II. usually the same as the crankshaft speed or the camshaft speed.
      A. I only
      B. II only
      C. Both I and II
      D. Neither I nor II

DATE COMPLETED ______________________ INSTRUCTOR'S SIGNATURE ______________________

unit 4

Engine Removal

In this unit, engine removal procedures are discussed in terms to cover a variety of engine sizes and manufacturer design differences.

After completing the preparation assignment, you will correctly answer a minimum of 7 out of 10 multiple choice questions in a written test, without reference material, on the following topics:

1. Engine removal preparation
2. Engine removal procedures

OBJECTIVE 4-A (written test)

Preparation:

Study pages 103 through 111.

ENGINE REMOVAL

Removal of the engine from a vehicle is necessary when the diagnosis reveals excessive internal wear or damage. To recondition the internal parts, the engine is first removed, then disassembled and inspected.

Because there are so many differences in removal procedures among the different manufacturers, the description of engine removal in this unit is purposely general. For the person with little or no experience in this work, the steps in Objective 4-A1 should prove helpful. This is especially true of those steps involving labeling and marking.

If engine work is needed, but can be completed without removing the engine, refer to Unit 14.

Note: Be certain that the work order is signed by the vehicle owner and that the teardown and assembly report sheet has all specification information filled in. Also, review the shop manual for engine removal procedures and discuss them with your instructor. As always, be safety conscious in all of your work.

OBJECTIVE 4-A1 (manipulative)

Working as a team with one other student and with the aid of Job Report Sheet 4-A1, you will remove the engine of an instructor-assigned automobile prior to its complete overhaul.

Time Recommended:

Twice the flat rate time allowed in *Motors, Chilton,* or *National Service Data* manuals.

Preparation:

1. Review pages 103 through 104.
2. Study pages 104 through 110.
3. Study manufacturer's or general repair manuals for the automobile and engine being serviced.
4. Complete the work order form and have the vehicle owner sign the work authorization statement.
5. Review the steps for completion of this job with your instructor prior to beginning work.
6. Steam clean the entire engine and under-hood areas (Fig. 4-1).
7. Locate a suitable storage space for all engine parts to be removed.

Tools and Materials Needed:

1. Steam cleaner
2. ½ or ⅜ drive socket set with handles
3. Assorted box/open-end combination wrenches
4. Assorted flare-nut (tubing) wrenches
5. Assorted screwdrivers, pliers, and miscellaneous hand tools
6. Shop towels and rags

FIG 4-1 Steam clean the entire under-hood area before beginning the engine removal process.

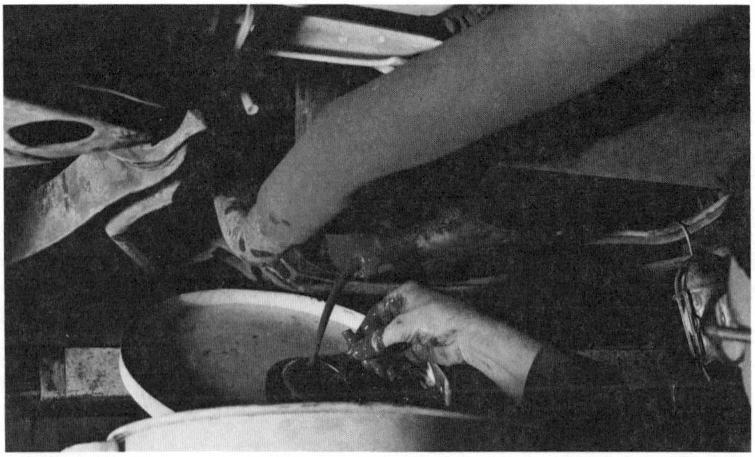

FIG 4-2 The first step is to drain the oil.

7. Mechanics wire
8. Tags or tape for making labels
9. Containers for nuts, bolts, and small pieces
10. Flashlight or trouble light
11. Floor jack or hoist
12. Engine hoist or chain hoist on "A" frame
13. Engine stand with proper mounting bolts

Procedure:

1. Drain the engine oil from the crankcase and all coolant from the cooling system (Fig. 4-2). Remove the oil filter.
2. After marking the position of the hood hinges, remove the bolts holding the hood (Fig. 4-3). Set the hood aside where it is out of the way.
3. Remove and store the battery.
4. Remove the air cleaner and the carburetor. Use a flare-nut wrench on the fuel line fitting (Fig. 4-4).
5. Attach the engine lift fixture to the carburetor mounting studs *or* use an engine sling with at least class 5 bolts. Bolts should enter the block at least 5 full turns (Fig. 4-5).
6. Remove the radiator with its hoses and shroud.
7. Remove the fan blade and belts.
8. Disconnect all engine wires and label or tag them with readable wording (Fig. 4-6). Locations of wires include:
   a. Alternator or generator
   b. Temperature sending unit
   c. Oil pressure sending unit
   d. Coil primary wires
   e. Starter motor
   f. Emission controls on engine

FIG 4-3 Place scratch marks around the hood hinges.

FIG 4-4 Use a flare-nut (tubing) wrench to loosen the fuel line.

FIG 4-5 Engine sling attached with a spacer.

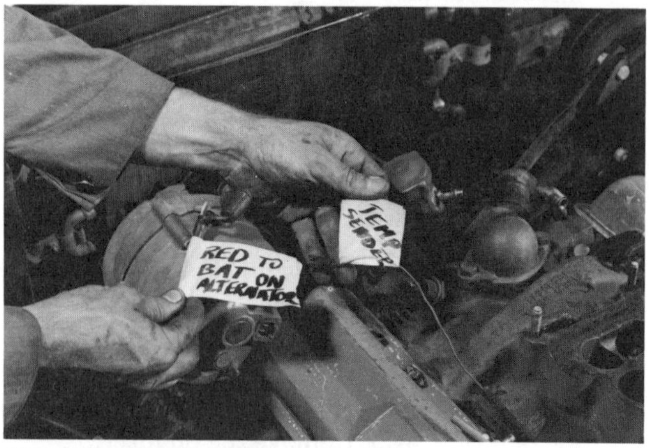

FIG 4-6 All wires, hoses, and cables should be labeled as they are detached from the engine.

FIG 4-7 Use mechanics wire to hold components out of the way.

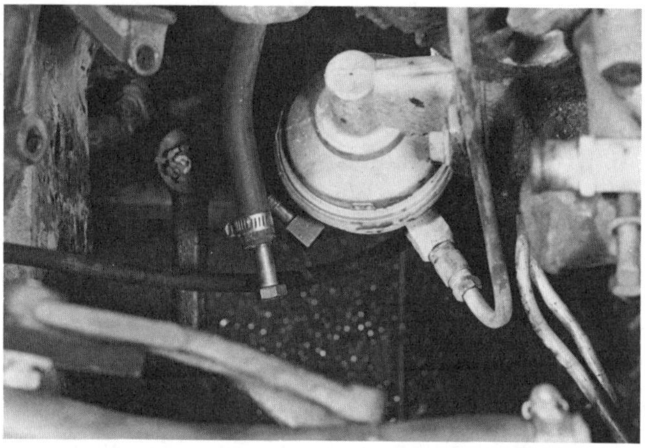

FIG 4-8 A bolt was used to plug the fuel line.

FIG 4-9 Using a rust penetrant will ease the removal of exhaust pipe nuts.

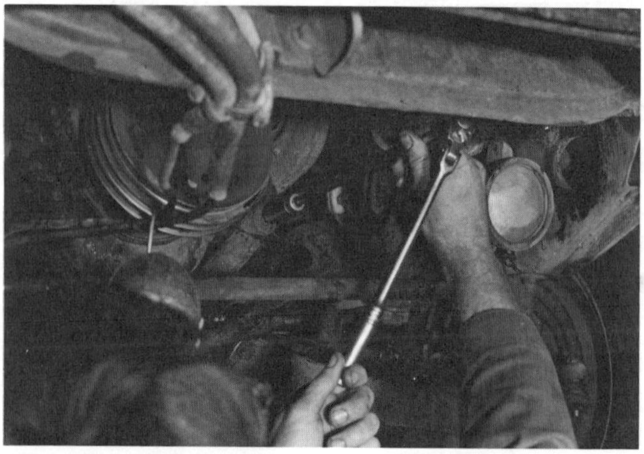

FIG 4-10 Removing front motor mount bolts.

9. Remove from the engine the alternator/generator power steering pump and air conditioner compressor. Note the location of all brackets.

Caution: Do not loosen fittings on air conditioner compressor or refrigerant will escape. Also, use mechanics wire to keep the A/C compressor secure and to hold power steering reservoir upright to prevent fluid spillage (Fig. 4-7).

10. Remove heater hoses attached to engine.
11. Remove ignition distributor.
12. Disconnect accelerator rod or cable and transmission linkage. Reinstall washers, bushings, clips, and pins so reinstallation time will be minimal.
13. Disconnect the fuel line at the pump coming from the tank. Plug the end of the line to prevent fuel spillage (Fig. 4-8).

Note: To complete the remaining steps, the vehicle will have to be raised. This can be accomplished by using a hydraulic hoist or lift (best) or a floor jack and then supporting the vehicle on safety (jack) stands.

14. Disconnect exhaust pipes from exhaust manifold (Fig. 4-9). In some instances, it is easier to remove exhaust manifolds from the engine, leaving them in the engine compartment.
15. Remove the bolts holding the front engine mounts (Fig. 4-10).

Note: In most cases, the transmission should remain in the vehicle when the engine is removed. Consult a repair manual or your instructor to be certain. If the transmission *is* to be removed, skip forward to step 19.

16. Remove flywheel cover pan or inspection plate.

Note: The designations A/T and M/T indicate whether the following steps are to be taken with automatic or manual transmission vehicles.

17. A/T: Remove converter to flywheel (flex plate–drive plate) bolts or nuts (Fig. 4-11).
    M/T: Disconnect clutch release linkage.
18. A/T: Making certain the engine and transmission are well supported, remove the transmission case (bell housing) to engine bolts.
    M/T: Supporting the engine and transmission, remove the transmission to clutch housing (bell housing) bolts. Install two guide bolts (Fig. 4-12). This prevents damage to the clutch disc as the transmission and engine are separated.

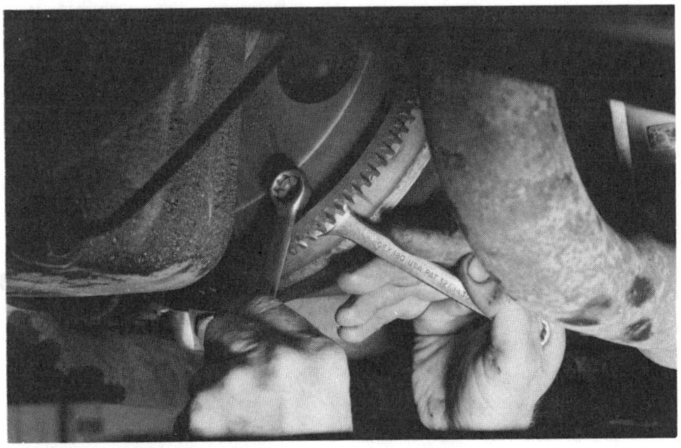

FIG 4-11 Removing torque converter-flywheel bolts.

FIG 4-12 Install guide bolts to prevent damage to the clutch disc.

19. If the transmission is to be removed (consult repair manual or your instructor), disconnect shift linkage and speedometer cable—with its gear, if necessary. Replace pins and clips in their respective places. Wrap the speedometer gear and the end of the cable with tape to prevent loss or damage.
20. Remove the drive shaft by loosening bolts or nuts on the yoke at the differential. Loosely reassemble bolts or nuts to maintain their original position. Label drive shaft with student mechanic's name and place it in the storage room.
21. Wrap the rear universal joint cups with tape to hold them in position (Fig. 4-13).
22. Install cup or dummy yoke over transmission output shaft to prevent oil spillage as the transmission is removed (Fig. 4-14).
23. Remove bolts holding the transmission mount in position. Support the transmission with a jack or temporary cross member while the mount is removed.
24. A/T: Remove oil cooler lines from the transmission.
25. Move engine hoist into position and attach to lifting fixture or sling (Fig. 4-15).
26. Raise engine (and transmission) enough to take their weight off the engine mounts. Inspect the entire engine for anything still attached to the vehicle. Remove where necessary.

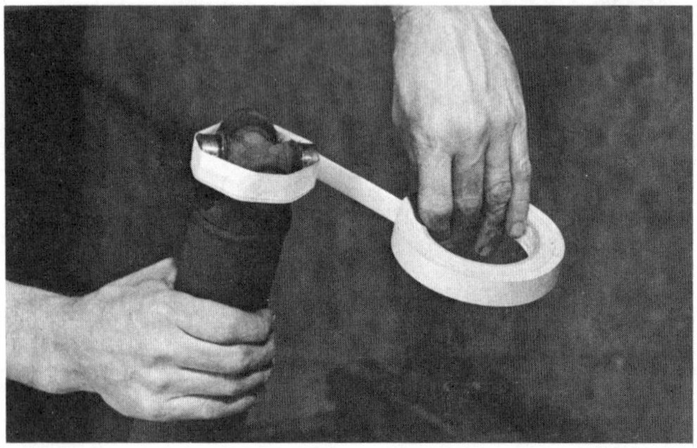

FIG 4-13 Using tape to hold the cups on the rear universal joint.

FIG 4-15 Engine hoist in position.

FIG 4-14 "Dummy" yoke installed on transmission output shaft to prevent oil spillage.

FIG 4-16 Engine being tilted upward in the front as it is being removed from the vehicle.

27. Pull forward on the engine as it is being lifted from the engine compartment. The engine should be tilted up in the front to make the removal operation easier (Fig. 4-16).
28. If the transmission is attached to the engine, lower both units to the floor and separate. Attach the engine stand adapter with bolts of the proper length and strength (Fig. 4-17).
29. Raise the engine alone and mount on engine stand (Fig. 4-18).
30. If the transmission is to remain in the car, support it to maintain the same position it would have with the engine attached. Placing a temporary cross member beneath the transmission or tying it to an overhead cross member often works well (Fig. 4-19).
31. Clean up any oil spills. Collect all loose parts and store in the trunk of the car.
32. Replace the hood of the vehicle and tighten the original mounting bolts finger tight (Fig. 4-20).
33. Complete Job Report Sheet 4-A1 and turn it in to your instructor.

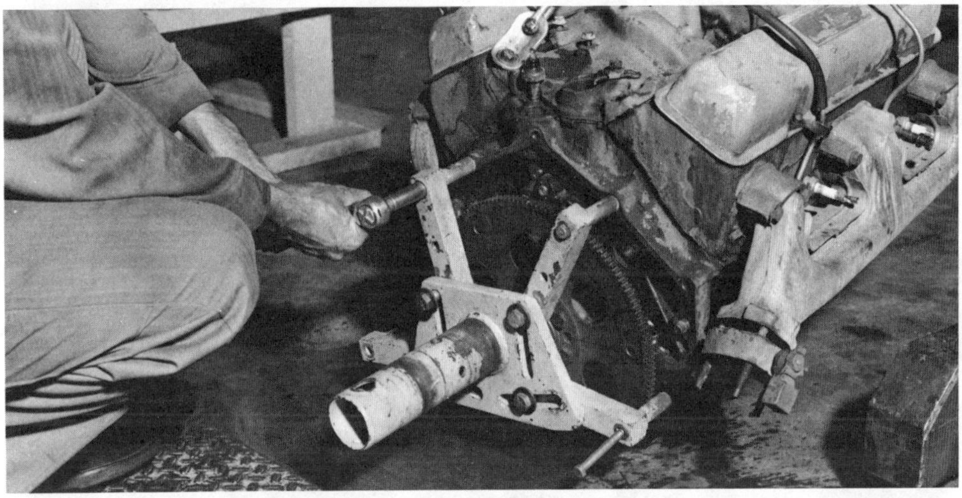

FIG 4-17 Attaching the engine stand adapter while the engine rests on the floor.

FIG 4-18 Engine mounted on its stand.

FIG 4-19 A wood cross-member used to support the transmission until the engine is replaced.

FIG 4-20 Replace the hood before moving vehicle to its storage area.

ENGINE REMOVAL

Student name ______________________________ Date ______________

Make of car ______________ Model ______________ Year ______________

Time started ______ Time finished ______________ Total time ______________

Flat rate time ______________________________

Specifications:

Refer to shop manuals for specifications and directions not covered in Objective Sheet 3-A1.

Checkup Questions:

1. Did engine removal include the transmission? ______________________________

   If not, how is the transmission supported? ______________________________

   ____

2. Are all wires labeled according to their proper terminals? ______________________________

   ____

3. How was the vehicle raised to perform removal operations on its underside? ______________

4. Where are parts being stored? ______________________________

   ____

5. Was engine removal area thoroughly cleaned up after work was completed? ______________

DATE COMPLETED ______________________ INSTRUCTOR'S SIGNATURE ______________________

Name ______________________________ Date ______________

DIRECTIONS

Select the best answer to each question and circle the letter at the left.

1. Labeling engine parts prior to their removal
   I. assists in their replacement.
   II. should be done after the engine is removed.
   A. I only
   B. II only
   C. Both I and II
   D. Neither I nor II
2. A suitable storage space for engine parts is
   I. one that can be locked.
   II. the trunk area of the automobile.
   A. I only
   B. II only
   C. Both I and II
   D. Neither I nor II
3. Marking the position of the hood hinges will
   I. keep the right and left sides identified.
   II. assist in replacing the hood when the engine is reinstalled.
   A. I only
   B. II only
   C. Both I and II
   D. Neither I nor II
4. The type of wrench recommended for use on the fuel line fittings is
   I. box end.
   II. open end.
   A. I only
   B. II only
   C. Both I and II
   D. Neither I nor II
5. The easiest method of preventing fuel spillage when the line is removed from the pump is to
   I. cover the opening with tape.
   II. plug the hose.
   A. I only
   B. II only
   C. Both I and II
   D. Neither I nor II
6. Recommended procedures for engine removal include
   I. keeping all fasteners in one container to prevent their loss.
   II. draining the transmission oil to prevent spillage.
   A. I only

B. II only
C. Both I and II
D. Neither I nor II

7. When removing an engine with its transmission,
   I. guide bolts should be installed.
   II. the speedometer gear should be removed.
      A. I only
      B. II only
      C. Both I and II
      D. Neither I nor II

8. As the engine is being lifted,
   I. it should be tilted upward in the front.
   II. a brief check for anything still attached should be made.
      A. I only
      B. II only
      C. Both I and II
      D. Neither I nor II

9. When the transmission is to remain in the car,
   I. it should be supported in some manner.
   II. the drive shaft should be removed.
      A. I only
      B. II only
      C. Both I and II
      D. Neither I nor II

10. After the engine is removed,
   I. the hood should be stored on top of the car.
   II. place all loose parts on the seats.
      A. I only
      B. II only
      C. Both I and II
      D. Neither I nor II

DATE COMPLETED ______________________ INSTRUCTOR'S SIGNATURE ______________________

unit 5

Engine Disassembly and Cleaning

This unit describes general procedures for disassembling an engine that has been removed from the vehicle. Also included are materials, equipment, and procedures for cleaning engine parts.

OBJECTIVE 5-A (written test)

After completing the preparation assignment, you will answer a minimum of 7 out of 10 multiple choice questions in a written test, without reference material, on the following topics:

1. Upper engine disassembly
2. Lower engine disassembly

Preparation:

Study pages 115 through 137.

ENGINE DISASSEMBLY

Disassembly is the process of systematically removing each part of the engine until each is an individual part or subassembly. These parts should be visually inspected for irregular wear or damage as the disassembly process continues.

The objective procedure sheets (5-A1 and 5-A2) give general steps for disassembly of an engine. These procedures are based on the average disassembly operations for an overhead valve (I-head), cast iron block engine. Consult a shop manual for reference.

Precautionary steps include the following:

1. Be certain the oil is drained and the oil filter is removed.
2. Drain all coolant from the block.
3. Do not remove the cylinder heads until the engine is cool, or warpage may result.
4. Be prepared to clean up minor liquid spills as soon as they happen.
5. Mark all duplicate parts with steel stamps or a center punch.
6. Make notes and/or drawings on the teardown sheet to assist later in the assembly process.

OBJECTIVE 5-A1 (manipulative)

Working as a team with one other student and with the aid of Job Report Sheet 5-A1, you will disassemble the upper portion (valve train and manifolds) of an engine from an instructor-assigned automobile prior to complete overhaul.

Time Recommended:

Twice the flat rate time allowed in *Motors, Chilton,* or *National Service Data* manuals.

Preparation:

1. Review pages 115 through 116.
2. Study pages 116 through 118.
3. Study manufacturer's or general repair manuals for the engine being serviced.

Tools and Materials Needed:

1. ½ or ⅜ drive socket set with handles
2. Assorted box/open-end combination wrenches
3. Assorted screwdrivers, pliers, punches, and hammers
4. Shop towels or rags
5. Containers for nuts and bolts labeled with locations from which they are removed (Fig. 5-1)
6. Steel number and letter stamps.

FIG 5-1 Containers for nuts and bolts labeled with their locations.

FIG 5-2 Using an air ratchet to remove the intake manifold.

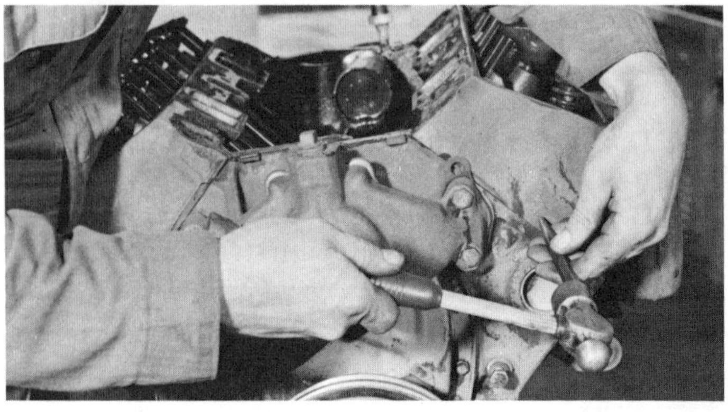

FIG 5-3 Marking the cylinder heads, right or left, as viewed from the driver's seat.

Procedure:

Note: The following steps are general guidelines for engine disassembly. For specific instructions on procedure, consult a shop manual. If at any time during the disassembly operations any sign of cracked or broken parts or damaged surfaces is noticed, consult your instructor for further directions. Save *everything* removed from the engine until your instructor approves of its disposal.

1. With the engine properly mounted on the engine stand, have your instructor check it before removing the engine lift from the fixture or sling. The fixture or sling should then be returned to its proper place.
2. Remove the intake manifold, inspecting the mating surfaces for water, vacuum, and/or exhaust leaks (Fig. 5-2). Make certain there was a good seal between the manifold and head.
3. Remove the exhaust manifolds, inspecting for cracks or warpage.
4. Remove the fuel pump and fuel lines.
5. Remove the water pump. Look at the bottom side of the pump for seal leakage.
6. Label R or L on appropriate cylinder head before each is removed (Fig. 5-3).
7. Remove cylinder heads carefully to avoid damage to the pushrods (Fig. 5-4). In some instances, the rocker arms or rocker arm assemblies must first be removed. Inspect combustion chamber surfaces for cracks or other irregularities (Fig. 5-5).
8. Carefully examine head gaskets for leakage or incorrect application on prior installation (Fig. 5-6). A blown head gasket will be burned through between cylinders and show a black color.
9. If the engine has nonadjustable valve clearance (see your instructor), keep

FIG 5-4 Loosening the cylinder head bolts by pulling towards yourself.

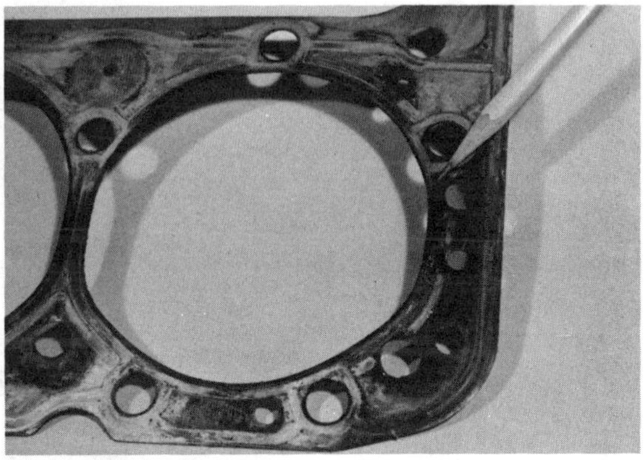

FIG 5-6 Evidence of a blown (coolant loss) head gasket. Note very dark area.

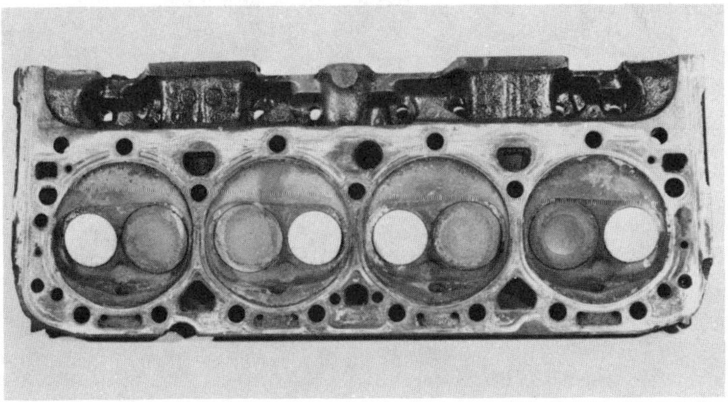

FIG 5-5 Examine the combustion chamber surfaces.

pushrods in exact order so that reinstallation is not complicated. Carefully inspect all parts for cracked, broken, or otherwise irregular surfaces.

10. With all the upper engine parts close at hand, have your instructor inspect your engine disassembly at this point.

Note: A preliminary check of cylinder wall wear and taper can now be made. To accomplish this task, study the portions of Unit 6 discussing telescoping gauges and micrometers and cylinder bore gauges.

11. Rotate the engine crankshaft until 2 pistons are at Bottom Dead Center.
12. Wipe the cylinder clean.
13. Expand the telescoping gauge to its maximum length in the area of the cylinder wall *just below* the ring ridge (Fig. 5-7).
14. Lock the gauge and measure with a micrometer (Fig. 5-8). Record this figure for the appropriate cylinder at A on Job Report Sheet 5-A1 (Teardown and Assembly).
15. From the figure at A, subtract the original bore. This will provide a rough estimate of the cylinder wear.
16. When using a cylinder bore gauge, make one reading just below the ring ridge and another reading as far down the cylinder wall as possible. The difference in the two readings will be the taper. The amount of wear will be the first reading (upper bore) minus the original bore size.
17. Rotate the engine crankshaft as necessary so all cylinders can be measured.
18. Consult with your instructor about your measurements and general engine condition.

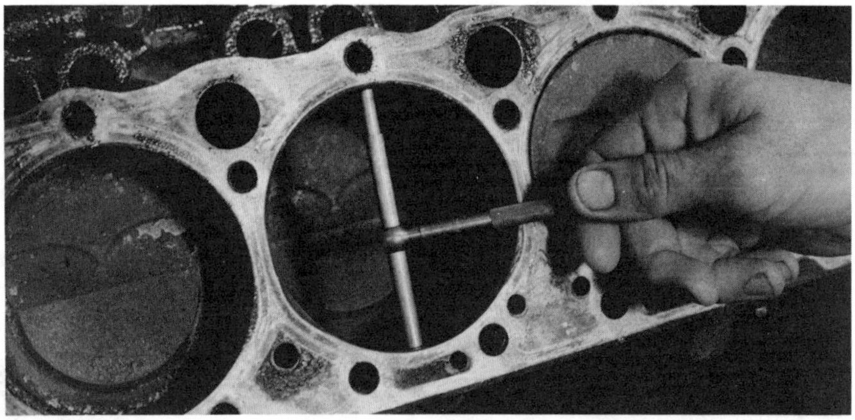

FIG 5-7 Expand the telescoping as shown just below the ring ridge.

FIG 5-8 Measure the expanded telescoping gauge.

JOB REPORT SHEET 5-A1

UPPER ENGINE DISASSEMBLY

Student name ______________________________ Date ______________

Make of car ______________ Model ______________ Year ______________

Time started ______ Time finished ______________ Total time ______________

Flat rate time ______________________________

Specifications:

Diagram factory cylinder numbering pattern.

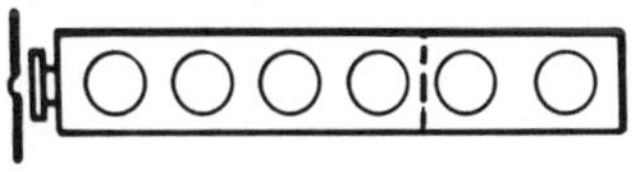

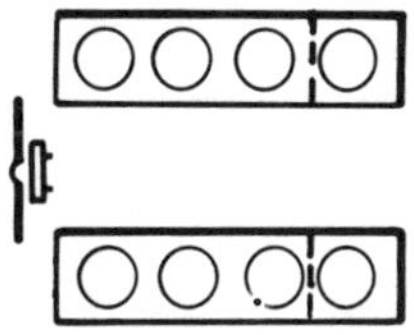

Checkup Questions:

1. Were any broken bolts or studs found? ______________________________

   If yes, where? ______________________________

2. Were all cylinder head bolts the same length and size? ______________________________

   If not, show location of all cylinder head bolts by completing this diagram:

3. Do pistons show any "wash" areas caused by excessive oil in the combustion chamber? ______________

4. Is there any sign of piston contact with the cylinder head? ______________________________

5. Do any combustion chambers show rust buildup due to cracks? ______________________________

6. Are all exhaust valves the same color? ______________________________

7. Results of preliminary cylinder wall tape measurement:

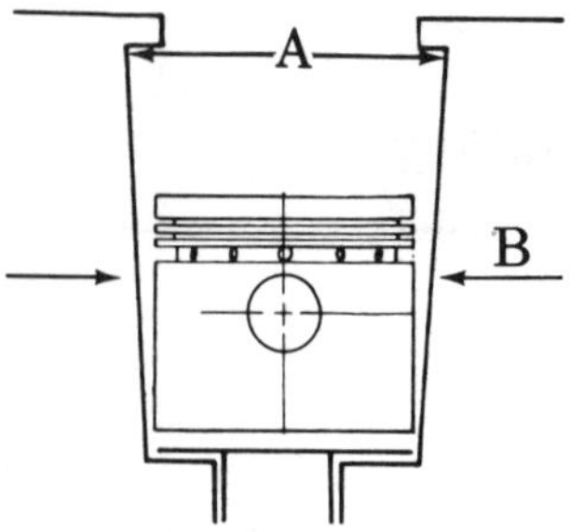

Cylinder #

	A	–	(cylinder bore)	=	Taper
1	______	–	______	=	______
2	______	–	______	=	______
3	______	–	______	=	______
4	______	–	______	=	______
5	______	–	______	=	______
6	______	–	______	=	______
7	______	–	______	=	______
8	______	–	______	=	______

DATE COMPLETED ______________ INSTRUCTOR'S SIGNATURE ______________

OBJECTIVE 5-A2 (manipulative)

Working as a team with one other student and with the aid of Job Report Sheet 5-A2, you will disassemble the lower portion (camshaft, crankshaft, and bearings) of an engine from an instructor-assigned automobile prior to complete overhaul.

Time Recommended:

Twice the flat rate time allowed in *Motors, Chilton,* or *National Service Data* manuals.

Preparation:

1. Review pages 115 through 118.
2. Study pages 121 through 125.
3. Study manufacturer's or general repair manuals for the engine being serviced.

Tools and Materials Needed:

1. ½ or ⅜ drive socket set with handles
2. Assorted box/open-end combination wrenches
3. Assorted screwdrivers, pliers, punches, and hammers
4. Shop towels or rags
5. Containers for nuts and bolts
6. Number and letter stamps
7. Vibration damper (torsional damper) puller with assorted bolts and step plate

Procedure:

Note: The following steps are a general guideline for engine disassembly. For specific instructions, use procedures as outlined in a shop manual. If, at any time during the disassembly operations, any sign of cracked or broken parts or damaged surfaces is noticed, contact your instructor for further direction. Save *everything* removed from the engine until your instructor approves of its disposal—EVERYTHING!

1. Remove vibration damper (torsional damper), using the proper puller and step plate (Figs. 5-9 and 5-10).

FIG 5-9 Hold the damper while turning the wrench.

FIG 5-10 Removing the vibration damper with a puller and connecting size step plate. (Oldsmobile Division, GMC)

2. Remove front cover (timing cover), as shown in figure 5-11. In some cases, the oil pan will have to be removed first.
3. Remove the oil pan. Examine interior of the oil pan for foreign matter. Wipe up any oil spills.
4. Remove the oil pump assembly (pump, screen, and shaft) as shown in figure 5-12.
5. Check all main bearing caps and connecting rods to be certain they are numbered before removal (Fig. 5-13). If the numbers cannot be seen or they appear irregular, mark each main and rod, upper and lower halves, with a center punch or number stamp according to the manufacturer's designated cylinder numbering pattern (Fig. 5-14).

Note: Step 4 must be done so that main and rod caps will be reassembled in exactly their original places. Some manufacturers recommend measuring the timing-chain slack before it is removed. Consult a repair manual. In most cases, the chain and timing gears are replaced as a routine procedure.

6. Remove the timing chain, if so equipped, by removing the timing gear from the camshaft (Fig. 5-15).
7. Remove any ring ridge (the area above the top piston ring travel in the cylinder) as shown in figure 5-16. Because there are many differences in the operation of ridge reamers in use today, no specific instructions will be given. Consult with your instructor for ridge reaming operations.

Caution: Do not remove any cylinder wall material below the ring ridge, for this can cause compression loss if the cylinders are not bored (Fig. 5-17).

8. With mains and rods properly marked, unbolt one rod at a time and push out of the cylinder from the bottom. Avoid damage to the crankshaft journals by covering rod bolts with hose or spark plug boots (Fig. 5-18). Keep rod caps and bearings in correct order.
9. Remove the main bearing caps and the crankshaft (Fig. 5-19).

FIG 5-12 Removing oil pump and screen assembly.

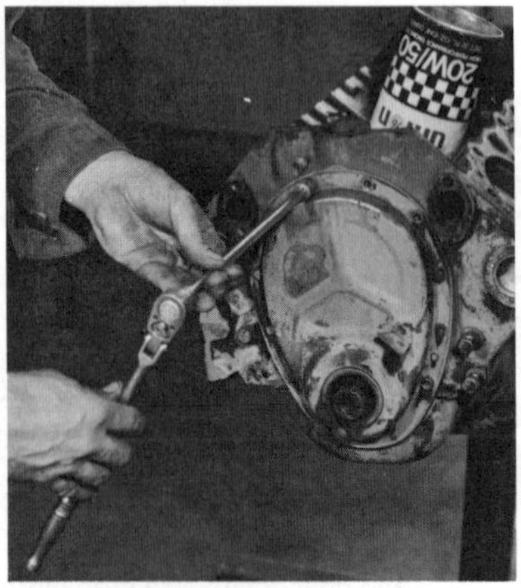

FIG 5-11 Removing timing chain cover.

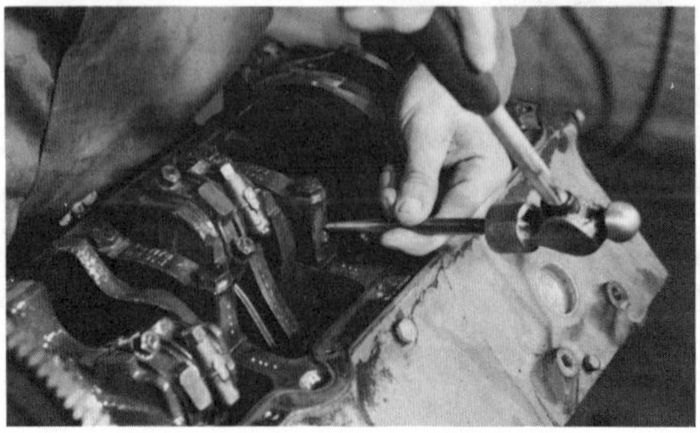

FIG 5-13 Marking main bearing caps with a center punch.

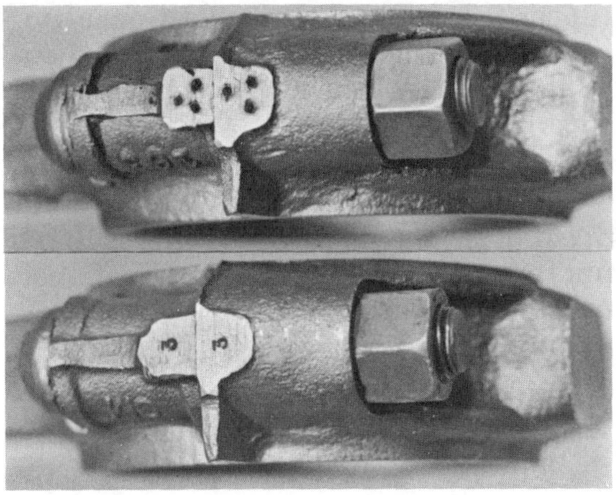

FIG 5-14 Two methods of marking connecting rods.

FIG 5-15 Removing timing chain and sprocket.

FIG 5-16 Cutting the ring ridge.

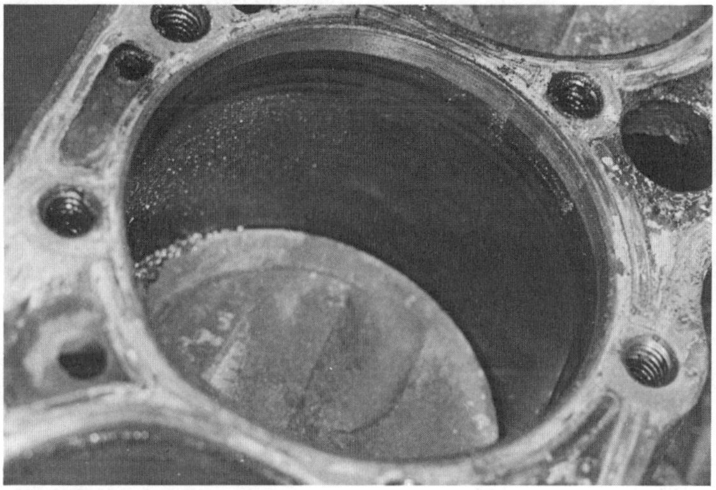

FIG 5-17 Ring ridge properly removed.

FIG 5-18 Using rubber hose over connecting rod bolts to protect crankshaft journals.
(Oldsmobile Division, GMC)

FIG 5-19 Removing the crankshaft with flywheel attached.

10. If an analysis of the bearings is to be made, mark each bearing half with its proper number and its position (upper or lower half). Example: Connecting rod number three should have two bearing halves; one marked 3U and one marked 3L. Clean each bearing half with fresh solvent and label it with a felt tip marker on the back side (Fig. 5-20).
11. If the engine is equipped with a fuel pump pushrod driven from the camshaft, remove it now.
12. Push or pull the lifters far enough out of the way to remove the camshaft. If some lifters stick when you attempt to remove them, wait until the camshaft is removed and then push them out from the camshaft side.

Note: If the camshaft and lifters are to be used again (unlikely), the lifters must be kept in order so that each one is reassembled to the same camshaft lobe (Fig. 5-21).

13. Remove the camshaft carefully by installing a long bolt in one of the holes on the front end and pulling gently forward. Support the rear end of the camshaft as it is being removed to reduce the chances of damage to the lobes and journals (Fig. 5-22).
14. Remove the rear main bearing oil seal from the cylinder block and rear main bearing cap.

FIG 5-20 Marking each bearing insert for later analysis.

FIG 5-22 Removing the camshaft by pulling forward while supporting the end.

FIG 5-21 Removing the lifters and maintaining their order for proper reuse.

FIG 5-23 Removing the oil gallery plugs.

15. Make a close inspection of all engine block surfaces to be certain this casting is free of cracks or other damage which would prevent its reuse.
16. Remove oil dip stick and its tube. The tube is normally pressed into place.
17. Remove all oil gallery plugs (Fig. 5-23).
18. Remove all core (freeze) plugs with a rocker pry bar.
    a. The first step is to drive the pointed end into the softest area of the plug, usually near the lower edge (Fig. 5-24).
    b. Next, pry outward until the plug is free (Fig. 5-25).
19. Remove all other nonferrous metals from the block. This is especially important if the cleaning is to be done in an alkaline (caustic) solution hot tank. These parts normally include camshaft bearings, distributor housing bushings, and oil pressure sending units. For camshaft bearing removal, refer to page 124.
20. Remove the oil deflector from the intake manifold (Figs. 5-26 and 5-27).
21. Have your instructor check out the block before you place it in the hot tank.
22. Complete Job Report Sheet 5-A2 and turn in to your instructor.

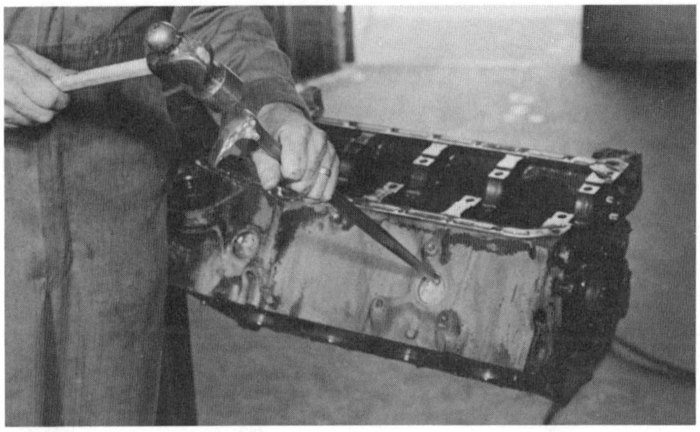

FIG 5-24 First step in removing core plugs—puncturing them with a pointed end pry bar.

FIG 5-25 Pry outward to remove the plug.

FIG 5-26 Remove the oil deflector from the intake manifold using a rocker type pry bar. Save all attaching pins.

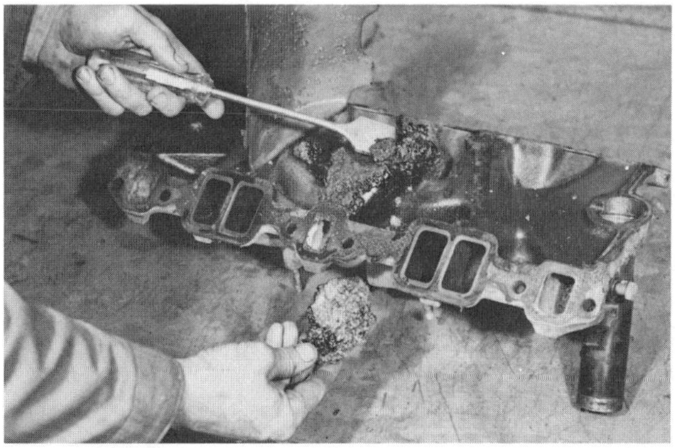

FIG 5-27 Remove the heavy carbon buildup before cleaning in the hot tank.

LOWER ENGINE DISASSEMBLY

Student name ______________________ ______________________ Date ______________________

Make of car ______________ Model ______________________ Year ______________________

Time started ______ Time finished ______________________ Total time ______________________

Flat rate time ____

Specifications:

Diagram factory cylinder numbering pattern.

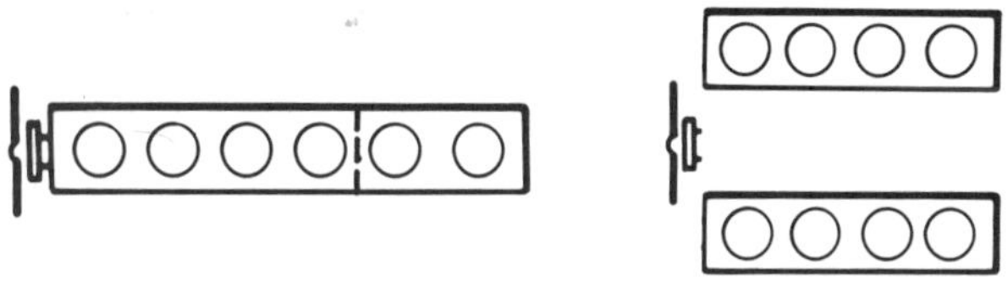

Checkup Questions:

1. What is the approximate depth of the groove on the timing cover seal surface of the vibration damper? ______________________
2. How much side-to-side movement can be found on the timing chain? ______________________
3. Were main and rod bearing caps previously marked? (Yes/No) ______________________

   If no, what did you use to mark them? ______________________
4. Are lifters to be used again? (Yes/No) If yes, are they kept in proper order? (Yes/No) ____________
5. Are timing gears and chain to be replaced (Yes/No) ______________________
6. Are any block repairs necessary? (Yes/No) ______________________

   If yes, describe them. ______________________

   ____

DATE COMPLETED ______________________ INSTRUCTOR'S SIGNATURE ______________________

CLEANING

The primary objective in cleaning engine parts is to remove all the foreign matter. This means stripping all metal surfaces of any dirt, oil, grease, gaskets, paint, and, in some cases, rust and scale. Competent engine repair is possible only if the parts involved are completely clean. Since a variety of materials is removed, many different cleaning methods and chemicals are used by engine rebuilders. In the information that follows both the chemicals and cleaning devices will be covered. The chemicals most often used for cleaning engine parts are considered dangerous when proper handling precautions are not observed. Safety procedures concerning these chemicals and methods will be noted.

CLEANING CHEMICALS

Proper cleaning of engine parts involves the correct application of certain chemicals. The following chemical compounds are most commonly found in engine repair facilities:

Chemical or Process	Common Material Form	Type of Machine
Alkaline-water	Caustic liquid	Hot tank, spray tank
Peening	Glass beads	Closed booth
Petroleum distillates	Stoddard solvent	Solvent wash tank
Chlorinated	Liquid carburetor cleaner	Cold tank or bucket

All chemicals used for cleaning are either alkaline (caustic) salts or acid compounds. One way chemists classify or identify chemicals is the pH scale, which runs from 0 to 14. Chemicals that are alkaline range from 7, the neutral point, to 14. Acids range from 0 to 7 on the scale.

Figure 5-28 shows the rating of certain chemicals on a pH scale. With few exceptions, the oils and greases found on or in engines are slightly acid, approximately 5.6 on the pH scale. Calcium carbonate (9.4) and lime (12.4) are common undesirable chemical compounds found along with iron oxide (rust) in the scale in the water jackets or passages in the block. When excessive, this scale can create overheating problems even after the most competent engine overhaul. Overheating occurs because the scale, an excellent insulator, does not allow the transfer of heat from the hot engine parts to the circulating coolant.

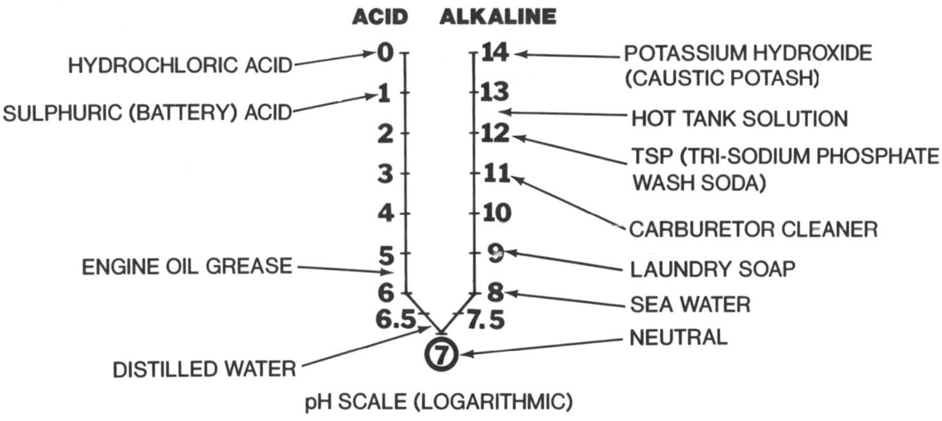

FIG 5-28 pH scale with common engine soils listed.

The chemical reaction of cleaning is best accomplished by subjecting parts to a solution of their opposite pH factor. As an example, oils and greases (5.6) will be thoroughly cleaned by using a chemical with a pH of 10 or 12. A typical chemical in this range is trisodium phosphate, T.S.P., with a pH of 12.0.

Normal household soaps and detergents, as well as materials used in a steam cleaner or hot tank, are alkaline. This means they have a pH factor somewhere between 7 and 14. Since most household and automotive cleaning involves the removal of oily (acid) substances, the alkaline compounds at the opposite side of the pH scale (7–14) work very well.

When cleaning alkaline substances from an object, an acid (pH 0–7) of some type must be used. The most common alkaline cleaning job for an acid in the shop is descaling the steam cleaner. In this operation, hydrochloric (muriatic) acid is pumped through the heating coils of the steam cleaner to remove the buildup of calcium carbonate and other minerals. Whenever water of average mineral content flows through piping, especially if it is heated as in a tea kettle or steam cleaner, the minerals attach themselves to the pipes. Acid must then be used to remove these alkaline substances to again allow the free flow of water through the coils.

CLEANING DEVICES

The cleaning of engine parts, which is a chemical reaction, is speeded up by heat and agitation. A steam cleaner is a common automotive cleaning device which heats water mixed with an alkaline solution. The heated "soapy" water is directed to the dirty engine parts so as to blast away or agitate the oil and grease materials. The alkaline solution reacts chemically with grease to help dissolve it. The heat of the solution also speeds up the chemical reaction of dissolving grease. Also, the force of the hot water and steam physically agitates the grease, helping to speed up the grease-dissolving action.

A steam cleaner solution should be 10–15 percent steam and 85–90 percent hot water. For effective cleaning, this solution should contain approximately 2 percent of a good quality steam cleaning compound (soap). For general use, the steam

FIG 5-29 Steam cleaner foam showing adequate amount of compound in solution.

FIG 5-30 Typical hot tank.

cleaner should have a *steady* flow of water and steam from the nozzle. The nozzle should not jump or pulse. Also, when the nozzle is directed at the floor, some evidence of bubbles or foam should be seen, indicating enough compound is being used (Fig. 5-29).

Note: *Always* wear gloves and eye protection when using the steam cleaner (Fig. 5-29).

Another common cleaning device for engine parts is the hot tank (Fig. 5-30). It is a large tank (200–400 gallons) to which heat (gas or electricity) can be applied. The cleaning results from the chemical reaction of the materials and heat. Some hot tanks also have an agitating device which reduces the total cleaning time. For effective cleaning, a solution of approximately pH 12 is necessary. The solution in the hot tank should be kept at approximately 180° to 200°F. (82° to 93°C.).

Place only engine parts made of ferrous metals (iron and steel) in the hot tank solution. Remove oil or temperature sending units, bypass adapters on the oil filter housings, bronze distributor bushings, etc. First make every effort to remove as much loose foreign material as possible from parts. Steam clean, scrape, or wash as many of them as possible. Under the very best conditions, parts must remain in the hot tank two to three hours.

Note: Wear suitable eye protection when inserting or removing parts.

Another cleaning device becoming more popular is the "jet cleaner" or spray cleaner (Fig. 5-31). This device is named for its method of spraying hot chemicals through "jets" onto the engine parts within its cabinet. It is much like a home dishwasher. This machine has the benefits of a hot tank, since it usually uses a heated, mild, alkaline solution; but it can also clean nonferrous metals such as aluminum, brass, and bronze. It cleans very rapidly (20–30 minutes) compared to a conventional hot tank.

Another type of engine cleaning device is called the high-pressure washer. This machine blends hot or cold water with a detergent (mild alkaline) cleaning chemical and pumps it to a high pressure (300–900 psi). This high-pressure liquid is

FIG 5-31 Spray cleaning machine.
(Storm-Vulcan)

FIG 5-32 Comparison of a glass beaded surface (right) and the ordinary hot tank method.

directed at soiled engines or parts with a hand-held nozzle. The combination of the detergent and the blasting force of the water cleans the parts.

Note: Eye protection should be used when operating this machine. This device has the advantages of low solution purchase price and operation cost and easy portability. It does not normally clean the inside parts of an engine as well as most engine technicians desire, however.

Peening is the process of air blasting different types of materials on a metal surface. This is sometimes referred to as dry parts cleaning. The material most often used is beads of glass that are 4 to 16 thousandths of an inch in size. Other materials used include walnut shells, steel shot, and silica sand. Glass beads have the advantage of cleaning all types of metals, both hard and soft, of all foreign matter including rust. It does this without upsetting or eroding any of their surfaces like some of the previously mentioned materials. Figure 5-32 shows a cast iron cylinder head cleaned by this method and an adjacent area cleaned by the hot tank method. Glass beading provides the only practical means of cleaning parts made of aluminum or other soft metals, since most alkaline solutions will eat away the metal. Also, scraping and wire brushing will scratch these soft, smooth surfaces.

Glass beading is done in a closed booth or cabinet so that the dust created can be controlled and the beads can be cleaned and reused as long as possible. The best designs for this equipment have reclaiming devices to separate the foreign matter from the beads and allow their reuse. The glass beads eventually break down into dust which the reclaimer gathers. Figure 5-33 shows a typical peening machine with a reclaimer on the left.

The operator views the work through a safety glass window while holding the part in one hand and the beading nozzle in the other. Both hands are protected by heavy rubber gloves. A foot pedal controls the flow of glass beads.

Glass beading is not only fast; it is also thorough. Cylinder head intake and exhaust parts cannot be cleaned as effectively with any other type of machine or materials.

Any parts to be cleaned must be totally free of oil, grease, and other wet substances, or the glass beads will stick to them. When cleaning parts such as valves or crankshafts, the polished surfaces such as the stems and journals should be masked off or otherwise protected to preserve the original finish. The temptation to clean complicated castings, such as Corvair or Volkswagen engine cases, should be avoided unless the technician has a foolproof method of removing all the glass beads before the engine is assembled.

Petroleum distillates are a group of naphtha-base liquids used primarily as enamel paint thinners and for the dry cleaning of clothes. The only formula within that group that is considered safe in engine part cleaning operations is Stoddard Solvent. This cleaning solvent is blended to have a high and narrow boiling range. This means it will support combustion but will not burn by itself at normal shop temperatures. Although prolonged contact with this solvent is not recommended, a mechanic generally has no ill effects from limited exposure to the skin.

Stoddard Solvent should be used in a cold parts washer where it is filtered and recirculated (Fig. 5-34). A typical unit in widespread use is a machine which features a combination of cleaning solvent delivery methods, including a flexible metal hose, pistol grip flush nozzle, and a jet manifold to keep the solvent in constant circulation. In case of a fire, a fusible link melts and closes the lid, thereby preventing further spreading of the fire. Safety precautions should be observed:

1. Always wear eye protection.
2. Close the lid after each use.

3. Remove all parts immediately after cleaning.
4. Do not use wire brushes, rags, or paper towels as cleaning aids.

Chlorinated hydrocarbon is a name given to any of those materials normally used for cleaning carburetors and other nonferrous metals (Fig. 5-35). The cleaning action of this material takes place as the parts rest submerged in the liquid. Devices are available which will lightly agitate the parts as they soak, thereby reducing the cleaning time somewhat. This material requires no added heat. It is easily washed off by a stream of fresh water.

Note: Do not use Stoddard Solvent to wash off parts coming out of this solution, for they will be coated with an oily film.

This cleaning material is normally sold in six-gallon buckets containing five gallons of cleaner and a parts basket. It has a high relative cost, along with an objectionable odor similar to that of creosote, but it has been used for many years because of its convenience. This material has a very low vaporization point, so a layer of dissimilar material that floats on top is added to containers of this material to reduce the evaporation and smell.

Engine parts that can be effectively cleaned with this material include piston and connecting rod assemblies, valves, and any part of a rocker assembly. Cleaning of painted surfaces should be discouraged since the paint does not freely dissolve, but comes off the part in large pieces which tend to stick to anything that is later cleaned in the same solution.

FIG 5-33 Typical glass bead machine with a reclaimer.
(Zero Manufacturing Co.)

FIG 5-34 Solvent recirculating cleaning tank.
(Graymills Corporation)

FIG 5-35 Typical cold parts cleaner.
(Graymills Corporation)

CHECKLIST FOR ENGINE RECONDITIONING

The following checklist is provided as a quick reference of reconditioning procedures commonly performed on a cast iron, overhead valve engine. All the steps listed are discussed in one or more of the units of this text. If you reach an operation that has not been covered by your objectives, instructor demonstrations, and classroom discussion, consult with your instructor before proceeding further.

Keep the checklist along with Job Report Sheet 3A-1, Teardown and Assembly, and the student copy of the work order on the engine stand at all times.

☐ 1. Strip the cylinder block to the bare casting.
  a. Remove core plugs (freeze plugs).
  b. Remove cam bearings.
  c. Remove oil gallery plugs.

☐ 2. Boil all ferrous parts in hot tank after steam cleaning:
  a. Cylinder block—stripped
  b. Cylinder heads—stripped
  c. Manifolds—intake and exhaust
  d. Valve cover
  e. Timing cover
  f. Oil pan
  g. Miscellaneous steel or iron parts—*no camshafts or crankshafts*

☐ 3. Check camshaft and crankshaft for wear. See items number 22, 23, and 25.

☐ 4. Flush out all passages with high pressure water or steam clean after hot tank.

☐ 5. Run wire brushes through oil galleries to clear all foreign matter.

☐ 6. Check cylinder heads for cracks and warped surfaces. Warped surfaces can be machined smooth. Cracked parts may be repaired or replaced.

☐ 7. Obtain pistons for boring if cylinder wall taper is over 0.010 in. (0.254 mm).
  a. If cylinder walls have less than 0.010 in. (0.254 mm) taper, new rings can be installed.
  b. Clean cylinder walls with lacquer thinner to cut glaze. Repeat as often as necessary to remove all traces of glaze. Use stone deglazer if necessary.
  c. Remove old rings and check grooves for wear. Regroove when necessary (+0.006 in., or, 0.152 mm).
  d. Clean oil return holes and ring grooves in piston.
  e. Check ring end gap in lower cylinder wall (at least 0.003 in., or, 0.0762 per millimetre of cylinder bore).
  f. Install rings on piston and check for free rotation in the grooves.
  g. Skip to number 10 on checklist.

☐ 8. Bore cylinders to within 0.0015 in. (0.0381 mm) piston size. Wet hone to within 0.0005 in. (0.0127 mm) of specified clearance.

☐ 9. Wet hone the remaining 0.0005 in. (0.0127 mm) to obtain a cross-hatch pattern finish to remove tool marks. Use 29-300 grit stone.

☐ 10. Clean all threaded holes with a tap and inspect for damage. Blow any foreign matter from the holes with air, guarding your eyes.

☐ 11. Scrape the machined surfaces with a sharp scraper and draw file all flat surfaces. Use a long screwdriver to break off any encrustations on the water jacket surfaces.

☐ 12. Check alignment of deck surfaces and main bearing bore.

☐ 13. Blow out all oil passages with air.

☐ 14. Remove the main bearing caps and clean under the mating surfaces.

☐ 15. Check lifter bores. Use a new lifter to check if the bore is more than 0.003 in. (0.0762 mm) oversize.

☐ 16. Wash and thoroughly flush the cylinder block by steam cleaning. The cylinders will need to be washed with a brush and soap to remove all grit. Be certain to flush the oil galleries and water jackets with plenty of water. Finally, coat the cylinder walls with a light oil to prevent rusting.

☐ 17. Replace oil gallery plugs. Seal with RTV Silicone.

☐ 18. Replace core plugs. Seal with RTV Silicone.

☐ 19. Check all oil passages with compressed air.

☐ 20. Check cam lobes and journals against specifications for wear. Replace cam if lobes are worn 0.006 in. (0.152 mm) or more.

☐ 21. Check the cam bearing bores for roundness.

☐ 22. Before installing the cam bearings, slip them onto each respective camshaft journal for a fit check.

☐ 23. Install the camshaft bearings. Always check the oil passages with air after installing bearings to check oil hole alignment.

☐ 24. Install cam and check the fit.

☐ 25. Mike crankshaft and regrind as needed. If crank is to be sent out to be ground, remove all keys, gears, dowels, and pilot bushing or bearing.

☐ 26. Clean out oil passages in crank.

☐ 27. Check woodruff key under the crank gear for wear.

☐ 28. Install new crank gear as needed.

☐ 29. Remove the main bearing caps and inspect for burrs. Replace and torque to specifications.

☐ 30. Mike main bearing bores for concentricity (within 0.0005 in., or, 0.0127 mm).

☐ 31. Mike each bearing shell crown wall thickness with a ball mike (check against specifications).

☐ 32. Install bearings in the block and caps (watch for dowels or grooves) and torque caps to specifications.

☐ 33. Mike I.D. (inside diameter) of installed bearing with a telescoping gauge and outside micrometer or dial bore gauge.

☐ 34. Subtract the O.D. (outside diameter) of the journal from the I.D. measured in number 33 for each bearing and check this reading against oil clearance specifications. Plastigage can be used instead of numbers 33 and 34.

☐ 35. Install the rear main bearing seal. Cut rope type flush or stagger neoprene ends.

☐ 36. Lube bearings with white grease, chassis grease, or oil and lay crank.

☐ 37. Before caps are tightened, align insert thrust surfaces.

☐ 38. Torque main cap bolts. Check crank for free floating fit.

☐ 39. Check end play with dial indicator or feeler gauge.

☐ 40. Check all rings for depth and side play and freeness in their respective grooves.

☐ 41. Check each ring in respective cylinder for end gap (0.003 in., or, 0.0762 per millimetre of cylinder bore).

☐ 42. Install pin bushings in rods and bore to fit the pin if full floating type.

☐ 43. Before installing new pistons on rods, check I.D. of connecting rod bearing surface with caps torqued in place (see instruction sheet on servicing connecting rods).

☐ 44. Assemble pistons, rods, and rings (follow specifications on direction of piston on rod). Oil the pin surfaces as they are assembled.

☐45. Install cam gear and time to crank gear. Have instructor check.

☐46. Install front cover. Before tightening the cover, install front pulley to align seal. Then tighten cover.

☐47. Examine oil seal surface on the front pulley to make certain it is not grooved.

☐48. Install pistons and rods with ring gaps properly aligned.

☐49. Check each connecting rod with plastigage. Alternate method: Check each insert crown against specifications with ball mike. Use same procedure as with main bearings to determine oil clearance. Check rod end play (side clearance).

☐50. Make a final check to see that all pistons and rods are in their correct number order with notches or arrows facing forward.

☐51. Disassemble oil pump and check side and end clearance.

☐52. Assemble oil pump after priming or packing with grease.

☐53. Check oil pressure relief valve and install oil pump drive shaft.

☐54. Install oil pump.

☐55. Replace oil pan after hammering its gasket surface level.

☐56. Mike valve stems for wear (wear area nearest valve head).

☐57. Check valve guides for wear and install new or knurl the old ones.

☐58. Reface the valves and grind valve seats.

☐59. Valve heads should have a head thickness of at least 1/32 in., or, 0.413 mm (margin).

☐60. Check valve springs for height, squareness, and tension.

☐61. Assemble valves in head with proper seals. Check the installed height. Add shims to the springs as necessary to correct for installed height.

☐62. Install lifters (new ones needed with new camshaft) and pushrods. Lube with a moly or E.P. grease.

☐63. Install heads and manifolds following torque sequence and specifications.

☐64. Install valve covers and other parts necessary for installation.

☐65. Install old set of spark plugs and cover all other openings in the engine with masking tape.

☐66. Paint engine with factory colors.

☐67. Cover with trash bag until installed.

Name ______________________________ Date ______________

DIRECTIONS:

Circle the best answer to each question.

1. Recommended disassembly procedures include
   I. constant inspection for irregular wear or breakage indications.
   II. marking all parts for positions as assembled.
   A. I only B. II only C. Both I and II D. Neither I nor II

2. The teardown and assembly sheet (Job Report Sheet 3-A1) can be used as a
   I. checkoff list during disassembly.
   II. place to make drawings or notes to assist in the assembly process at a later date.
   A. I only B. II only C. Both I and II D. Neither I nor II

3. The following can be used to label engine parts:
   I. chalk.
   II. center punch and hammer.
   A. I only B. II only C. Both I and II D. Neither I nor II

4. The exhaust manifolds
   I. need to be labeled R or L.
   II. should be checked for cracks.
   A. I only B. II only C. Both I and II D. Neither I nor II

5. Preliminary wear and taper checks should be made
   I. on all cylinders.
   II. and the results discussed with your instructor.
   A. I only B. II only C. Both I and II D. Neither I nor II

6. The diameter of the cylinder just below the ring ridge is
   I. the original bore size.
   II. the area of least wear.
   A. I only B. II only C. Both I and II D. Neither I nor II

7. The tools necessary to remove the vibration damper include
   I. damper puller and bolts.
   II. assorted wrenches and step plates.
   A. I only B. II only C. Both I and II D. Neither I nor II

8. If the camshaft is to be reused, the
   I. lifters must be reground.
   II. pushrods must be shortened.
   A. I only B. II only C. Both I and II D. Neither I nor II

9. The crankshaft should be protected during disassembly by
   I. spraying with oil or coating with grease.
   II. using rubber hose or spark plug boots on the connecting rod bolts.
   A. I only B. II only C. Both I and II D. Neither I nor II

?

What about assembly - should we put boots on the con rod bolts

10. If a bearing wear analysis is to be made,
I. place them in the hot tank for cleaning.
II. label each half with its number and position.
A. I only B. II only C. Both I and II D. Neither I nor II

11. Oil gallery and core plugs
I. should be removed prior to hot tank cleaning.
II. can be removed with a rocker pry bar.
A. I only B. II only C. Both I and II D. Neither I nor II

12. Chemicals used for cleaning engine parts are most often
I. acids.
II. in the upper range (10–12) on the Ph scale.
A. I only B. II only C. Both I and II D. Neither I nor II

13. A steam cleaner
I. produces a solution that is 85–90% steam.
II. uses an acid solution with the steam.
A. I only B. II only C. Both I and II D. Neither I nor II

14. Strong alkaline solutions can damage
I. brass, copper, bronze, and aluminum.
II. nonferrous metals.
A. I only B. II only C. Both I and II D. Neither I nor II

15. Peening with glass beads will clean a metal part
I. without eroding its surface.
II. that cannot be cleaned with an alkaline solution.
A. I only B. II only C. Both I and II D. Neither I nor II

16. Before glass beading is done on any engine part, make certain
I. it is free of oil or grease.
II. the part is of the type your instructor has approved for this type of cleaning.
A. I only B. II only C. Both I and II D. Neither I nor II

17. Glass bead cleaning is not considered complete until
I. all glass beads are removed from the part while it is within the cabinet.
II. the machine itself has been properly cleaned.
A. I only B. II only C. Both I and II D. Neither I nor II

18. Some form of eye protection should be worn
I. only when alkaline solutions are being used for cleaning engine parts.
II. whenever eye hazards exist in the shop.
A. I only B. II only C. Both I and II D. Neither I nor II

19. The solvent tank should
I. be kept free of burnable materials.
II. have its lid (cover) closed after each use.
A. I only B. II only C. Both I and II D. Neither I nor II

20. Cleaning of painted surfaces should not be done
I. in a hot tank solution.
II. in carburetor cleaner solutions.
A. I only B. II only C. Both I and II D. Neither I nor II

DATE COMPLETED ____________ INSTRUCTOR'S SIGNATURE ____________________

unit 6

Measurements and Inspection

When the parts of an engine have been thoroughly cleaned, measurements are taken on each piece. This is done to determine whether it needs replacement, machining, or no further work. This unit covers the use of measuring tools and general inspection procedures.

OBJECTIVE 6-A (written test)

After completing the preparation assignment, you will answer a minimum of 7 out of 10 multiple choice questions in a written test, without using reference material, on the following topics:

1. English and metric systems of measurement
2. Measuring tools
3. Engine parts list

Preparation:

Study pages 141 through 157.

The importance of accurate measurement by a modern auto mechanic cannot be overemphasized. As the car goes through constant development, components and assemblies are manufactured to finer and finer tolerances. Most of the automotive parts produced today are designed to be used in a number of applications and to provide interchangeability for faulty, worn, or damaged pieces. Careful and exact measurement by service technicians are required before making recommendations as to reuse or replacement of parts.

Automotive manufacturers use many different precision-measuring tools. The student should be able to do the following:

1. Understanding the English and the metric system
2. Angular measurement
3. Measuring tools
   a. Steel rule
   b. Feeler gauge
   c. Telescoping hold gauge
   d. Micrometer
   e. Vernier caliper
   f. Dial indicator
   g. Torque wrench
4. Tables and conversion charts

ENGLISH VS. METRIC SYSTEM

The English system of measurement in inches and pounds is still the most common system used throughout the automotive industry of the United States. However, the rest of the world uses the metric system of measurement.

The metric system has 10 as its multiplier just like our system of money. As an example, 10 millimetres equal 1 decimetre, and 10 decimetres make 1 metre. Listed below are the common distance and weight measurements used in automotive service work:

English	*Metric*
12 inches = 1 foot	10 millimetres = 1 centimetre
3 feet = 1 yard	10 centimetres = 1 decimetre
5280 feet = 1 mile	10 decimetres = 1 metre
16 ounces = 1 pound	1000 metres = 1 kilometre
2000 pounds = 1 ton	100 centigrams = 1 gram
	1000 grams = 1 kilogram

A fuller table of English-metric equivalents may be found in the Appendixes, inside the back cover of this text.

The Metric Act of 1975, which became a law in December of that year, establishes a national policy for the United States to convert to the metric system; but there is no timetable included in the law. An understanding of the metric system can be of great value because of import vehicles requiring service and repair (Table 6-1).

MEASURING TOOLS

STEEL RULE

The simplest measuring tool for automotive service work is the 6 in. steel rule, which is accurate to 1/32 in. This rule may be divided in increments of 1/16 to 1/64 in. The most popular steel rule has inches divided into 1/32 in. on one edge and millimetres on the other edge (Fig. 6-1). The rule marks both for measuring systems makes learning the metric system somewhat easier. The 6 in. rule with the dual scales measures to approximately 152.4 millimetres (1 inch = 25.4 millimetres, so 6 inches = 152.4 millimetres).

On the backside of the rule is a decimal equivalent chart for the conversion of fractions to decimals and vice versa. Conversion without a chart is accomplished by dividing the numerator by the denominator.

Examples: ¼ inch = 250 thousandths (0.250), or 1.000 ÷ 4 = 0.250
⅛ inch = 125 thousandths (0.125), or 1.000 ÷ 8 = 0.125

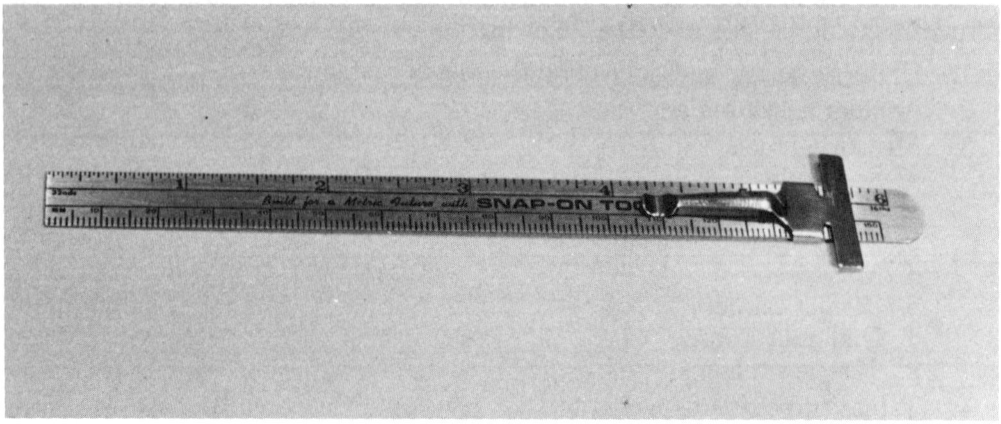

FIG 6-1 Six-inch (154 mm) steel rule with a sliding clip.

TABLE 6-1

METRIC SYSTEM—SI

The International System of Units (Systeme International d'Unites) officially abbreviated "SI" in all languages – the modern metric system

QUANTITY	EXAMPLES OF APPLICATIONS	METRIC UNIT	SYMBOL
Length	Dimensions Tire rolling circumference Turning circle/ radius Braking distance	metre	m
	Greater than 999 metre	kilometre	km
	Dimensions	millimetre	mm
	Depth of surface finish	micrometre	um
Area	Glass & Fabrics Brake & Clutch linings Radiator area etc.	square centimetre	cm^2
	Small areas	square millimetre	mm^2
Volume	Car Luggage Capacity	cubic metre	m^3
	Engine Capacity	litre	l
	Vehicle fluid capacity	cubic centimetre	cm^3
Volume Flow	Gas & Liquid	litre per second	l/s
Time Interval	Measurement of elapsed time	second minute hour day	s min h d
Velocity	General use	metre per second	m/s
	Road speed	kilometre per hour	km/h
Acceleration & Deceleration	General use	metre per second squared	m/s^2
Frequency	Electronics	hertz kilohertz megahertz	Hz kHz mHz
Rotational Speed	General use	revolution per minute revolution per second	rpm rps
Mass	Vehicle mass Legal load rating	megagram	t
	General use	kilogram	kg
	Small masses	gram milligram	g mg
Density	General use	kilogram per cubic metre gram per cubic centimetre kilogram per litre	kg/m^3 g/cm^3 kg/l
Force	Pedal effort Clutch spring force Handbrake lever effort etc.	newton	N
Moment of Force (Torque)	Torque	newton metre	N-m
Power, Heat Flow Rate	General use Bulbs Alternator output Engine performance Starter performance	watt kilowatt	W kW

QUANTITY	EXAMPLES OF APPLICATIONS	METRIC UNIT	SYMBOL
Celsius Temperature	General use	degree Celsius	^{o}C
Thermodynamic Temperature	General use	kelvin	k
Electric Current	General use	ampere milliampere microampere	A mA μA
Potential Difference (Electromotive Force)	General use	kilovolt volt millivolt microvolt	kV V mV μV
Electric Resistance	General use	megohm kilohm ohm	MΩ kΩ Ω
Electric Capacitance	General use	farad microfarad picofarad	F μF pF
Fuel Consumption	Vehicle performance	litre per 100 kilometre	l/100 km
Oil Consumption	Vehicle performance	litre per 1000 kilometre	l/1000 km
Stiffness	Linear stiffness	kilonewton metre	kN/m
Tire Revolutions	Tire Data	revolution per kilometre	rev/km
Pressure	Tire Coolant Lubricating oil Fuel pump delivery Engine compression Manifold Brake line (hydraulic) Car heating & ventilation Barometric pressure	kilopascal	kPa
Luminous Intensity	Bulbs	candela	cd
Accumulator Storage Rating	Battery	ampere hour	A-h

U.S.A./METRIC COMPARISON

QUANTITY	USA	METRIC – SYMBOL	
Length	Inch-Foot-Mile	Metre	m
Weight (mass)	Ounce-Pound	Kilogram	Kg
Area	Square inch/Foot	Square Metre	m^2
Volume-Dry	Cubic inch/Foot	Cubic Metre	m^3
-Liquid	Ounce-Pint-Quart-Gallon	Litre	l
Velocity	Feet Per Second	Metre per Second	m/s
Road Speed	Miles Per Hour	Kilometre per Hour	km/h
Force	Pound-Force	Newton	N
Torque	Foot-Pounds	Newton metre	N-m
Power	Horsepower	Kilowatt	kW
Pressure	Pounds Per Square Inch	Kilopascal	kPa
Temperature	Degrees Fahrenheit	Degrees Kelvin and Celsius	K ^{o}C

FEELER GAUGE

The feeler gauge is often used to check the gap between two parts, such as an oil pump gear end clearance (Fig. 6-2). It can measure a minimum distance in a slot, such as the end gap of a piston ring. Each feeler strip is labeled with its thickness in thousandths of an inch or millimetres or both (Fig. 6-3). The strips are usually made from high quality steel ground to a consistent thickness. This tool should be kept dry to prevent rust.

STRAIGHTEDGE

The straightedge is simply a rigid piece of steel with a precision-ground flat, straight surface (Fig. 6-4). When checking the flatness of a large engine part, such as a cylinder head or block, the surface of the piece is first cleaned. Then, placing the ground edge on the head surface, the engine repairman can attempt to freely pass a feeler gauge of a given thickness between the straightedge and the surface. If the gauge passes, the part is warped and may require resurfacing or reboring.

This tool should be treated with care by avoiding hard contact with other metals. Store in a moisture-free area and let it hang to prevent warpage.

TELESCOPING GAUGES

Telescoping gauges are sold in sets and labeled as to the limits of their measurement. This gauge is most often used to measure the inside of an opening. For engine repair work, this includes measuring the diameters of cylinder walls, main bearing housing bores, camshaft housing bores, and valve guide sizes. The head of the telescoping gauge expands across the opening and is then locked to stop its movement (Fig. 6-5). A micrometer is then used to determine the exact distance to which the gauge expanded (Fig. 6-6).

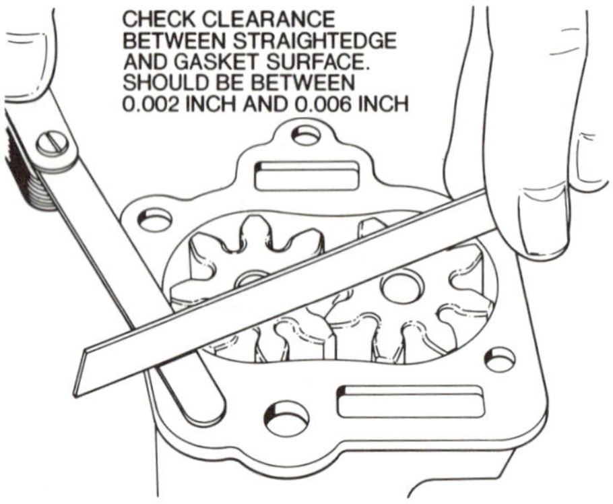

FIG 6-2 Measuring oil pump end clearance with a feeler gauge and small straightedge.
(Buick Motor Division, GMC)

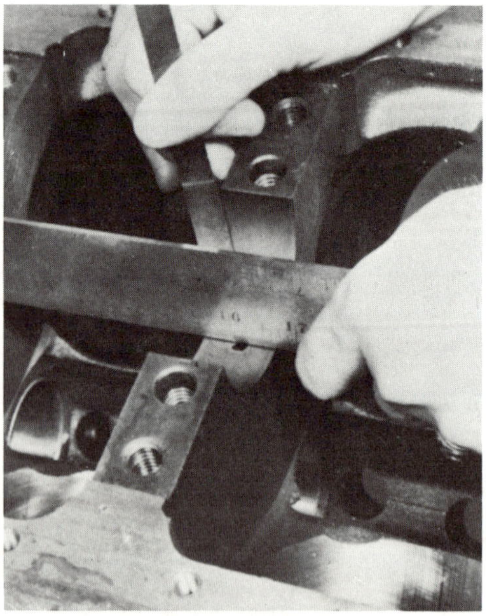

FIG 6-4 Using a straightedge to check main bearing housing bore alignment.
(Chevrolet Motor Division, GMC)

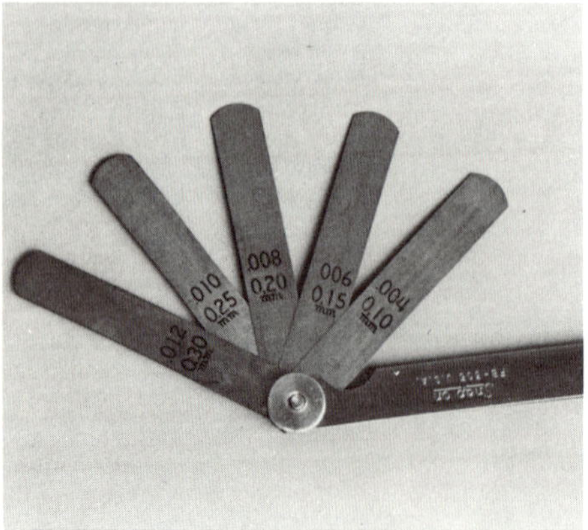

FIG 6-3 Feeler gauge set marked for inches and millimetres.

Small hole and split ball gauges serve the same purpose as the telescoping gauge, but they are usually used to measure small holes such as valve guides.

MICROMETER

Micrometers are the most widely used tools of the automotive mechanic. They measure accurately to one-thousandth of an inch (0.001 in.). Some models with a vernier scale are designed to read ten-thousandths of an inch (0.0001). The micrometer consists of a frame, anvil, spindle, sleeve, and thimble (Fig. 6-7). The sleeve is sometimes called a hub or barrel. The part to be measured is placed between the anvil and the spindle faces. The spindle is moved toward the anvil by rotating the thimble until the spindle and anvil contact the part. The dimension is then found by reading the sum of the figures on the sleeve and thimble.

Micrometers are available in many sizes. The most common size is the one-inch micrometer, which provides measurement up to one inch. A two-inch micrometer measures between one and two inches, and so forth.

The beveled edge of the micrometer thimble is divided into 25 equal parts, each line representing one-thousandth of an inch (0.001 in.). Rotating the thimble from one line to the next moves the thimble one-thousandth of an inch (0.001 in.). Each complete turn of the thimble moves the spindle 25 thousandths of an inch (0.025 in.).

To read the micrometer in thousandths (see example, Figure 6-8),

a. Multiply the number of vertical divisions visible on the sleeve by 0.025.
b. Add this to the number of thousandths indicated by the number of the thimble that is closest to the long horizontal line on the sleeve.

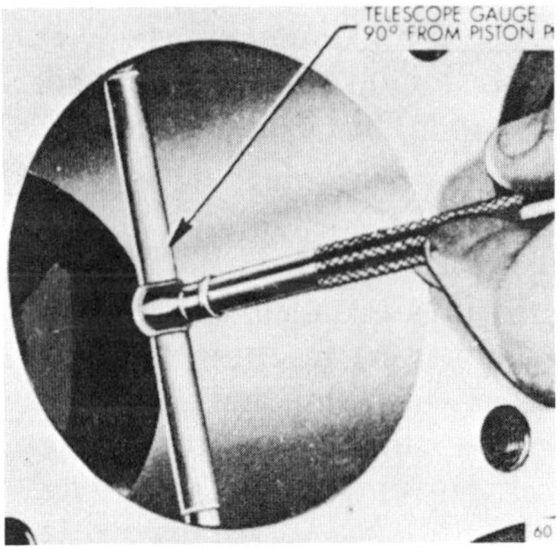

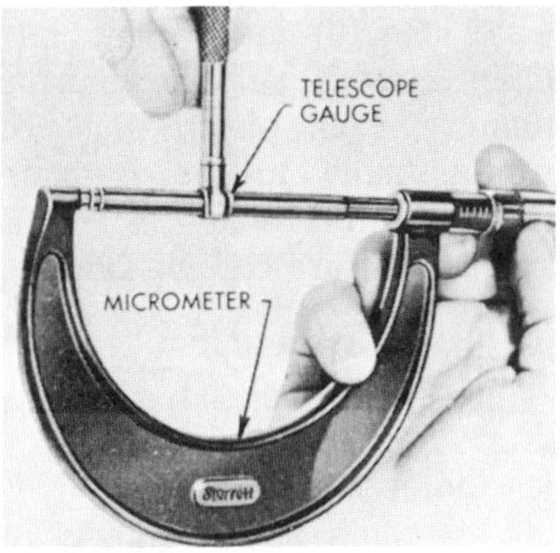

FIG 6-5 Expand the telescoping gauge to its maximum diameter.
(Buick Motor Division, GMC)

FIG 6-6 Use an outside micrometer to measure the telescoping gauge.
(Buick Motor Division, GMC)

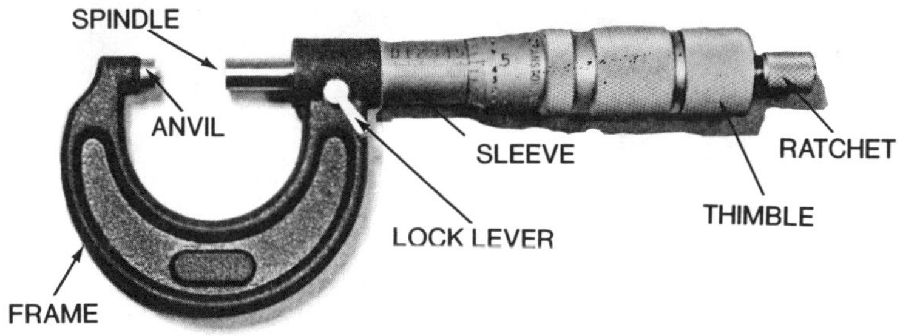

FIG 6-7 Typical 1 inch (0–1 inch) micrometer with parts named.

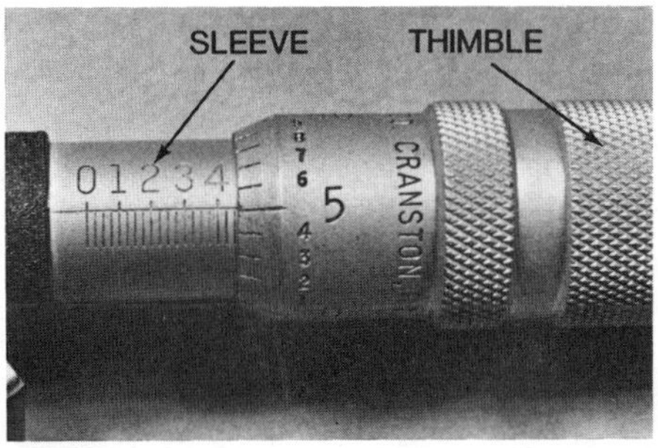

FIG 6-8 Micrometer readings are taken from the sleeve and thimble.

One way to remember how to read the micrometer is to compare it to making change from a ten-dollar bill. Count the numbers on the sleeve as dollars, the vertical line on the sleeve as quarters, and the numbers on the thimble as cents. Add up the change you would receive and put a decimal point in front of the numbers instead of a dollar sign.

Readings in ten-thousandths of an inch (0.0001 in.) can be obtained on micrometers with a vernier scale on the top of the sleeve (Fig. 6-9). Since this micrometer allows more accurate measurements, an additional reading must be made. When reading the ten-thousandths micrometer, first note the thousandths reading as with the ordinary micrometer. Then note which line on the top of the sleeve aligns exactly with any of the lines on the thimble. Figure 6-10 shows the sleeve number 4 is the one most exactly lined up. If the regular thousandths reading was 0.304 and the sleeve vernier reading was 4, the total measurement would be 0.3044.

The metric micrometer uses the same names for its parts, but the indications are different (Fig. 6-11). One revolution of the spindle moves it a distance of 0.5 mm. The sleeve is graduated in mm from 0 to 25 so that it takes two spindle revolutions to move it 1 mm. The thimble is graduated in 50 divisions with every fifth line numbered. Rotating the thimble from one graduation to the next moves the spindle only 1/100 mm. Two graduations of movement would equal 2/100 mm and so on.

Micrometers are available which have both English and metric readouts (Fig. 6-12).

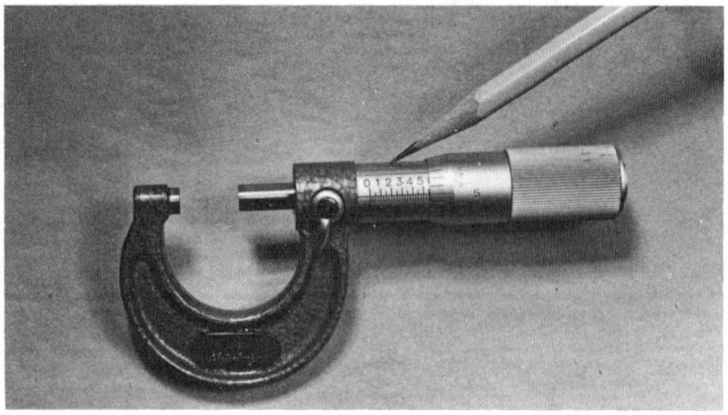

FIG 6-9 A micrometer with a vernier scale on top will measure in ten-thousandths (1/10,000) of an inch.

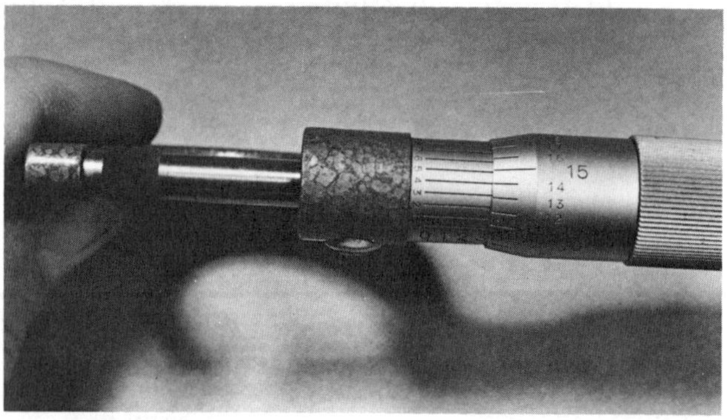

FIG 6-10 The sleeve number 4 is the closest to alignment with any of the thimble marks.

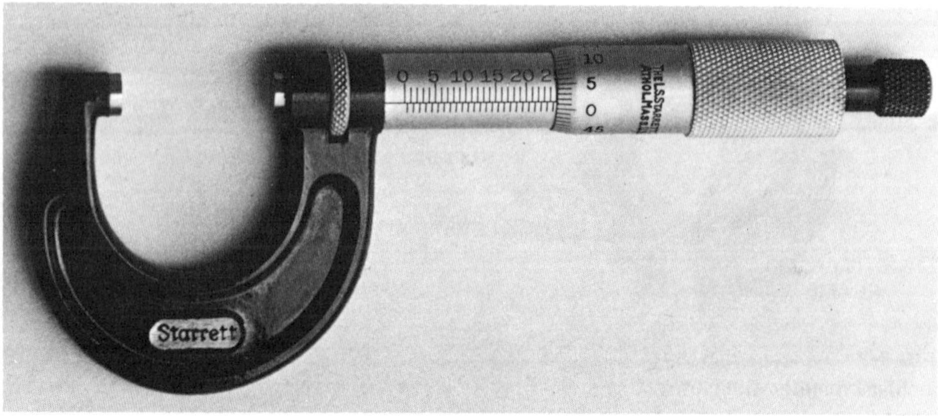

FIG 6-11 Metric micrometer.

Helpful hints when using micrometers:

1. Keep all micrometers clean by storing in a dust-free container and wiping clean after each use.
2. Avoid excessive tightening of the thimble. This can damage internal threads.
3. Check micrometer accuracy often by using a micrometer standard.
4. Do not store micrometers with the anvil and spindle in contact.
5. Hold the micrometer with your little finger through the opening in the frame. This allows one hand to measure while the other hand holds the part to be measured (Fig. 6-13).

VERNIER CALIPER

A vernier caliper is a common measuring tool used by the automotive machinist (Fig. 6-14). It offers accurate indications of one-thousandth of an inch (0.001 in.) for inside, outside, and depth measurements and is usually faster than using a micrometer.

This device consists of two parts: stationary and movable. The stationary part is the beam, with English or metric graduation of approximately six to eight inches in length, a hooked part of which forms one half of the measuring jaws. The movable part holds the Vernier scale and the other half of the jaws. Each graduation mark on the beam indicates twenty-five thousandths of an inch (0.025 in.). Each graduation mark on the Vernier scale indicates one-thousandth of an inch (0.001 in.).

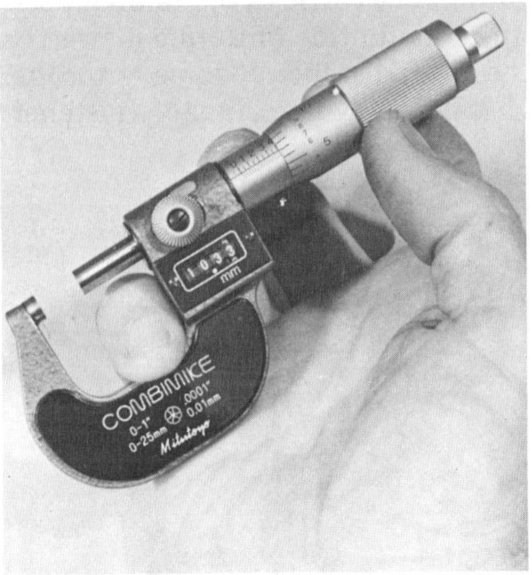

FIG 6-12 Combination English-metric micrometer.

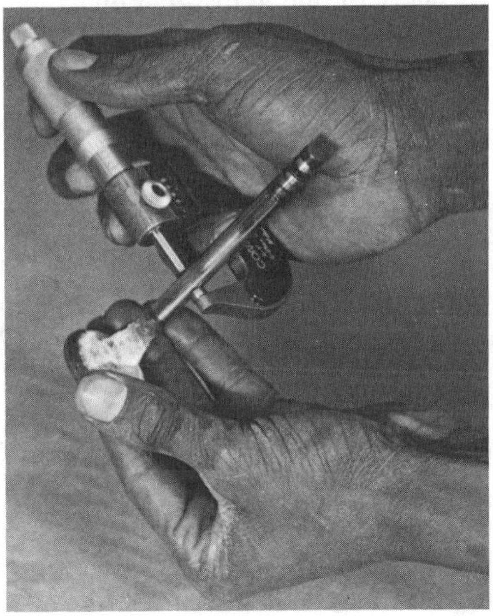

FIG 6-13 Correct holding position for a micrometer. The little finger is looped through the frame opening.

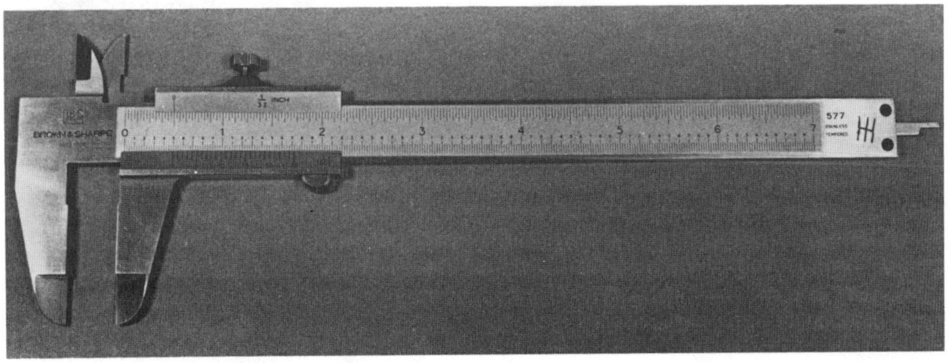

FIG 6-14 Vernier caliper.

To read the vernier caliper, bring the two jaws to the point to be measured. First note the position of the zero of the Vernier scale (movable part). In the example (Fig. 6-15), the zero is between 1.100 and 1.125, so the first figure to read is 1.100. The second step is to look on the vernier scale and find the number which most exactly lines up with one on the beam scale. In the example, 7 lines up best with one of the thousandths on the beam. By adding the two numbers, the total reading is obtained: 1.100 plus 0.007 equals 1.107.

DIAL INDICATOR

Another measuring tool used frequently by the automotive machinist is the dial indicator. Dial indicators come in a variety of designs, two of which are used in engine repair (Fig. 6-16). The indicator on the left in Figure 6-16 is often used to measure out-of-roundness or the amount of movement between two parts, which is called *clearance*. The one pictured has a screw-base attachment. The instrument on the right, used to measure cylinder size, including taper and out-of-roundness, is frequently referred to as a dial bore gauge.

The units of measurement are usually thousandths (0.001) of an inch for the indicator on the left and five ten-thousandths (0.0005) for the dial bore gauge.

The range of measurement for the dial indicator used in engine work should be at least one-half (½) inch, preferably one (1) inch. For the dial bore gauge, different adapters can be attached to give a range of 2 in. to 8 in. Fig. 6-17 .

TORQUE AND TORQUE WRENCHES

Nearly every threaded fastener on a modern automobile carries a tightening specification commonly known as torque. The simplest definition of torque is twist or turning effort. Torque is based on the law of the lever: distance × force = torque. Torque is the twist effort about one point, which is normally the center of a fastener.

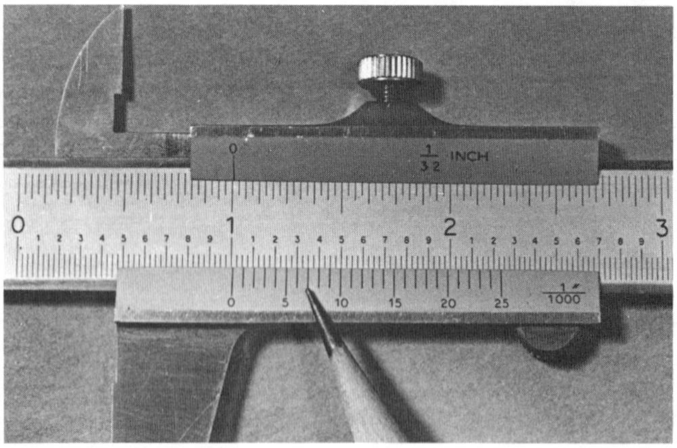

FIG 6-15 Reading the vernier scale.

FIG 6-16a Dial indicator with a screw base.
(Oldsmobile Division, GMC)

Torque, as related to fasteners, is really the resistance offered in the tightening process and must not be confused with tension, which is the amount of pull in a straight line. One of the classifications of a spring is its tension.

As nuts or bolts are tightened to a specific torque, they provide a certain amount of tension force to the pieces being held together. This is done to keep the parts aligned as originally assembled and to maintain a tight seal. To provide the proper tension between parts, the threads of both fasteners must be free of all foreign matter or damage. An important step in completing any engine repair job is to clean the threaded surfaces of all fasteners, a process known as thread chasing. This allows the torque wrench to bring all assembled pieces together with the proper tension.

Torque specifications are listed in units of foot-pounds (ft-lbs) and inch-pounds (in-lbs) for domestic vehicles while the Newton-metre (N•m) and metre-kilogram (mkg) are the common units for foreign vehicles. An English-metric conversion table is provided on page 37.

The importance of correct torque cannot be overemphasized. If a nut or bolt is not tightened enough, it may loosen and fall from the vehicle, causing serious personal injury. On the other hand, a fastener that is overtightened may break off, causing lost time in the repair or replacement of an expensive part.

Gasket leakage is often the result of minor undertorque while distortion of metal parts is the most common overtorque problem. The parts being assembled—not the fasteners—often dictate the amount of torque necessary for good sealing and resistance to disassembly by vibration. When in doubt, consult manufacturer's specifications for proper torque. If no specifications are available, the maximum torque capacities of the common sizes of bolts can be found on page 47.

Torque wrenches are available in many different designs. Capacity of torque, drive end size, and method of indication are all features to consider in buying and using a torque wrench.

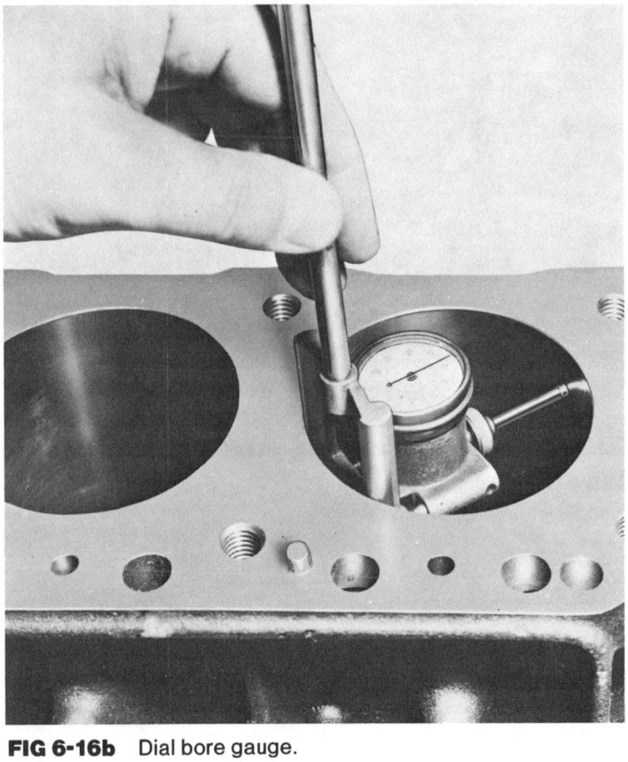

FIG 6-16b Dial bore gauge.
(Oldsmobile Division, GMC)

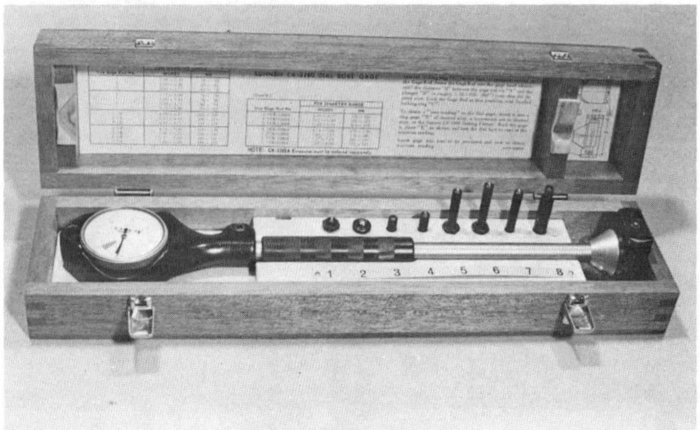

FIG 6-17 Dial bore indicator with adapters.

Wrenches with small torque capacity are graduated in inch-pound units and have a 1/4 or 3/8 drive end. One hundred fifty inch-pounds is a common average capacity for small torque applications.

Large capacity torque wrenches have a 3/8 or 1/2 drive end and are graduated in foot-pounds. Two hundred foot-pounds would accommodate most large torque requirements.

These wrenches have different means of indicating when the proper torque has been reached. The flexible beam wrench uses a semi-rigid pointer on a scale; the dial indicator wrench has a moving needle to indicate amounts of torque (Fig. 6-18). Another type, possibly the most popular in automotive service work, clicks or breaks slightly when the preset torque is reached (Fig. 6-19).

Helpful hints when using torque wrenches:

1. All good torque wrenches are precision instruments and should be used in accordance with manufacturer's recommendations.
2. Always follow manufacturer's specifications on torque requirements.
3. Use a torque wrench within its designed limits. If extensions are to be used to change the effective length of a torque wrench, consult your instructor about calculating the torque.
4. Do not use a pipe or other materials on the handle of a torque wrench for easier turning.
5. While pulling on the handle of the wrench, use your other hand for support by pushing on the drive end (Fig. 6-20). This will help prevent slippage of the wrench from the nut or bolt and will give more accurate torque settings.

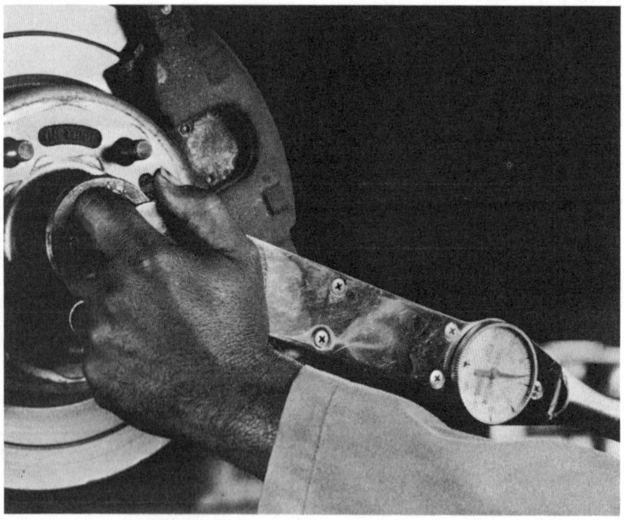

FIG 6-18 Dial indicator torque wrench.

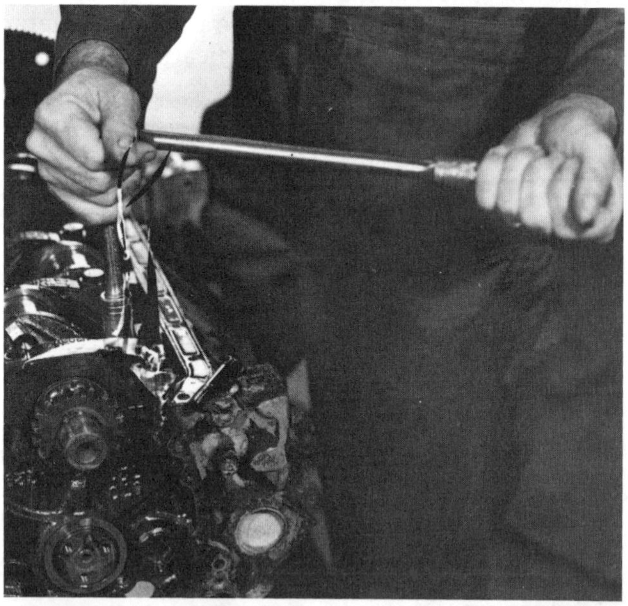

FIG 6-20 Pull the wrench handle while supporting the drive end.

FIG 6-19 Click torque wrench. (Proto)

6. Damaged fasteners can cause torque inaccuracy. Replace all nuts and bolts with damaged threads. Replace bolts with surface damage on the underside of the head.
7. If oil is used as a thread lubricant, increase torque values by 20 percent.

COMPONENT FAILURE ANALYSIS

During disassembly and after the engine parts have been cleaned, all components should be inspected. Inspection is a process of examining each piece for wear and damage in order to determine through measurement and judgment, what parts should be discarded or reused. Even pieces which are to be used again may require some machine work. Measurement operations on pieces with normal, average wear will be covered in later units related to each part.

When components fail during engine operation, the engine technician should be able to give the owner some reasons why they failed. Among the reasons for component failure are lack of care in operation and maintenance, failure to properly maintain the engine, or both.

The following figures (6-21 through 6-34) show typical engine part failures and their causes. This listing is by no means complete. For problems not treated here, consult your instructor, who can assist you in determining component failures. Also, review the information in Unit 3, Diagnosis Procedures.

FIG 6-21

FIG 6-22

FIG 6-21 Cocked valve spring. This valve is an example of the damage a cocked valve spring can cause. Notice the scuffing at the upper left and lower right portions of the stem. A spring cocked badly enough to cause this scuffing pattern would also force the valve face to seat unevenly. The uneven contact can be seen on the valve face, which was partially "cleaned up" on a grinder before the picture was taken. Valve burning or breaking is a likely result of uneven seating. Also, the scuffed stem could cause the valve to stick open. (Dana Corporation)

FIG 6-22 Poor seating. Even though the valve face contacts its seat around the entire perimeter of the face, it still may not be seating properly. The correct seating pattern is in the middle of the valve face (see valve at top). The lower valve was seating on the inner edge of the face, and valve burning might soon have resulted because of the excessive amount of material protruding into the combustion chamber. (Dana Corporation)

FIG 6-23

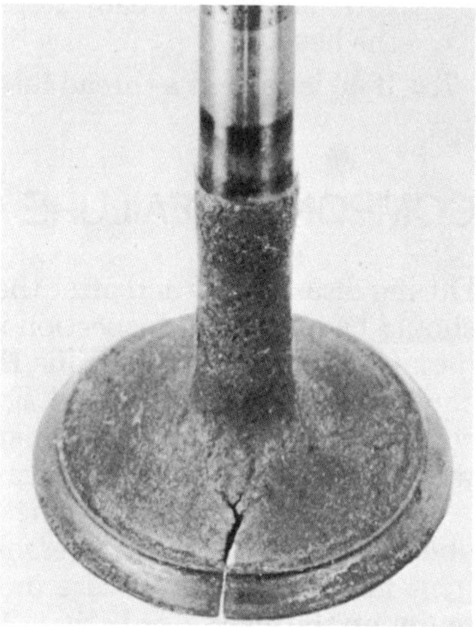

FIG 6-24

FIG 6-25

FIG 6-23 Face burning.
Although valve faces are made of materials that can withstand very high temperatures, there is a limit to how much heat they can take. Valve burning is usually progressive, and quite often starts when a flake of carbon comes off a heavily carboned valve. A channel is formed through which hot gases can pass, raising the temperature of the valve face and burning it. As the valve burns, the leakage area is increased, causing a larger area of the valve to be affected. Other common causes of valve burning are an incorrect seat angle, a loose or cracked seat insert, and abnormally high combustion chamber temperatures.
(Dana Corporation)

FIG 6-24 Hoop stressing.
Hoop stressing is a very common cause of valve failure. It starts because the center of a valve head cools more slowly than the outer edge. As the outer edge cools, it tries to contract. But the center of the valve, still hot, is expanded. Something has to give, so a crack in the outer edge of the valve face is the result. Like valve burning, this condition is caused by valve temperatures much higher than the valve was designed to withstand. The valve shown here has a hardened face. As a result, the face material only cracked, but the softer head material was burned and eroded by the escaping gases.
(Dana Corporation)

FIG 6-25 Scoring.
Both piston and rings are scored on the piston shown above. Scoring results when metal-to-metal contact occurs between rubbing surfaces and the temperature of one of these surfaces reaches the melting point of the material. Scuffing is a lighter form of damage than scoring and has the same causes: (A.) improper break-in; (B.) improper combustion; (C.) lack of cylinder lubrication; (D.) coolant leakage into the cylinder; (E.) overheating due to a faulty cooling system, and (F.) incorrect or insufficient bearing or piston clearance.
(Dana Corporation)

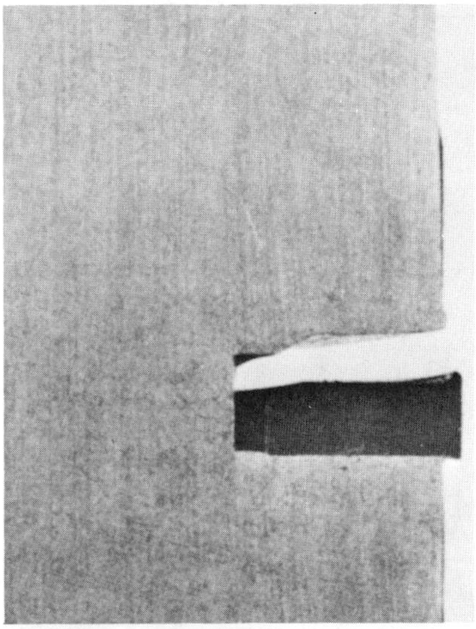

FIG 6-26

FIG 6-27

FIG 6-28

FIG 6-26 Top ring and groove side wear.
More than 90 percent of all aluminum pistons removed for reringing have excessively worn top grooves as above. This area wears most because it is exposed to maximum combustion temperature and pressure and all airborne abrasives that enter the engine. Before installing new rings, always check with a Perfect Circle Wear Gauge or a 0.006 inch feeler gauge and a new ring. Correct excessively worn top grooves quickly and easily by remachining the grooves with a PC Manulathe to accommodate a steel 0.025-inch top groove spacer above the top compression ring.
(Dana Corporation)

FIG 6-27 Connecting rod misalignment.
Note the wear pattern on this piston. Contact with cylinder wall shows on bottom of the skirt at left and at ring lands on the right. There is also a diagonal wear pattern extending across the skirt of the piston. This uneven wear is due to a bent or twisted connecting rod or cylinder bored at an angle to the crankshaft. When connecting rods are misaligned, rings do not have proper face contact with cylinder wall, pistons wear rapidly and unevenly, oil consumption is excessive and the engine is more prone to scuff or score. Always check rod alignment.
(Dana Corporation)

FIG 6-28 Detonation.
This piston has burned completely through the piston lands as a result of detonation. Note the broken oil ring. Detonation finally results in severe piston burning, accelerated top groove wear, piston ring breakage, scoring, sticking and, ultimately, complete engine failure. Detonation is one form of abnormal combustion where the temperature and pressure in the combustion chamber become excessive. Detonation is caused by: (A.) lean fuel mixture; (B.) fuels of too low an octane; (C.) over advanced ignition timing; (D.) lugging; (E.) carbon deposits on pistons and cylinder heads, and (F.) excessive milling of cylinder heads (which increases compression ratio). (Dana Corporation)

FIG 6-29

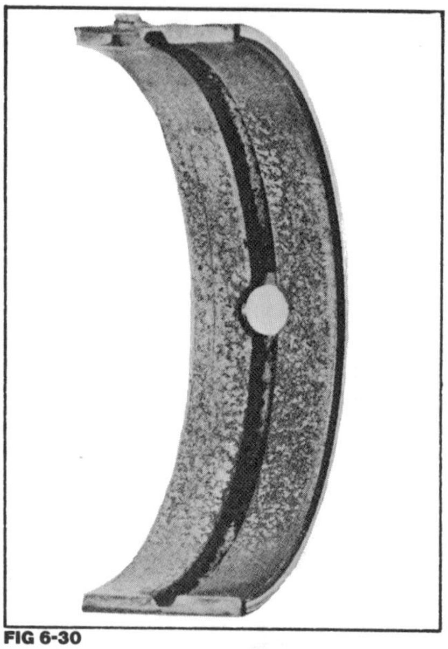

FIG 6-30

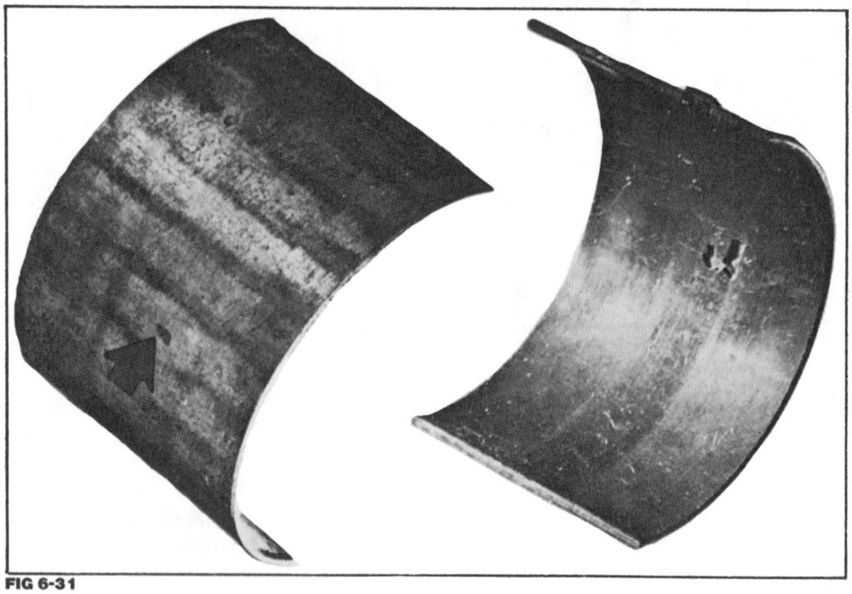

FIG 6-31

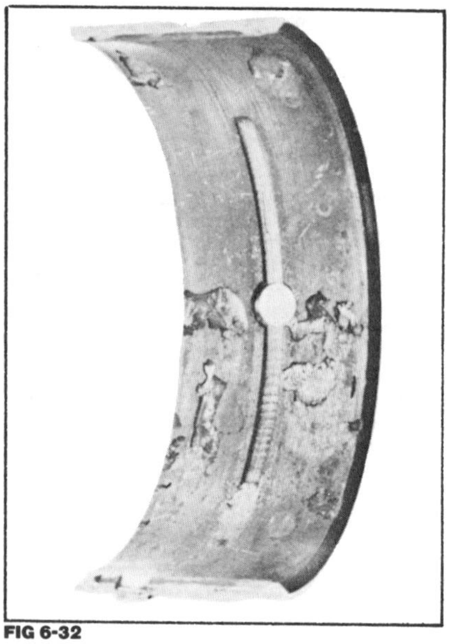

FIG 6-32

FIG 6-29 Preignition.
The excessive heat and pressure resulting from preignition burned a hole in the head of this piston. The pressure drove the top ring down with sufficient force to bend the intermediate ring land and shear the lower ring land. Preignition is the igniting of the fuel charge prior to the regular ignition spark. Detonation, wild ping and power loss are all caused by preignition, resulting in major damage to pistons, rings and valves. Causes of preignition are (A.) carbon deposits that remain incandescent; (B.) spark plugs of too high a heat range or plugs with cracked or broken ceramic cores; (C.) detonation or the conditions leading to it, and (D.) sharp edges in the combustion chamber.
(Dana Corporation)

FIG 6-30 Dirt is the most common cause of bearing damage. Dirt can embed itself in the soft bearing to cause wear and decreased life of both the bearing and its journal. This can be prevented by cleaning the engine thoroughly during bearing installation and by proper maintenance of both air and oil filters.
(Dana Corporation)

FIG 6-31 A particle of dirt was left (see arrow) during bearing installation. It remained behind this bearing on its cap to push the bearing inward at one point, thus increasing localized pressure and heat. This damage started because of improper cleaning of a bearing. Dirt was allowed to remain and create an area of excessive heat that led to this damage.
(Dana Corporation)

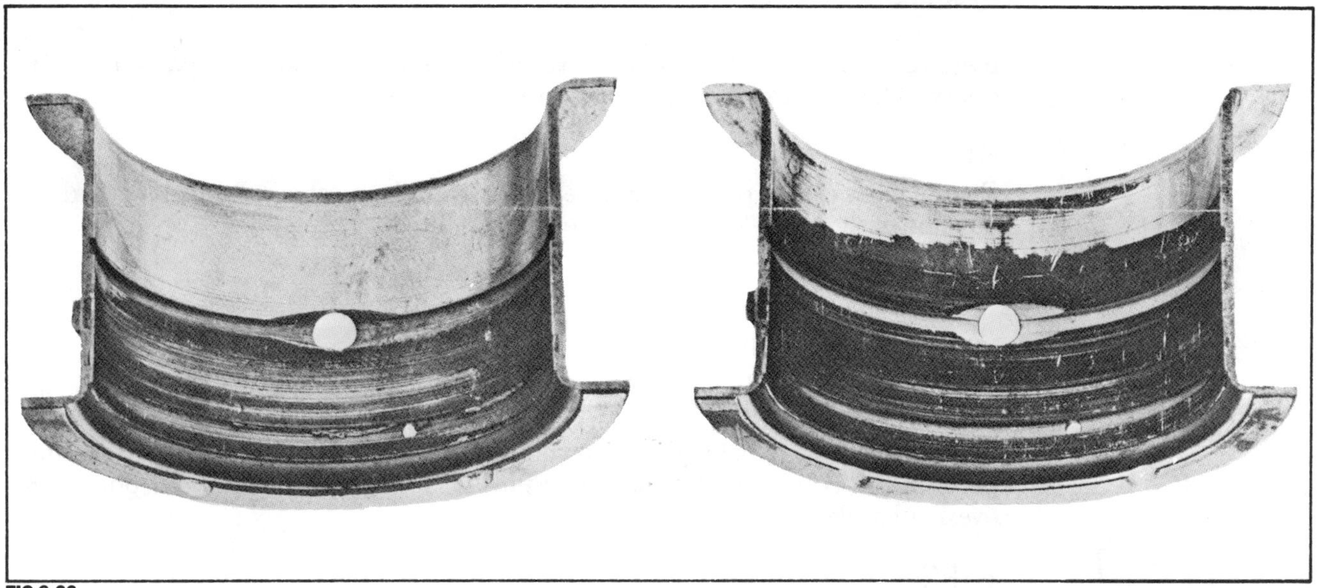

FIG 6-33

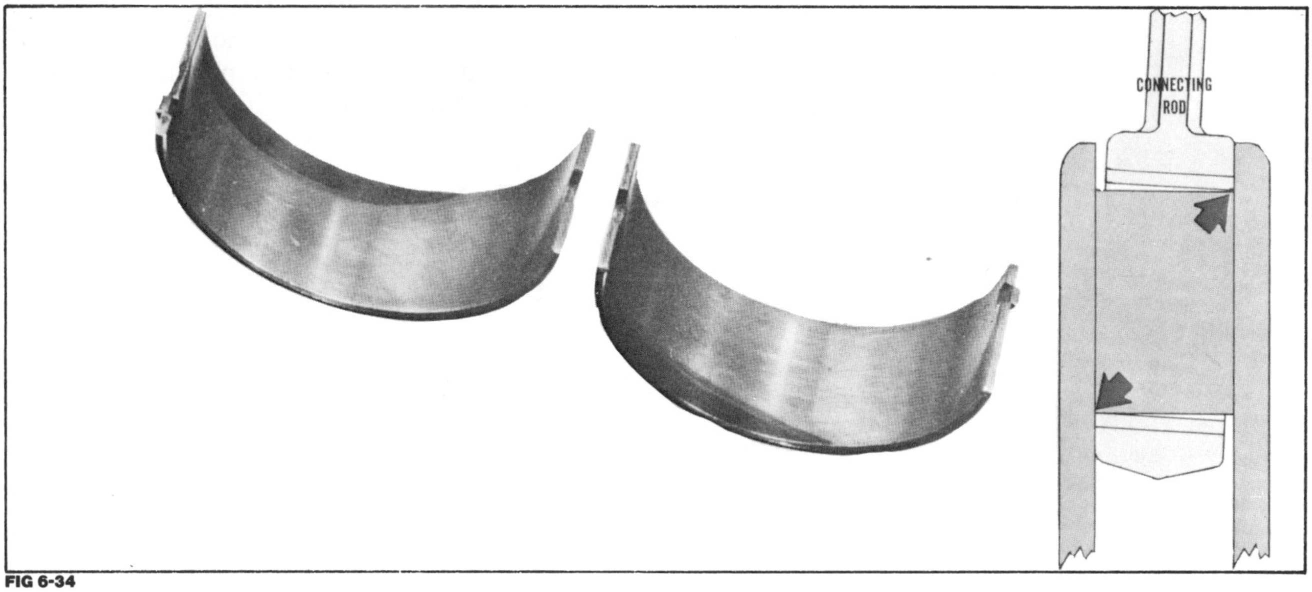

FIG 6-34

FIG 6-32 Fatigue of the metal has caused breaking away and voids on the surface of the bearing. This occurs from excessive loading of the engine and the resulting heat. Fatigue starts with fine radial cracks on the bearing's surface. They gradually move inward toward the bond line where the bearing surface and back are joined together. These cracks expand, turning parallel with the bond line, to ultimately meet other cracks. This releases small sections of bearing metal which crowd the journal, become overheated, melt and hasten fatigue to complete bearing failure.

(Dana Corporation)

FIG 6-33 Tapered journals allow areas of excessive clearance between the journal and bearing, which distribute more wear on one edge of the bearing, as shown. Before installing new bearing inserts, always check the journals for wear or taper with a micrometer or with Plastigage.

(Dana Corporation)

FIG 6-34 Bent connecting rods can cause concentrated wear on the bearings. This angular loading from a bent rod will cause the excessive wear shown on one edge of the upper bearing insert and the opposite edge of the lower bearing insert. When this wear pattern is found, check the connecting rod for alignment.

(Dana Corporation)

ENGINE PARTS

Regardless of the condition of the engine being serviced, some new parts and materials will be required before the assembly operation can begin.

KITS

Engine parts can be purchased in several ways. The most economical method is to buy one of the three kinds of engine kits:

1. Crank kit
2. Overhaul or recondition kit
3. Master or complete kit

The crank kit contains a reground crankshaft with matching main bearings and connecting rod bearings. The overhaul kit includes the crank kit plus piston rings, cam bearings, rod bushings, and a timing gear or chain. The master kit includes parts in the crank and overhaul kits plus valve lifters, camshaft, oil pump, and overhaul gasket set.

PARTS LIST

To reduce the "down" time of an engine and avoid delays in obtaining replacement engine parts, it is important to begin making a parts list as soon as possible. Figure 6-35 shows an example of an engine parts list, a list of the common parts needed to complete the average job of reconditioning an engine. Be certain to fill in all necessary information concerning sizes and quantities before placing an order.

Fig. 6-35 AUTOMOTIVE TECHNOLOGY Engine Parts List

Student Name ______________________________ Date ______________

Make of Car ______________ Year ________ C.I.D. or LITRE ______________ No. of Cyl. ______________

Quan.	Description "A"	Price	Quan.	Description "B"	Price
	Pistons (size)			Thermostat (Degree)	
	Piston Rings (size)			Coolant (Qts)	
	Rod Bearings (size)			Motor Mounts R L	
	Main Bearings (size)			Fan Belts	
	Cam Bearings			Air Filter	
	Piston Pins (size)			Fuel Filter (size)	
	Connecting Rod(s)			Fuel Hose I.D.	
	Valves (int) (ex)			Spark Plugs (no.)	
	Valve Springs			Points	
	Valve Guides			Condenser	
	Valve Lifters			Distributor Cap & Rotor	
	Push Rods			Ignition Coil	
	Valve Seals - See gasket set			Ignition Resistor	
	Rocker arm(s)			Spark Plug Wires	
	Rocker arm shaft(s)			Carb. Kit (No.)	
	Timing Chain & Gears			Nuts, Bolts	
	Camshaft			Paint, Eng. Enamel	
	Oil Pump			Battery Cables	
	Gaskets (Overhaul Head, Pan)			Trash Bags (Lg.)	
	Crankshaft Seal(s)			Core (freeze) Plugs:	
	Oil (quarts)			Block dia	
	Oil Filter			Camshaft dia	
	Heater Hose (size)			Oil Gallery dia	
	Radiator Hose(s)				
	Radiator Cap (lbs)			Total "B"	
				Total "A"	
				Tax	
	Total "A"			Grand Total	

These parts are to be used by students for instructional purposes only.
Parts will not be resold.

Authorized by ______________________________ Date ______________

TEST 6-A

Name ______________________ Date ______________

DIRECTIONS:

Circle the best answer to each question.

1. One millimetre is equal to
   I. 1/100 of a metre.
   II. 0.03937 inches.
   A. I only B. II only C. Both I and II D. Neither I nor II
2. Angular measurement
   I. uses both English and metric units.
   II. is the space or movement in an arc.
   A. I only B. II only C. Both I and II D. Neither I nor II
3. The straightedge
   I. is used with a feeler gauge to check for warpage or flat surfaces.
   II. should be hung up when not in use.
   A. I only B. II only C. Both I and II D. Neither I nor II
4. The telescoping gauge is used
   I. to measure circular openings only.
   II. to measure any opening with parallel sides.
   A. I only B. II only C. Both I and II D. Neither I nor II
5. The micrometer is designated by its
   I. maximum opening measurement.
   II. units of measurement.
   A. I only B. II only C. Both I and II D. Neither I nor II
6. The measurement of a micrometer is the sum of the figures on
   I. the sleeve and the spindle.
   II. the spindle and the frame.
   A. I only B. II only C. Both I and II D. Neither I nor II
7. A vernier micrometer
   I. has a label on the frame to indicate this fact.
   II. measures in units of 0.0001.
   A. I only B. II only C. Both I and II D. Neither I nor II
8. The accuracy of a micrometer can be
   I. altered by excessive tightening of the thimble.
   II. checked by using a micrometer standard.
   A. I only B. II only C. Both I and II D. Neither I nor II
9. A vernier caliper
   I. measures in inches or millimetres.
   II. can be used for inside, outside, or depth measurements.
   A. I only B. II only C. Both I and II D. Neither I nor II
10. A dial indicator can be used to measure
    I. clearance (end play) and out-of-roundness.
    II. the diameter of a bearing journal.
    A. I only B. II only C. Both I and II D. Neither I nor II

11. A torque wrench measures
   I. the tension between two or more pieces being tightened.
   II. resistance to turning of a fastener.
      A. I only B. II only C. Both I and II D. Neither I nor II
12. When using a torque wrench,
   I. it is permissible to use an extension on the handle to make the turning easier.
   II. pull on the handle while applying pressure on the drive end with the other hand.
      A. I only B. II only C. Both I and II D. Neither I nor II
13. Careful observation of a disassembled part
   I. can give clues as to the reason for the failure or excessive wear.
   II. eliminates the need for any measurement.
      A. I only B. II only C. Both I and II D. Neither I nor II
14. Valve face burning is caused by
   I. incorrect stem materials.
   II. overcooled combustion chambers.
      A. I only B. II only C. Both I and II D. Neither I nor II
15. Scoring of a piston
   I. is the advanced stages of scuffing.
   II. involves metal transfer between the cylinder and the piston.
      A. I only B. II only C. Both I and II D. Neither I nor II
16. Top ring groove wear
   I. is most often caused by airborne abrasives.
   II. should be checked before installing new rings.
      A. I only B. II only C. Both I and II D. Neither I nor II
17. Detonation (abnormal combustion)
   I. can cause the piston rings to break.
   II. usually damages the center of the piston.
      A. I only B. II only C. Both I and II D. Neither I nor II
18. A bent connecting rod
   I. often wears a diagonal pattern on the piston skirt.
   II. causes irregular wear on the connecting rod bearing.
      A. I only B. II only C. Both I and II D. Neither I nor II
19. Engine parts
   I. can be purchased most economically on an individual basis.
   II. in a complete kit include a new camshaft and lifters.
      A. I only B. II only C. Both I and II D. Neither I nor II
20. A replacement parts list
   I. is helpful in making an estimate of the cost of repairs.
   II. should be completed as soon as possible after the engine is assembled.
      A. I only B. II only C. Both I and II D. Neither I nor II

DATE COMPLETED ______________ INSTRUCTOR'S SIGNATURE ______________________

unit 7

Cylinder Heads and Valve Service

This unit covers the fundamentals of overhead valve cylinder heads and common service operations performed on them by engine technicians. The objectives and procedures are presented in the order which they would normally follow in a shop.

OBJECTIVE 7-A (written test)

After completing the preparation assignment, you will correctly answer a minimum of 14 out of 20 multiple choice questions in a written test, without using reference material, on the following topics:

1. Cylinder head design and materials
2. Cylinder head disassembly
3. Valve guide wear measurements
4. Crack detection
5. Warpage checks
6. Surfacing operations

Preparation:

Study pages 161 through 181.

CYLINDER HEADS

The cylinder head assembly of the modern overhead valve engine is a group of durable, precision parts to provide for and control the combustion process (Fig. 7-1).

The common design used today consists of a single casting of grey iron or aluminum with one intake and one exhaust valve per cylinder. The area surrounding the valves, known as the combustion chamber, can have many irregular shapes (Fig. 7-2). The springs, located opposite the combustion chamber, have the job of closing the valves once they have been opened by the camshaft.

The combustion chamber design is said to be the most influential factor in determining the power output and efficiency of an engine. The fuel must enter the chamber in the correct form (gas) and amount (air-fuel ratio). It must be burned without abnormal combustion and leave as exhaust gases for an engine to run at its maximum capacity. For many years, combustion chamber shapes were chosen to offer the maximum power output and long-term operation. Today, the top priority in their design is reduced emissions and improved mileage.

Since 1971, many design modifications have been made in automotive engines: higher operating temperatures (195°F. or 90.6°C.), retarded ignition timing, leaner air-fuel mixtures, exhaust gas recirculation (EGR), and catalytic converters. These changes have reduced the undesirable emissions; however, they significantly increase the overall temperatures, especially the cylinder head area.

MANUFACTURING MATERIALS

The materials used in the manufacture of cylinder heads are grey iron and aluminum. The iron head is made by pouring molten iron into a complex mold of sand. This process is referred to as sand casting. The aluminum head is made by pouring liquid aluminum into a steel mold called a die or by using the sand casting method. This method of making complicated parts is called die casting. When the poured metal has cooled into a solid piece, it is machined to the designer's specifications with smooth surfaces and threaded holes.

The following service information is meant for the average cast iron cylinder head. For those makes not covered, consult a manufacturer's or general repair manual. The sequence of events is fairly constant, regardless of the make of engine.

V-8 ENGINE HEAD AND MANIFOLD (350)

1. COVER
2. REINFORCEMENT
3. GASKET
4. NUT
5. ARM—Valve Rocker
6. CAP
7. SPRING
8. VALVE—Exhaust
9. VALVE—Inlet
10. SPARK PLUG
11. WASHER
12. BOLT
13. MANIFOLD ASSY.—Exhaust
14. GASKET
15. STUD
16. GASKET UNIT
17. EXHAUST GASKET
18. HEAD ASSY.—Cylinder
19. BOLT
20. ROD
21. SHIELD
22. STUD
23. BALL
24. CAP

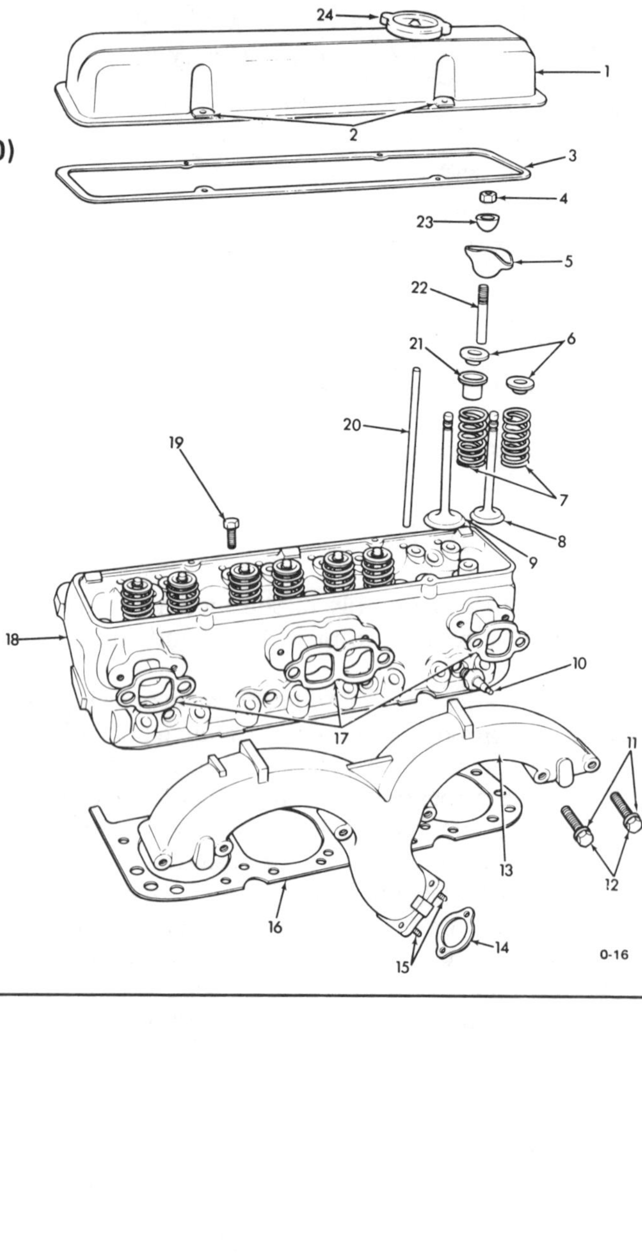

FIG 7-1 Engine cylinder head and exhaust manifold assembly.

INITIAL CLEANING

The first step in servicing a cylinder head is to steam clean it, if possible. This will remove much of the sludge (hardened oil). It will also partially clean the combustion chambers, allowing any large cylinder head cracks to be discovered. Should a crack appear, consult with your instructor.

Do not hot tank a cylinder head before the valve guide wear is measured. The manufacturers of valve guide knurling equipment recommend that knurling be done while the guide has its "natural" lubrication (oil film left from former operation).

CYLINDER HEAD DISASSEMBLY

If, during disassembly of a cylinder head, the valve springs will not compress as they should, the sludge-varnish buildup in the retainer and locks (keepers) is holding them together. To overcome this problem, use a piece of pipe and a soft-face hammer to break them loose (Fig. 7-3).

VALVE GUIDE WEAR MEASUREMENT

As an engine operates at normal highway speeds, the valves open and close as many as 70,000 times per hour while temperatures in the combustion chamber are as high as 4000°F. (2200°C.). Under these conditions, any valve guide and stem will wear, so it is important to check them. If the valve guides are not within the wear limits, valve action, while the engine is running, will be poor. The accuracy of the valve seat grinding process depends upon having a guide with minimum wear, since the grinding stone must run on an arbor placed into the guide.

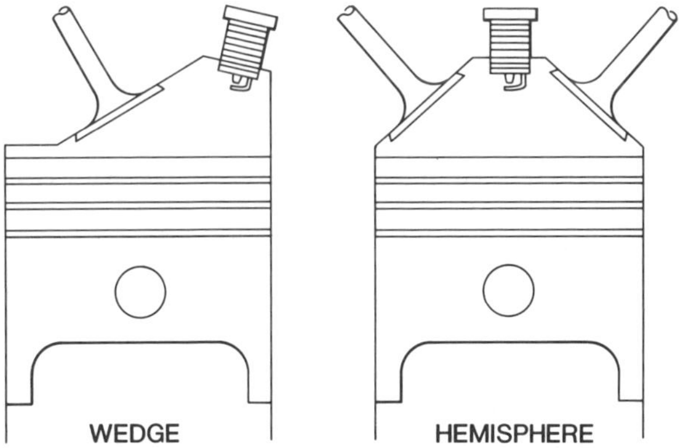

FIG 7-2 Common combustion chamber shapes.

FIG 7-3 Loosening valve locks (keepers) by striking retainer.

OBJECTIVE 7-A1 (manipulative)

Working as a team with one other student and with the aid of Job Report Sheet 7-A1, you will steam clean, disassemble, and measure the valve guide wear of a cylinder head with a minimum of four combustion chambers from an instructor-assigned engine.

Time Recommended:

Twice the flat rate time allowed in *Motors, Chilton,* or *National Service Data* manuals; approximately 20 minutes per combustion chamber.

Preparation:

1. Review pages 161 through 163.
2. Study pages 164 through 165.
3. Study manufacturer's or general repair manuals for the engine being serviced.
4. Attend lectures and demonstrations on cylinder head service.

Tools and Materials Needed:

1. Teardown sheet—valve stem and guide specifications
2. Steam cleaner
3. Valve spring compressor
4. Telescoping gauge (5/16 in. or 7.92 mm min.) and micrometer
5. Valve guide cleaning brush
6. Cylinder head holding stand
7. Valve rack
8. Soft-faced hammer and short piece of ¾ in. pipe

Procedure:

1. Steam clean all surfaces of the cylinder head.
2. Remove the rocker arms if not previously done.
3. Position the head on the bench with the valve springs facing away from you.
4. Adjust the jaws of the valve spring compressor to fit the diameter of the spring retainer (Fig. 7-4). Adjust the length of stroke of the compressor (Fig. 7-5).
5. Compress the valve spring and remove the locks (Fig. 7-6). If the valve spring cannot be compressed, the retainer and locks can be loosened by striking, as shown in Figure 7-3.
6. Release the valve spring very carefully. Keep all parts of your body away from the spring in case the jaws slip off the retainer.
7. Remove the valves from the cylinder head. *Keep them in order* so they may be assembled in the same guide from which they were removed. Use the valve rack (Fig. 7-7). If a valve cannot be removed easily, it probably has a mushroomed tip. Carefully file off the mushroomed edge before removing.
8. Clean the valve guides with a brush.
9. Expand a telescoping gauge or a split-ball gauge in the valve guide on the combustion chamber side (Fig. 7-8).

10. Mark the cylinder head with each measurement; then record the measurements on Job Report Sheet 7-A1.
11. Measure the valve stem with a micrometer at the location shown in figure 7-9.
12. Record the measurements and complete Job Report Sheet 7-A1. Consult with your instructor before proceeding further.

FIG 7-4 Adjusting width of spring compressor jaws.

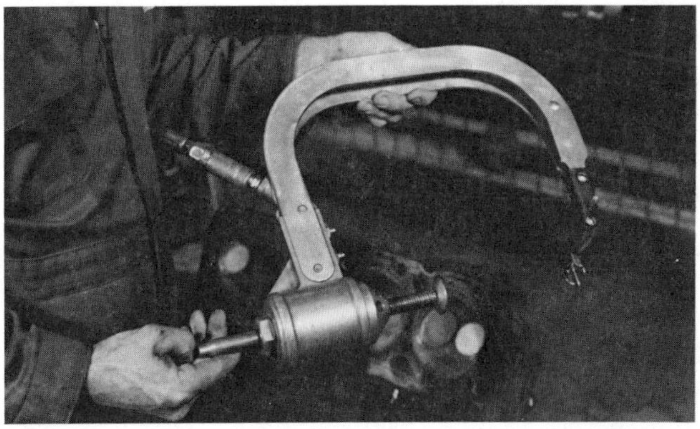

FIG 7-5 Adjusting the length of a stroke on the spring compressor.

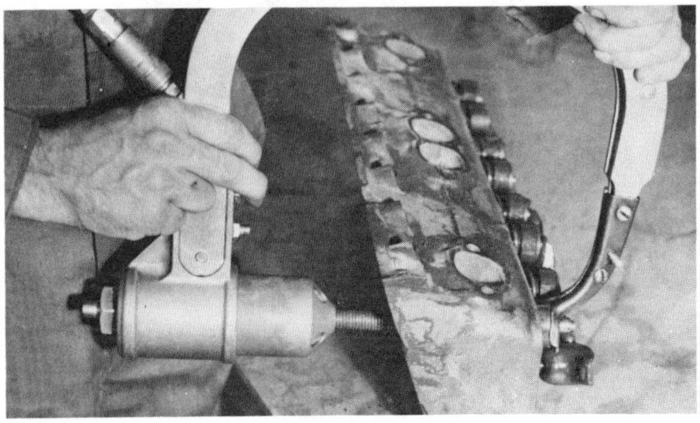

FIG 7-6 Compressing a valve spring.

FIG 7-7 Keeping valves in order by using a rack.

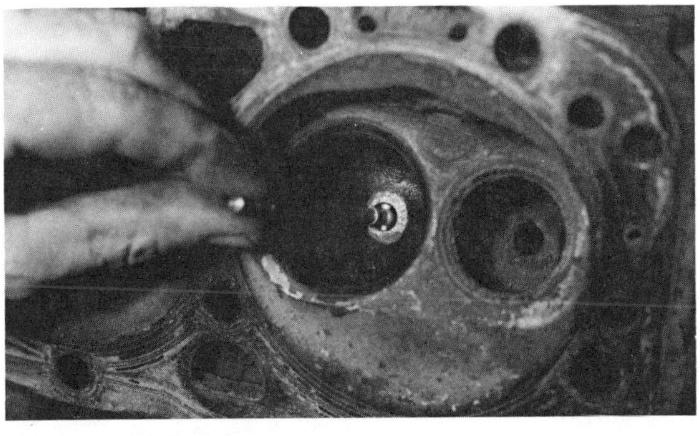

FIG 7-8 Measuring valve guide wear from the combustion chamber.

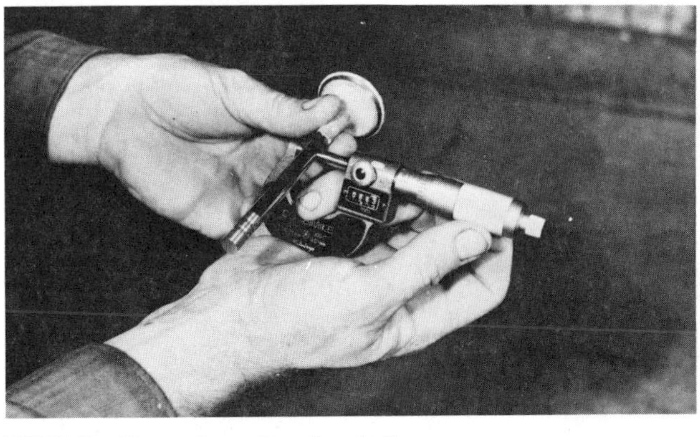

FIG 7-9 Measuring valve stem in the guide wear area near the head.

JOB REPORT SHEET 7-A1

CYLINDER HEAD DISASSEMBLY AND MEASUREMENT

Student Name ____________________ Date ____________

Make of Car ____________ Model ____________ Year ____________

Time Started ____________ Time Finished ____________ Total Time ____________

Flat Rate Time ____________________

Specifications:

Indicate inches or millimetres.

Valve stem diameter:

Intake ____________ Exhaust ____________

Valve guide diameter (bore):

Intake ____________ Exhaust ____________

Valve guide clearance:

Intake ____________ Exhaust ____________

Checkup Questions:

1. What type of valve seals were installed? ____________
2. Were any of the valve tips mushroomed? (Yes/No) ____________
3. What would cause the mushrooming of a valve? ____________

   ____
4. Did any of the guides show excessive wear? ____________

   Which ones? ____________
5. Why should the valve stems be measured near the head end of the valve? ____________

   ____

DATE COMPLETED ____________ INSTRUCTOR'S SIGNATURE ____________

Once it has been decided what service operations the valve guides must have, the cleaning process should begin. Refer to Unit 5 for information on cleaning materials and devices and to Objective 7-A2.

CRACK DETECTION

Cylinder heads will crack for several reasons. The most common cause of cracking is overheating the casting. This expands the metal in the area of greatest heat, usually the exhaust valve area of the combustion chamber. When the metal reaches its elastic limit through expansion, it finally begins to break. This break is a crack which very often allows coolant to leak into the combustion chamber.

Another way cracks develop in engine parts is through the expansion that takes place when water freezes in a confined space such as the water jackets. These cracks show up in the exterior part of the casting in most cases.

The materials and equipment used to detect cracks are of two types: chemical and magnetic (Figs. 7-10 and 7-11). Chemical detection is done with dye penetrants or nonmagnetic materials such as aluminum or magnesium. Magnetic particle detection is used for crack checks on iron and steel parts (see Objective 7-A2).

All cracks must be repaired before any valve service work takes place. Repairing cracks with tapered plugs will be covered in Objective 7-A3; chemical (epoxy) crack repair will be covered in Unit 9, Cylinder Block Preparation.

WARPAGE CHECKS

Engine castings become distorted (warped) through the heating and cooling cycles they routinely face. At times, stress through improper bolt torque will distort a casting. Overheating, for whatever reason, may distort a casting so that it does not properly fit its mated part. The use of a straightedge and feeler gauge set can be used to detect warpage.

When warpage becomes excessive, the part must be machined in one of several ways to return it to a straight, even surface. Milling, which uses a cutting tool to remove a minimum of material and restore the straightness of a piece, is most often used by automotive machinists. Objective 7-A4 covers the general use of a milling machine. Grinding is another method of restoring the surfaces of engine parts.

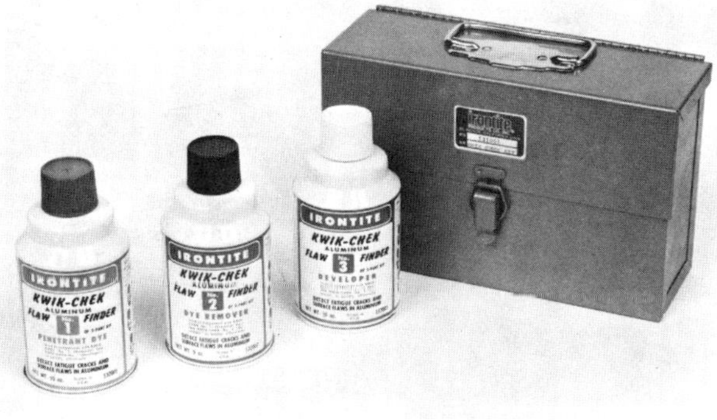

FIG 7-10 Chemical crack detection materials.

(Irontite Products Co., Inc.)

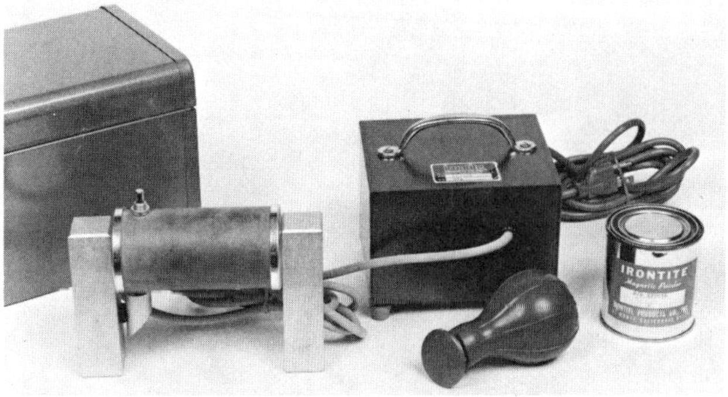

FIG 7-11 Magnetic crack detection kit.

(Irontite Products Co., Inc.)

OBJECTIVE 7-A2 (manipulative)

Working as a team with one other student and with the aid of Job Report Sheet 7-A2, you will thoroughly clean and check for cracks and warpage a cylinder head from an instructor-assigned engine.

Time Recommended:

Twice the flat rate time allowed in *Motors, Chilton,* or *National Service Data* manuals; approximately 10 minutes per combustion chamber.

Preparation:

1. Review pages 161 through 169.
2. Study pages 170 through 171.
3. Study manufacturer's or general repair manuals for the engine being serviced.
4. Attend lectures and demonstrations on cylinder head service.

Tools and Materials Needed:

1. Cleaning equipment, depending on methods chosen
2. Magnetic crack detection kit or dye penetrant
3. Straightedge and feeler gauge set

Procedure:

1. Clean the bare cylinder head by one or more of the following methods as recommended by your instructor (refer to Unit 5 on cleaning):
   a. Hot tank (cast iron only)
   b. Spray cleaner (iron and aluminum, depending on solution)
   c. Rotary wire brush with a drill motor (Fig. 7-12).
   d. Cold tank (carburetor cleaner)
   e. Peening (glass beading)

FIG 7-12 Wire brush being used with a drill motor to clean combustion.

FIG 7-13 Glass beaded cylinder head.

The most thorough job will be done by cleaning in two steps. First, place the head in one of the first four cleaning materials or processes. Next, glass bead all of the combustion chamber surfaces (Fig. 7-13).

2. Visually inspect the combustion chamber surfaces for irregularities and/or cracks. After glass beading, most cracks are quite easily seen, especially in aluminum alloy beads. Dye penetrant chemicals are often used in checking for cracks in an aluminum alloy cylinder head. Follow the chemical manufacturer's instructions for best results.
3. When checking iron heads, attach the magnets to the head adjacent to one combustion chamber (Fig. 7-14).
4. Lightly sprinkle magnetic powder over the combustion chamber area.
5. Cracks will appear as a light-colored line (Fig. 7-15). If a crack is detected in a cast iron head, consult with your instructor before proceeding. Crack repair will be covered in Objective 7-A3.
6. With a large file (10–12 in. mill, smooth), lightly draw file the block side of the cylinder head (Fig. 7-16). This will remove small nicks or gasket materials that would cause an incorrect reading with the straightedge.
7. Lay the straightedge end to end on the head (Fig. 7-17). Attempt to insert a 0.006 in. (0.152 mm) feeler gauge anywhere under the straightedge. If this can be done without raising the straightedge, the head needs to be surfaced (milled or ground).
8. Lay the straightedge side to side at various places. Use a 0.003 in. (0.076 mm) feeler for this check. The maximum warpage that is allowable in any 6 in. is 0.003 in.

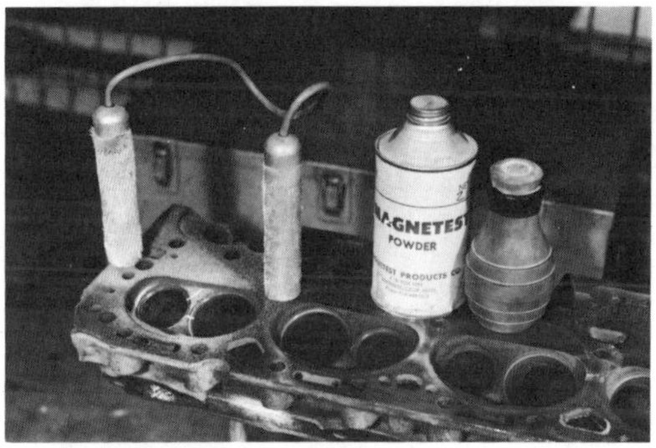

FIG 7-14 Magnets attached to one combustion chamber.

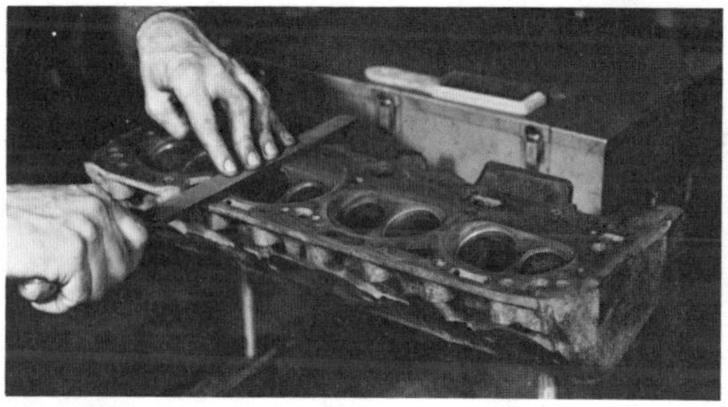

FIG 7-16 Draw-filing the gasket surface.

FIG 7-15 Cracks form light-line of powder as it is applied. (Irontite Products Co., Inc.)

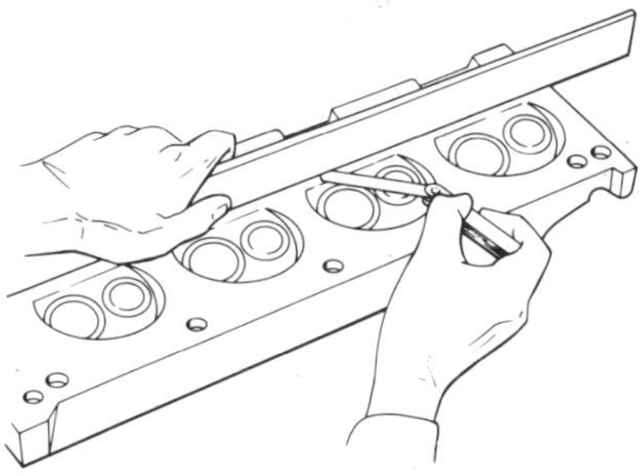

FIG 7-17 Using a straightedge and feeler gauge to check warpage. (Chrysler Corporation)

JOB REPORT SHEET 7-A2

CYLINDER HEAD CLEANING, CRACK DETECTION AND WARPAGE CHECKS

Student Name ______________________________ Date ______________

Make of Car ______________ Model ______________ Year ______________

Time Started ______________ Time Finished ______________ Total Time ______________

Flat Rate Time ______________________________

Specifications:

Warpage limits (general):

0.003 in. (0.076 mm) in any 6 in. (150 mm)

0.006 in. (0.152 mm) maximum overall

Checkup Questions:

1. What cleaning methods were used?

   a. ______________________________

   b. ______________________________

   c. ______________________________

   d. ______________________________

2. Were any cracks located? (Yes/No) ______________________________

   If so, did they reach the valve seat? ______________________________

3. What size feeler gauge would pass freely between the straightedge and the head surface? __________

DATE COMPLETED ______________ INSTRUCTOR'S SIGNATURE ______________

CRACK REPAIR

When cracks in cast engine parts are detected, a decision must be made whether the part is to be repaired. This decision will depend on the cost and availability of a replacement part versus the repair cost.

Crack repair of aluminum parts is usually accomplished by welding or chemical methods. Either method requires equipment and materials not found in most automotive repair shops.

Cast iron is repaired frequently by the use of cast iron taper plugs (threaded) or epoxy-metallic adhesives (Fig. 7-18). The threaded pins are used in cylinder heads and in the upper cylinder areas of the block. If the crack extends into the valve seat, an insert will have to be installed after the crack has been repaired (Fig. 7-19).

The following procedures can also be used for the repair of cast iron parts other than cylinder heads.

Working as a team with one other student and with the aid of Job Report Sheet 7-A3, you will use taper, threaded plugs to repair a cracked cylinder head from an instructor-assigned vehicle engine.

OBJECTIVE 7-A3 (manipulative)

Time Recommended:

Consult your instructor: ——— minutes.

Preparation:

1. Review pages 161 through 175.
2. Study pages 175 through 176.
3. Study manufacturer's or general repair manual for the engine being serviced.
4. Attend lectures and demonstrations on cylinder head service.

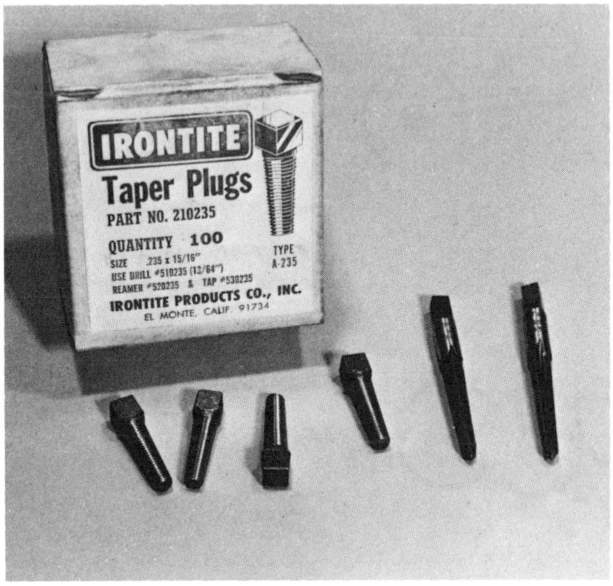

FIG 7-18 Taper plugs with special tap and reamer to repair cast iron cracks.

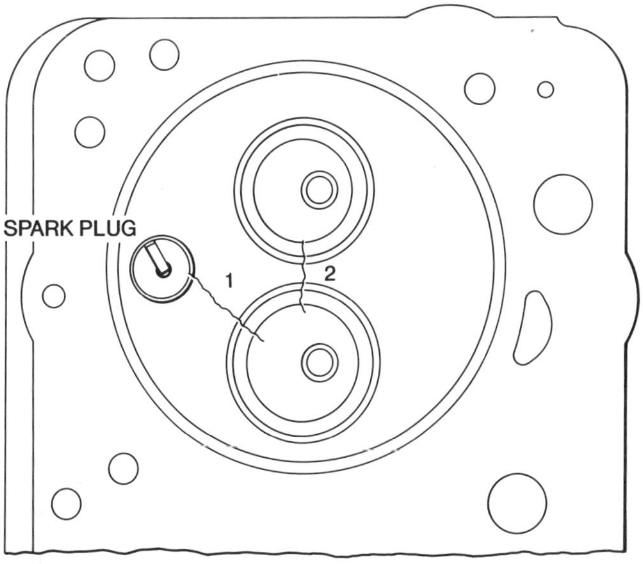

FIG 7-19 Common locations of cylinder head cracks.

Tools and Materials Needed:

1. Hammer
2. Center punch
3. Special tap and reamer
4. Taper plugs
5. Locking chemical (Loctite)
6. Hacksaw
7. Wrench
8. Drill motor and rotary file
9. Twist drills of various sizes
10. Drift punch

Procedure:

1. Determine the exact length of the crack and centerpunch each end (Fig. 7-20).
2. After deciding upon the size of plugs to be used, select the twist drill that matches the plug.
3. Drill a hole at one end of the crack. Be careful not to break the drill by applying excessive pressure to the drill motor, especially when the drill is breaking through into the water jackets.
4. Ream the hole with tapered reamer of the appropriate size.
5. Carefully thread the hole, using the special tap which matches the plug size.
6. Coat the threads of the plug with a sealer and turn it firmly into the hole. DO NOT TIGHTEN UNTIL BREAKAGE OCCURS.
7. Saw off the plug approximately 1/32 in. (0.794 mm) above the surface.
8. Using the steps above, begin "lacing" with additional plugs until the entire length of crack is covered with plugs (Fig. 7-21). This will provide an interlocking seal.
9. With all plugs in place, peen over the slightly protruding surfaces with an air hammer equipped with a blunt-nose tool.
10. Smooth the surface area of the plugs with a rotary file or stone.

FIG 7-20 Determine the ends of the crack, and centerpunch.

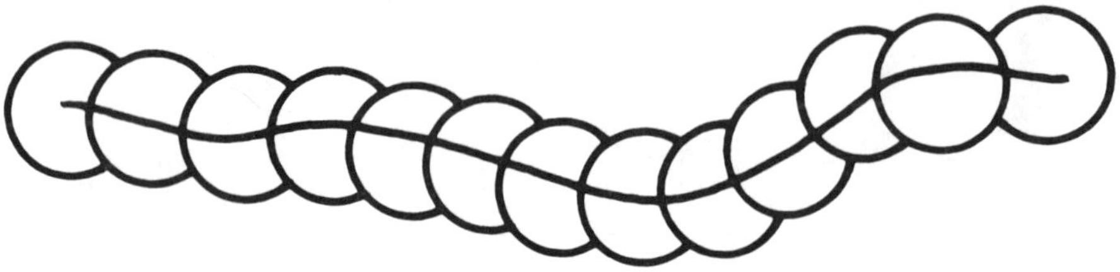

FIG 7-21 Crack interlaced with taper plugs.

JOB REPORT SHEET 7-A3

CRACK REPAIR

Student Name ____________________ Date ____________

Make of Car ____________ Model ____________ Year ____________

Time Started ____________ Time Finished ____________ Total Time ____________

Flat Rate Time ____________________

Specifications:

Drill size for plug tap being used ____________.

Checkup Questions:

1. Why is lacing necessary? ____________________

   ____

2. From what material are the taper plugs made? ____________________

3. List the manufacturer numbers for the following:

   a. Plug ____________________

   b. Reamer ____________________

   c. Tap ____________________

4. Describe the cause of the crack. ____________________

   ____

   ____

DATE COMPLETED ____________ INSTRUCTOR'S SIGNATURE ____________

SURFACING OPERATIONS

When the contact surfaces of engine castings become distorted, they must go through a machining operation known as *surfacing.* This process involves removal of minor amounts of metal from cylinder heads, block, or intake manifolds. Many different designs of machines are used for this purpose in automotive machine shops. One uses a milling cutter; another uses a large grinding wheel.

There is very little problem with the alignment of the surfaces of an in-line engine. With a V-type engine, if the cylinder heads are surfaced more than 0.010 in. (0.254 mm), the intake manifold bolt holes may not line up during the assembly phase.

After machining surface "A" on most OHV cylinder heads, machine the intake manifold side of the cylinder head, surface "B" (Fig. 7-22). The amount to be removed varies with the degree at which the cylinder head is machined, but both heads must have the *same* amount removed. The chart below shows the amount which must be removed from the intake manifold side of the head so the bolts will line up. To find the angle of the two surfaces, place the head with the combustion chamber side down. The angle from the vertical (90°) is the head angle:

5°	Amount removed at "A" × 1.1 = Amount to be removed at "B"
10°	Amount removed at "A" × 1.2 = Amount to be removed at "B"
15°	Amount removed at "A" × 1.4 = Amount to be removed at "B"
20°	Amount removed at "A" × 1.7 = Amount to be removed at "B"
25°	Amount removed at "A" × 2.0 = Amount to be removed at "B"
30°	Amount removed at "A" × 3.0 = Amount to be removed at "B"

For example, suppose that 0.015 in. (0.381 mm) was removed from surface "A" and that Surface "B" is at a 10° angle to "A." Then:

0.015	Amount for "A"
×1.2	Factor for 10° angle head
030	
015	
0.0180	Amount to be removed from "B"

OBJECTIVE 7-A4 (manipulative)

Working as a team with one other student and with the aid of Job Report Sheet 7-A4, you will surface the combustion chamber side and manifold side, if necessary, of a cylinder head from an instructor-assigned vehicle engine.

Time Recommended:

Twice the flat rate time allowed in *Motors, Chilton,* or *National Service Data* manuals. Consult your instructor.

Preparation:

1. Review pages 161 through 179.
2. Study pages 179 through 180.
3. Study manufacturer's or general repair manuals for the engine being serviced.
4. Attend lectures and demonstrations on cylinder head service.

Tools and Materials Needed:

Surfacing machine

Procedure:

No specific procedures will be given here because of differences in equipment design and brand names. Complete Job Report Sheet 7-A4 following the surfacing operations.

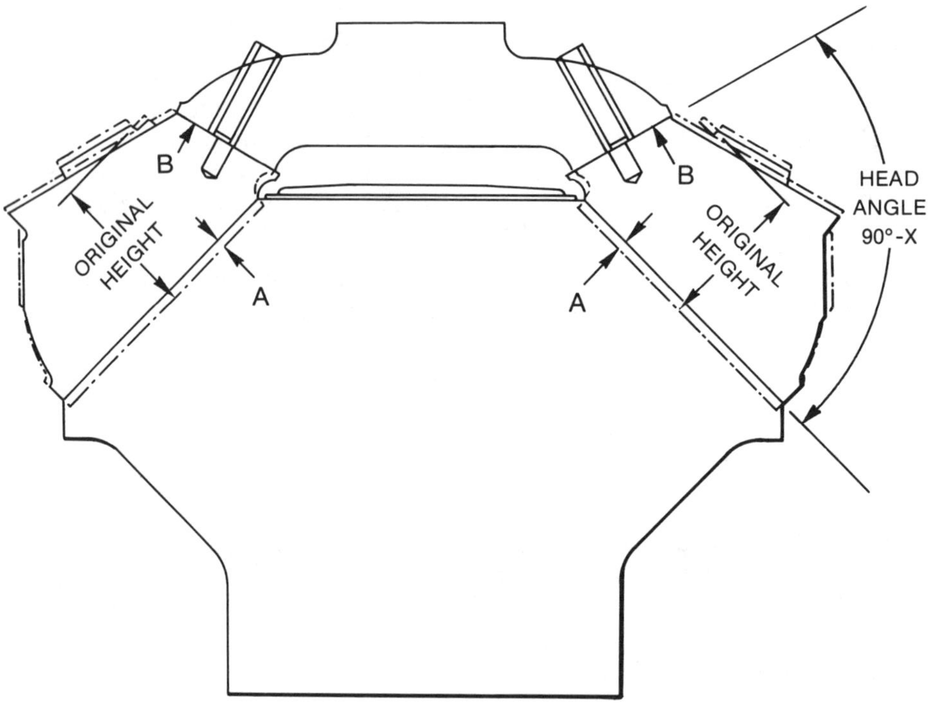

FIG 7-22 How to machine the intake manifold surface of a Vee-type OHV engine cylinder head.

(Van Norman Machine Co.)

CYLINDER HEAD RESURFACING

Student Name ______________________ Date ____________

Make of Car ____________ Model ______________ Year ____________

Time Started ____________ Time Finished ____________ Total Time ____________

Flat Rate Time ______________________________

Specifications:

Cylinder head angle (intake to block angle) ______________

Checkup Questions:

1. What is the make and model of surfacing machine? ______________

   ____

2. Were the cutters sharpened adequately? ______________

3. What was the total amount of material removed from the block surface? ______________

   from the intake manifold surface? ______________

4. How would excessive surfacing affect the valve stem end of the rocker arm when the head is installed?

   ____

   ____

DATE COMPLETED ______________ INSTRUCTOR'S SIGNATURE ______________

Name ______________________ Date ______________

DIRECTIONS:

Circle the best answer to each question.

1. In an overhead valve engine
   I. the camshaft opens and closes the valves.
   II. The valves are part of the combustion chamber.
   A. I only B. II only C. Both I and II D. Neither I nor II
2. Combustion chamber temperatures
   I. influence the speed of the engine.
   II. can increase with retarded ignition timing.
   A. I only B. II only C. Both I and II D. Neither I nor II
3. Before disassembly of the cylinder head
   I. the head should be hot tanked.
   II. the guides should be checked for wear.
   A. I only B. II only C. Both I and II D. Neither I nor II
4. Valve guide wear beyond acceptable limits
   I. causes increased operating temperatures.
   II. can be corrected with a valve spring compressor.
   A. I only B. II only C. Both I and II D. Neither I nor II
5. The valve spring compressor requires adjustment for
   I. the size of the spring retainer.
   II. length of stroke.
   A. I only B. II only C. Both I and II D. Neither I nor II
6. Compressing the valve spring
   I. should be done with the spring facing toward the operator.
   II. should take place after the locks are removed.
   A. I only B. II only C. Both I and II D. Neither I nor II
7. A valve rack should be used to
   I. keep the valves in order so each one is returned to its original position.
   II. determine which valves are worn beyond acceptable limits.
   A. I only B. II only C. Both I and II D. Neither I nor II
8. Valve stems
   I. with mushroomed ends must be filed before removal from the head.
   II. and guides should be knurled when necessary.
   A. I only B. II only C. Both I and II D. Neither I nor II
9. Measurement for wear of the valve parts
   I. is not necessary if a valve job is to be completed.
   II. determines the extent of work to correct excessive clearances.
   A. I only B. II only C. Both I and II D. Neither I nor II

10. The following tool can be used to measure valve guide wear:
    I. dial bore gauge.
    II. valve guide driver.
    A. I only B. II only C. Both I and II D. Neither I nor II

11. The greatest amount of wear on the valve stem occurs
    I. just above the retainer.
    II. just above the locks.
    A. I only B. II only C. Both I and II D. Neither I nor II

12. Cylinder heads develop cracks
    I. during cleaning operations.
    II. during disassembly operations.
    A. I only B. II only C. Both I and II D. Neither I nor II

13. Crack detection
    I. on aluminum cylinder heads requires magnetic powder and a magnet.
    II. with dye chemicals is used most often on iron and steel parts.
    A. I only B. II only C. Both I and II D. Neither I nor II

14. Warpage of a cylinder head
    I. makes it unusable for further work.
    II. is best determined through the use of dye penetrants.
    A. I only B. II only C. Both I and II D. Neither I nor II

15. The maximum limits of warpage are
    I. 0.003 in. (0.076 mm) overall.
    II. 0.060 in. (1.52 mm) in any 6 in. distance.
    A. I only B. II only C. Both I and II D. Neither I nor II

16. Draw filing the head gasket surface
    I. is necessary to remove enough metal to correct any warpage.
    II. is done to remove minor nicks and unwanted gasket material.
    A. I only B. II only C. Both I and II D. Neither I nor II

17. Crack repair of cast iron cylinder heads
    I. begins by determining the length of the crack.
    II. can be completed successfully with taper plugs.
    A. I only B. II only C. Both I and II D. Neither I nor II

18. Crack repair with taper plugs follows a step-by-step procedure
    I. requiring a drill motor and special tap.
    II. where plugs are overlapped to cover the crack.
    A. I only B. II only C. Both I and II D. Neither I nor II

19. Surfacing an in-line cylinder head
    I. requires milling of the manifold surface.
    II. can be accomplished only on a cutter mill.
    A. I only B. II only C. Both I and II D. Neither I nor II

20. Vee cylinder heads
    I. may require manifold surfacing if more than 0.010 in. (0.254 mm) is removed from the combustion chamber surface.
    II. made from aluminum cannot be resurfaced.
    A. I only B. II only C. Both I and II D. Neither I nor II

DATE COMPLETED ______________ INSTRUCTOR'S SIGNATURE ______________________

OBJECTIVE 7-B (written test)

After completing the preparation assignment, you will correctly answer a minimum of 14 out of 20 multiple choice questions in a written test, without reference material, on the following topics:

1. Valve service
2. Valve guide service
3. Valve spring testing
4. Valve seat service
5. Cylinder head assembly

Preparation:

Study pages 185 through 220.

VALVES AND THEIR SERVICE

In normal engine operation, the intake and exhaust valves absorb a great amount of punishment. Valve heads are exposed to high combustion pressures and extreme heat (4000° F. or 2200° C.). Miles of movement within the guide wear the stem areas, and corrosive combustion materials erode the faces of the valves. As a result, the valves, especially the exhaust valves, become pitted, burned, warped, and grooved. When any of these problems arise, the valves do not seal, thus reducing the operating efficiency and economy of the engine.

DESIGNS

Figure 7-23 shows the common valve designs in use today. Increased engine operating temperatures to reduce exhaust emission controls have required improved

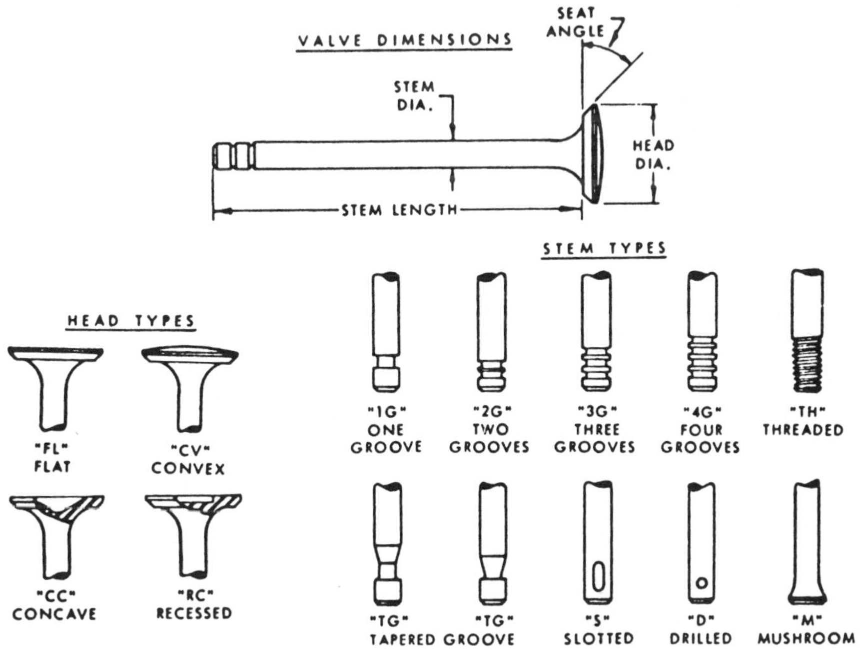

FIG 7-23 Engine valves. (NAPA)

valve materials. However, these materials have not reduced the need for valve service. On the contrary, as much or more valve service is being done today as ever before.

MATERIALS

Valves are made from several different alloys of steel containing large percentages of chromium and nickel so they can withstand high temperatures for hours of continued operation. The average running temperature of the valve head is around 1400°F. (760°C.).

The manufacturing process by which valves are made is called forging or drop forging. A piece of steel alloy is literally pounded into the rough shape of a valve. It is then ground and polished to specification. The head and tip are heat treated to harden these areas. Finally, the stem is plated, usually with chrome, to resist the wear that will take place in the guide. Very often the head areas of valves are plated with nickel or aluminum in an effort to reduce the effects of erosion from corrosive gases.

Valve service does not usually involve changing the shapes of the combustion chambers, although in high performance engine work, some designs are modified for better engine breathing. This will be discussed in Unit 15.

FACING AND TIPPING

If, after the valve stem is checked for wear, the valve margin appears to have at least 1/32 in. (0.794 mm) thickness, the valve face can be resurfaced. This is normally called facing or refacing. The tip end must also be resurfaced. This operation, which squares the end of the stem, is called tipping.

OBJECTIVE 7-B1 (manipulative)

Working as a team with one other student and with the aid of Job Report Sheet 7-B1, you will resurface the face and tip end of at least eight valves from an instructor-assigned automobile.

Time Recommended:

Twice the flat rate time allowed in *Motors, Chilton,* or *National Service Data* manuals. Approximately 4 minutes per valve.

Preparation:

1. Review pages 185 through 186.
2. Study pages 186 through 190.
3. Study manufacturer's or general repair manuals for the automobile being serviced.
4. Attend lectures and demonstrations on valve service.

Tools and Materials Needed:

1. Wire wheel and face shield or glass bead machine
2. Valve refacing machine

Procedure:

Valve Refacing

1. Keep all valves in order to ensure that they are reassembled into the same guides from which they were removed. If you have not thoroughly cleaned the valves, do so now. Use a wire wheel and face shield or glass bead machine.

Caution: Do not allow wire wheel or glass beads to contact the valve guide wear area of the stem.

Note: Any valve with less than 1/32 in. (0.787 mm) margin or excessive stem wear should be discarded.

2. Study the refacing machine instruction sheet to be certain you understand all operations and controls.
3. Adjust the chuck holding the valve to the proper angle. This is usually 44° or 45° for 45° valve seats and 29° or 30° for 30° valve seats (Fig. 7-24).
4. Dress the refacing wheel to a smooth surface with the diamond nib. Seek your instructor's help (Fig. 7-25).
5. Insert the valve into the chuck opening. Position the valve so that the unworn upper portion of the stem contacts the chuck jaws.

FIG 7-24 Worktable on valve facing machine showing angle settings for valve chuck.

FIG 7-25 Dressing refacer grinding stone.

6. Tighten the chuck and then adjust the sliding stop by pushing it against the valve stem (Fig. 7-26). This will allow the remaining valves of the same size to be quickly positioned. Do all the exhausts in one group and all the intakes in another group.
7. Bring the grinding wheel and the face of the valve close together before turning on the motors (Fig. 7-27).
8. Using the transverse lever, move the valve just off the stone face before starting the motors.
9. Start both motors. Check the valve head for excessive wobble (warp) as it turns and adjust coolant knob.
10. Bring the valve face and grinding stone together until a slight grinding noise is heard.
11. Move the valve face back and forth across the grinding stone with short strokes. NEVER allow the valve face to leave the stone while grinding away the imperfections.

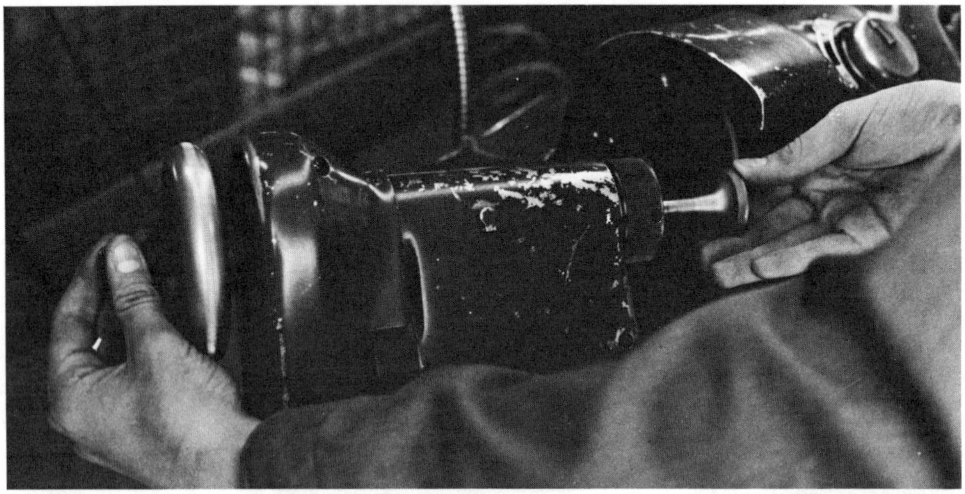

FIG 7-26 Adjusting the sliding stop inside the chuck.

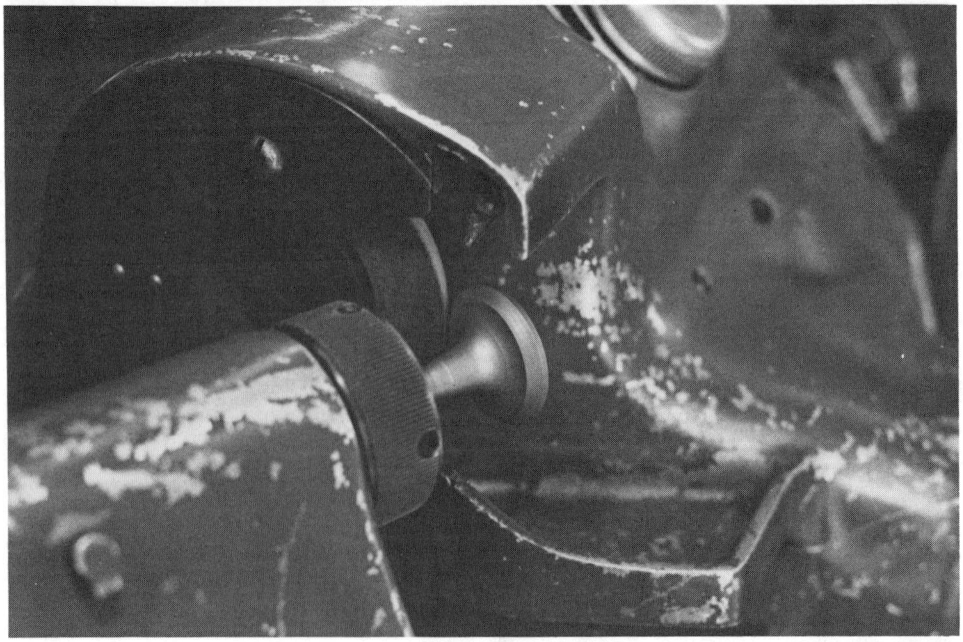

FIG 7-27 Position valve face close to grinding wheel.

12. Remove only enough material from the face to bring it to a smooth satin finish (Fig. 7-28).
13. Bring the valve face to the center of the wheel and hold it there momentarily. Then move the stone away from the valve face. DO NOT move the valve face away from the stone.
14. Shut off the machine and remove the valve. Reface the remaining valves of the same group, then the other group.

Valve Tipping

1. Lay the valve stem in the V groove so that the tip just contacts the grinding wheel.
2. Tighten the clamp to hold the valve in the V groove (Fig. 7-29).

FIG 7-28 Grind only enough off the face for a smooth finish. The valve should be moved back and forth across the stone surface.

FIG 7-29 Place valve stem in the rear groove and lightly tighten the clamp.

3. Turn on the motor and adjust the coolant so that a small stream will cool the end of the stem while it is being ground.
4. Turn the feed knob inward about 0.001 in. (0.0254 mm) and pass the valve stem over the grinding wheel several times to square the tip (Fig. 7-30).

Note: Remove only enough material to grind away slight imperfections. Excessive grinding (more than 0.030 in. or 0.762 mm) will remove the hardened portion of the tip, causing the valve to mushroom very quickly when reassembled and run. Valves with heavily pitted stem tips should be replaced.

5. Grind the tips of all remaining valves.
6. Clean up all tools, machinery, and areas used to complete this objective.

FIG 7-30 Grinding the valve tip to remove minor surface problems.

JOB REPORT SHEET 7-B1

VALVE REFACING AND TIPPING

Student Name ______________________ Date ____________

Make of Car ____________ Model ______________ Year ____________

Time Started ____________ Time Finished ____________ Total Time ____________

Flat Rate Time ______________________________

Specifications:

Valve face angle

Intake	Exhaust
°	°

Checkup Questions:

1. Why would a valve face be ground at 44° to meet a 45° seat? ______________________________

2. Why is it necessary to dress the grinding wheel? ______________________________

3. What is the proper place to tighten the chuck on the valve stem?

4. How much grinding should be done on the valve face? ______________________________

5. If the surface of the valve face appears smooth after grinding, what part of the grinding machine should be moved to separate the valve and the grinding wheel? ______________________________

DATE COMPLETED ____________ INSTRUCTOR'S SIGNATURE ____________

VALVE GUIDES AND THEIR SERVICE

As the engine operates, the valve is constantly being pushed down by the rocker arm and pulled up by the valve spring. As the rocker arm and spring push and pull, they apply side force to the valve stem to wear the guide into an out-of-round oval shape. This is often referred to as bell mouthing (Fig. 7-31). The greatest amount of wear of this type occurs on the end of the guide closest to the combustion chamber, where there is less lubrication. Most often the wear is greater in the exhaust valve guide because the exhaust valve is exposed to greater heat. When guide wear appears to be beyond specified limits, consult with your instructor before proceeding.

DESIGNS

Valve guide designs are of two types, the integral and the replaceable. The integral guide is a hole drilled into a cast iron head. The replacement guide is an iron or bronze sleeve of the correct inside diameter that is pressed into the head. The integral guide has the advantage of carrying away heat buildup in the valve stem more easily. The replaceable guide has the advantage of quick, inexpensive repair to return the guide to factory specifications. Under most conditions, both types of guides incur some wear.

To service the integral guide, one of four operations can be performed: (1) reaming the hole to an oversize for use with a valve having an oversize stem (Fig. 7-32); (2) reaming the hole to allow a replaceable guide to be installed; (3) knurling to reform portions of the metal within the guide; and (4) installing a bronze wall insert.

VALVE GUIDE CORRECTION

Valve guide knurling is one of two processes currently used to correct excessive guide clearance on cylinder heads with integral valve guides. The other method is bronze wall insert. The knurling process uses a tool (resizing arbor) to form a threadlike channel within the guide bore (Fig. 7-33). The metal closest to the

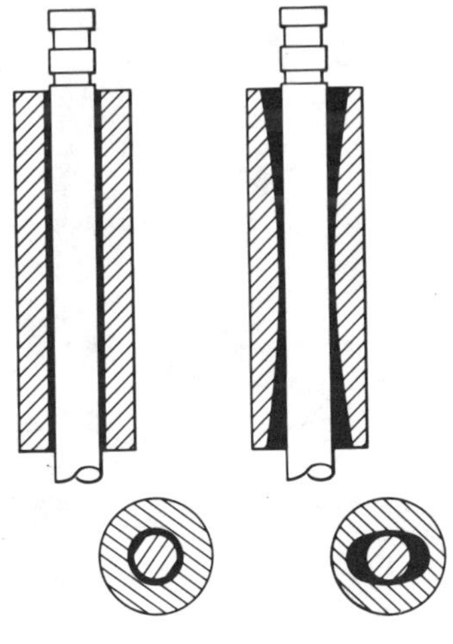

FIG 7-31 Typical valve guide wear is a result of rocker arm side thrust.

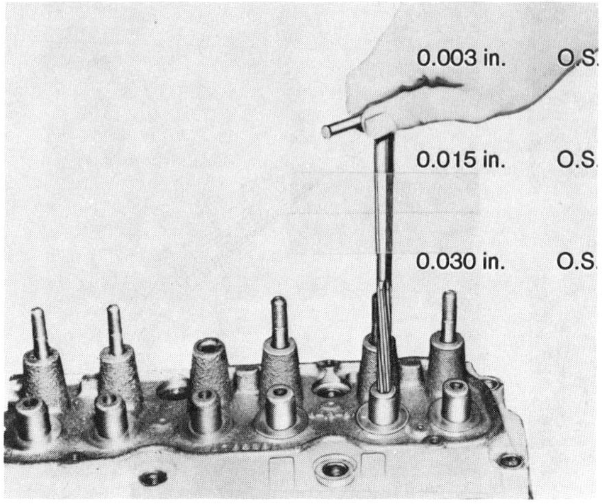

FIG 7-32 Reaming a valve guide to fit an oversize valve stem.

(Oldsmobile Division, GMC)

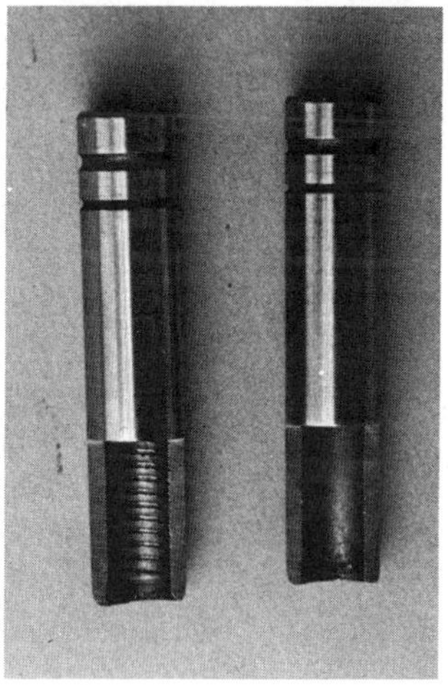

FIG 7-33 Knurled valve guide.

thread groove is raised above the original height of the bore, thereby reducing its diameter. Precision reaming is then used to establish the stem to guide clearance. This new clearance can be up to one-half less than the original clearance specification.

The bronze wall insert is installed after a course thread is cut into the guide bore. A broach is driven through the bronze insert opening to force it into position. Precision reaming is then done to bring about the desired clearance between valve stem and guide. One advantage of this method, as well as the knurling process, is that the grooves within the guide will hold lubrication, thereby reducing much ordinary guide wear.

VALVE GUIDE REPLACEMENT

For cylinder heads that have replaceable guides, removing guides that are worn beyond acceptable limits and replace them with new ones can be accomplished by driving (using a valve guide driver and striking with a hammer) or pressing (using a valve guide driver with an arbor or hydraulic press).

For engines with bronze guides (many imports), the best method of guide replacement is to top the old guide with threads in the spring end (Fig. 7-34) and then screw in a matching bolt at least five turns (Fig. 7-35). Then use a valve guide driver and press against the bolt rather than the soft edges of the guide. This will remove the guide without damaging the surface of the head that holds the guide in place (Fig. 7-36).

OBJECTIVE 7-B2 (manipulative)

Working as a team with one other student and with the aid of Job Report Sheet 7-B2, you will recondition a minimum of four valve guides, using a knurling system such as the K-line on an instructor-assigned cylinder head.

FIG 7-34 Cutting threads with a tap in the spring side of the guide.

FIG 7-35 Insert a bolt into the threaded guide.

FIG 7-36 Drive or press out the guide by pushing against the bolt.

Time Recommended:

Twice the flat rate time allowed in *Motors, Chilton,* or *National Service Data* manuals; approximately 6 minutes per guide.

Preparation:

1. Review pages 185 through 194.
2. Study pages 194 through 196.
3. Study manufacturer's or general repair manuals for the automobile being serviced.
4. Attend lectures and demonstrations on valve service.

Tools and Materials Needed:

1. Guide knurling set
2. Drill motor (¼ in. 1600–1800 rpm)

Procedure:

Caution: Do not attempt to knurl guides with more than 0.010 in. (0.254 mm) wear. Satisfactory results cannot be achieved. Do not attempt to knurl the chilled steel guides as used in International 450 and many import engines. These guides are too hard to be resized and must be replaced.

1. Clean all valve guides with a brush or expanding blade cleaning tool (Fig. 7-37). If guides have been knurled previously, notify your instructor. Normally guides cannot be knurled twice.
2. Attach the speed reducer in the kit to the drill motor.
3. Select the proper resizing arbor to fit the guide being knurled. The size is marked on the shank end (Fig. 7-38).

FIG 7-37 Cleaning the valve guide.
(Oldsmobile Division, GMC)

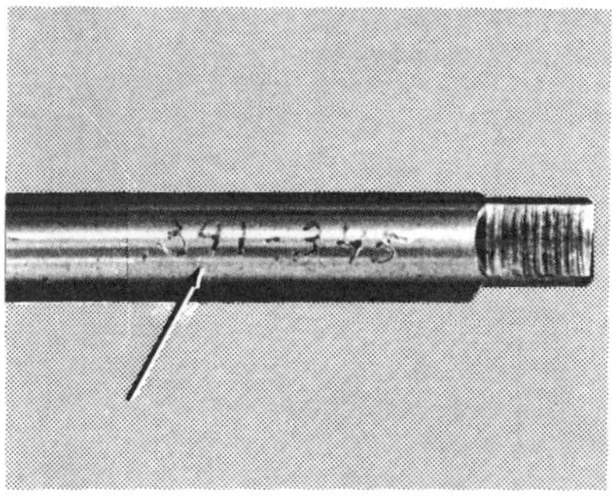

FIG 7-38 Resizing arbor with its size range listed on the drive end.

4. Lubricate the guide and resizing arbor with special lubricant provided.
5. Insert the pilot end of the arbor in the combustion chamber side of the guide.
6. Engage the speed reducer and arbor and begin the resizing operation. Be certain the arbor is aligned with the guide hold as the resizing takes place. Hold the motor in one hand and the speed reducer in the other (Fig. 7-39).
7. Turn the resizing arbor all the way through the guide.
8. Clean the guide bore with the brush furnished in the kit.
9. Select the proper reamer (Fig. 7-40), the one which is the same size as the original guide bore. A reamer that is 0.001 in. to 0.002 in. smaller than the original guide bore can be used if the valve does not bind in the guide when installed.
10. Ream the guide (Fig. 7-41). Lubrication is not necessary for this step.
11. Clean the guide bore with a brush and compressed air. Use eye protection.
12. Repeat steps 4–11 until all guides are resized.

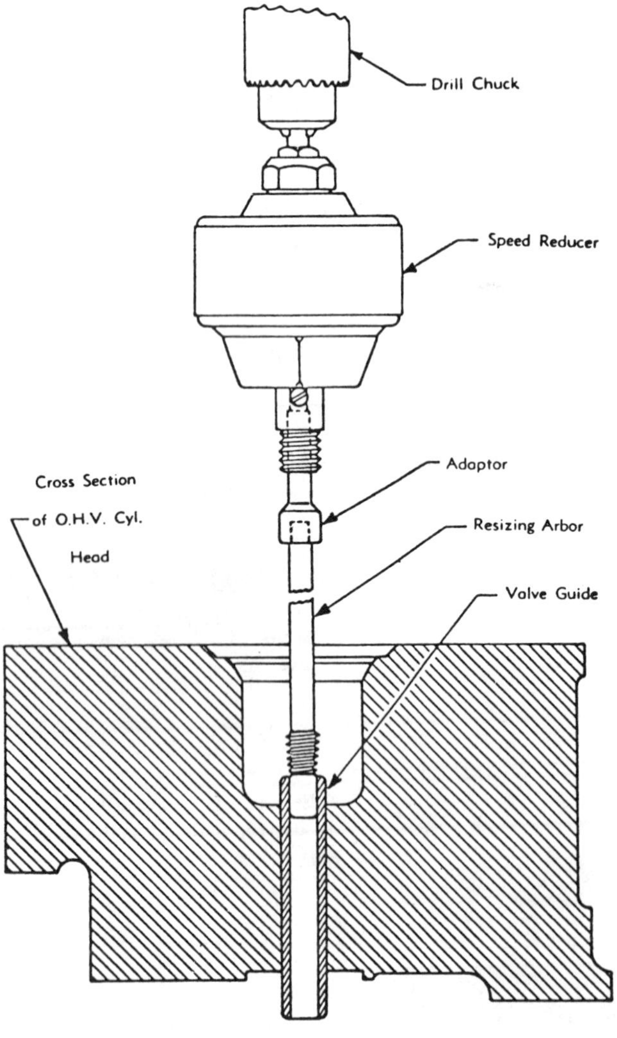

FIG 7-39 Assembly of drill motor, speed reducer, and resizing arbor. (K-Line Tool Company)

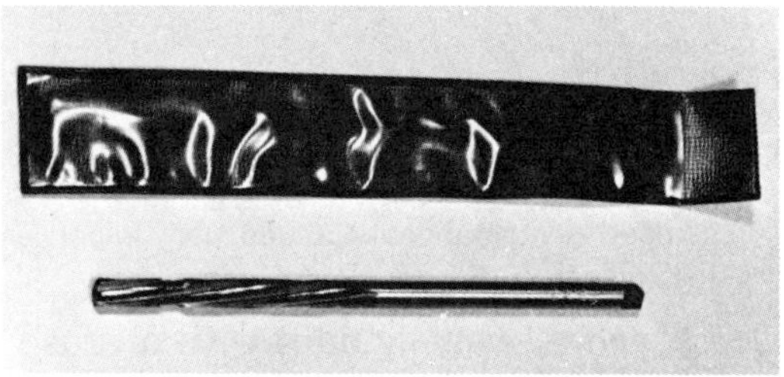

FIG 7-40 A valve guide reamer with its protective package.

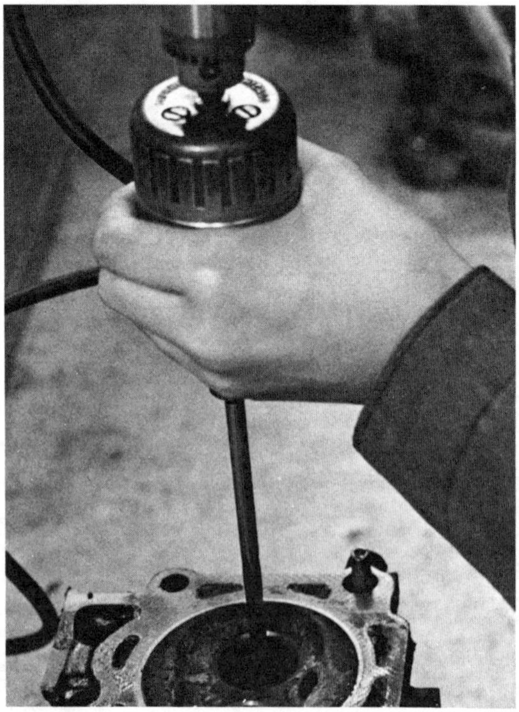

FIG 7-41 Reaming the valve guide.

JOB REPORT SHEET 7-B2

VALVE GUIDE KNURLING

Student Name ______________________ Date ____________

Make of Car ____________ Model ____________ Year ____________

Time Started ____________ Time Finished ____________ Total Time ____________

Flat Rate Time ______________________

Specifications:

Factory Valve Guide Specifications:

	Intake	Exhaust
Valve Guide (inside diameter)	______	__________
Valve Stem (guide area diameter)	______	__________
Stem-to-Guide Clearance	______	__________

Checkup Questions:

1. What type of lubrication is used during the resizing operation?

   ____

2. Is lubrication necessary during the reaming process? ______________________

3. How much can the stem-to-guide clearance be reduced through the knurling process? __________

4. Briefly fill in the steps in order to complete the knurling process:

   a. Clean the valve guide______________________________

   b. ______________________________

   c. ______________________________

   d. ______________________________

   e. ______________________________

   f. ______________________________

   g. ______________________________

DATE COMPLETED ____________________ INSTRUCTOR'S SIGNATURE ____________________

OBJECTIVE 7-B3 (manipulative)

Working as a team with one other student and with the aid of Job Report Sheet 7-B3, you will recondition a minimum of four valve guides using the Winona Bronze Wall process on an instructor-assigned vehicle engine cylinder head.

Time Recommended:

Twice the flat rate time allowed in *Motors, Chilton,* or *National Service Data* manuals; approximately 8 minutes per guide.

Preparation:

1. Review pages 185 through 196.
2. Study pages 199 through 200.
3. Study manufacturer's or general repair manuals for the automobile being serviced.
4. Attend lectures and demonstrations on valve service.

Tools and Materials Needed:

1. Bronze wall kit
2. Bronze wall bushings (one per guide)
3. Diagonal cutting pliers
4. Hammer

Procedure:

1. The cylinder head and guides must be absolutely clean and free from rust. Place the head on a stand with the combustion chamber side down. Use a stand that can be clamped to a bench. This is necessary because the turning force of the tap will tend to rotate the head.
2. Chuck the hex/square drive adapter into the power drive unit. Screw auxiliary pipe handle into the drill body.
3. Set the switch to the TAP (right hand rotation) position (Fig. 7-42).

FIG 7-42 Reversing switch placed in the Tap position.

FIG 7-43 Tapping the guide. Use no lubrication.
(Winona Manufacturing Company)

FIG 7-44 With the bushing wound on the inserting tool, screw it into the guide and break the lower tail.
(Winona Manufacturing Company)

4. Place the tap (pilot end) into the upper guide bore *without lubrication.*
5. Engage the power unit and tap the guide bore completely through the guide (Fig. 7-43).
6. Blow out or brush the chips from the guide bore. Use eye protection.
7. Select the correct size bronze wall bushing and install it on the inserting tool by fitting the closed tail into the slot on the bottom of the tool and the protruding tail into the slot at the top.
8. Wind up the bushing on the installing tool until it begins to tighten on the shaft.
9. Insert the bushing into the guide by turning it in until the end of the tool can just be felt at the opposite end of the guide at the valve port. This should leave the bushing recessed about 1/16 in. (1.54 mm) from the port end of the guide (Fig. 7-44).
10. Firmly grasp the top knob and turn the entire insert until the bottom tab breaks.
11. Unwind the excessive portion of the bushing above the guide.
12. With diagonal cutting pliers, cut off the bushing so that it extends 1/32 in. (0.794 mm) beyond the O.D. of the valve guide (Fig. 7-45).
13. Install the plastic (nylon) serrated bushing so the tail of the bronze wall bushing will engage one of the serrations. Use the clamp provided in the kit to hold the plastic bushing. When tightening the clamp, push downward on the plastic bushing to ensure that the bronze wall does not turn during the reaming operation.
14. Using the reaming oil provided, lubricate the guide bore and the bulb end of the broach. Drive the broach, bulb end first, completely through the guide. This presizes the bushing for finish reaming (Fig. 7-46).
15. Switch the power drive unit to the REAM position. This causes the chuck to turn toward the left (reverse).
16. Lubricate the reamer (correct size) and the guide bore before attaching the power unit.
17. Apply light downward pressure and turn on the motor. This will ream the guide to its correct size as the reamer passes through (Fig. 7-47).
18. Check the fit of the valve and guide.
19. Remove the plastic bushing and thoroughly clean the guide bore.

FIG 7-45 Cut off bushing, leaving a 1/32 inch (0.794 mm) tail.
(Winona Manufacturing Company)

FIG 7-46 Lube the bore and broach and drive it through the guide.
(Winona Manufacturing Company)

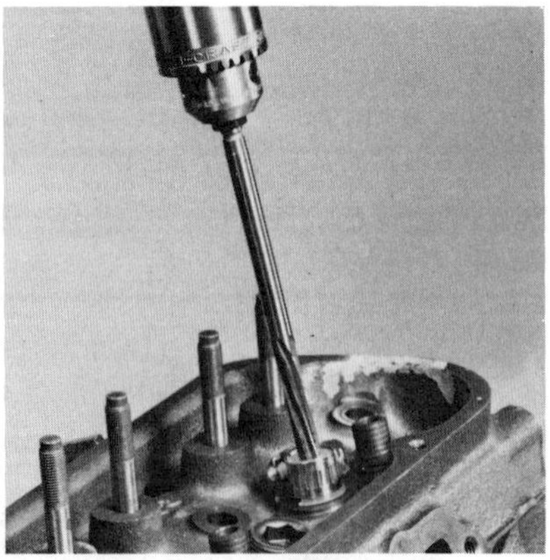

FIG 7-47 Switch motor to the Ream position and ream the guide.
(Winona Manufacturing Company)

JOB REPORT SHEET 7-B3

BRONZE WALL VALVE GUIDE REPAIR

Student Name ______________________________ Date ______________

Make of Car ______________ Model ______________ Year ______________

Time Started ______________ Time Finished ______________ Total Time ______________

Flat Rate Time ______________________________

Specifications:

Factory valve guide specifications. Indicate inches or millimetres.

	Intake	Exhaust
Valve Guide (inside diameter)	______	______
Valve Stem (guide area diameter)	______	______
Stem-to-Guide Clearance	______	______

Checkup Questions:

1. Why is lubrication not necessary when tapping cast iron guide bores? ______________________________

2. Why is broaching necessary? ______________________________

3. What would happen if the serrated plastic bushing was not firmly clamped to the tail of the bronze bushing during the reaming operation? ______________________________

4. What size reamer was used for final reaming? ______________________________

DATE COMPLETED ______________ INSTRUCTOR'S SIGNATURE ______________

SEALS

Valve seals are designed to prevent oil from entering the valve guides and being burned in the combustion chamber (intake) and in the exhaust gases (exhaust).

It has been long assumed that most oil consumption takes place as loss through intake valve guides. This occurs because the intake stroke creates an area of lesser pressure (vacuum) at the port end of the guide. This, combined with the washing effect of the incoming fuel, would cause intake guide oil consumption. A study conducted by a prominent maker of engine parts revealed that as much or more oil is lost through the exhaust valve guide as the intake. Hotter surfaces in the exhaust valve parts which make the oil flow more easily, greater guide clearance, and the drawing effect of the exhaust gas flow are the reasons for exhaust guide oil consumption.

There are several types of valve stem oil seals in use today, including the following:

1. Square ring groove seal ("O" ring)
2. Deflector (umbrella) seal
3. Positive seal

The ring seal is installed on the second groove of the valve stem. The locks are then installed on the first groove (closest to the tip). As the retainer meets the locks, the ring is compressed slightly, sealing the upper end of the valve stem (Fig. 7-48). A metal shield is often assembled with the retainer to further assist in keeping oil from entering the valve guide after running down the valve stem.

The deflector seal, frequently called an umbrella, is a rubber cup placed on the valve stem (cup side down) to shed the oil like an umbrella before it has a chance to enter the guide (Fig. 7-49).

Positive seals are usually affixed to the valve guide after some machining work is accomplished. As the valve moves up and down in the guide, a rubber or Teflon ring within the seal scrapes off the majority of the oil. However, enough of an oil film remains to properly lubricate the valve stem and guide. If seals of this type are to be installed, follow the manufacturer's directions enclosed with the seals.

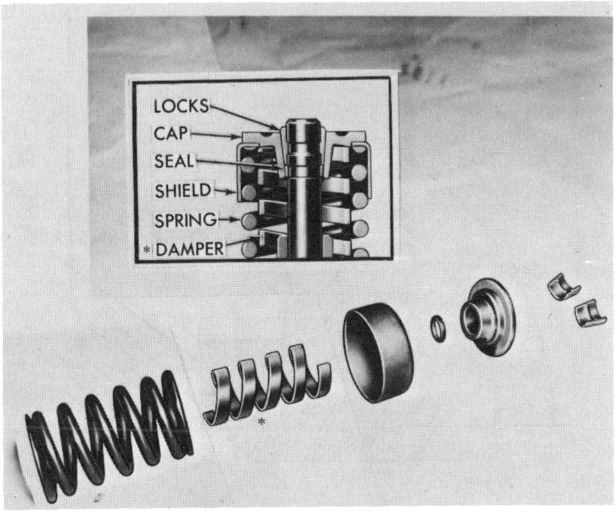

FIG 7-48 Ring valve stem seal location.

(Chevrolet Motor Division, GMC)

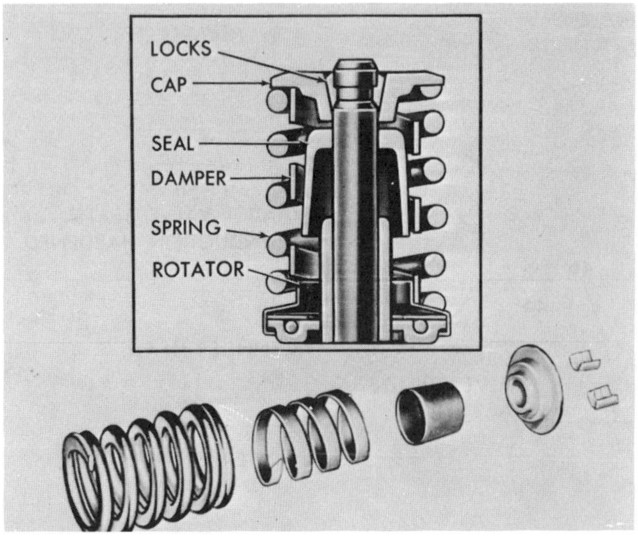

FIG 7-49 Deflector (umbrella) valve stem seal.

(Chevrolet Motor Division, GMC)

VALVE SEATS

The point of contact between the face of the valve and the head is called the valve seat. Valve seats are of two types: the integral and the insert.

INTEGRAL VALVE SEAT

The valve seats in most iron cylinder heads are referred to as integral because they are merely machined surfaces in the parts of the combustion chamber (Fig. 7-50). The iron used to cast the head itself is normally strong enough to hold its shape as a valve seat under average operating conditions.

Many engines that require unleaded fuel for emission control devices have valve seats that are heat treated (induction hardened) to reduce seat wear. Without tetraethyl and tetramethyl lead in the gasoline to cushion the face and seat contact of the valve, some engines experience accelerated wear to the seats, referred to as valve seat recession. This can be corrected and prevented by the use of valve seat inserts.

INSERT VALVE SEAT

Engines designed to run on L.P.G. (liquefied petroleum gas) or L.N.G. (liquefied natural gas) and others destined for heavy duty usage often have cylinder heads with valve seat inserts (Fig. 7-51). This is done to compensate for the lack of valve seat and face cushion which lead additives in gasoline provide. Valve seat wear is also accelerated by high combustion chamber temperatures, which are common in heavy duty service. Any engine with an aluminum cylinder head will have valve seat inserts, since the aluminum alloy used in casting this type of head will not withstand the operation requirements of a valve seat.

The valve seat inserts themselves are manufactured from several different materials chosen for their hardness, from plain cast iron, which is fairly soft, to an alloy called stellite, which is very hard. Some import engines have valve seat inserts made from bronze.

Two styles of grinding valve seats are commonly used today. The type used in most passenger car engines, called the interference valve seat (Fig. 7-52), has a valve face that is ½ to 1½ degrees less than the valve seat angle. In the initial stages

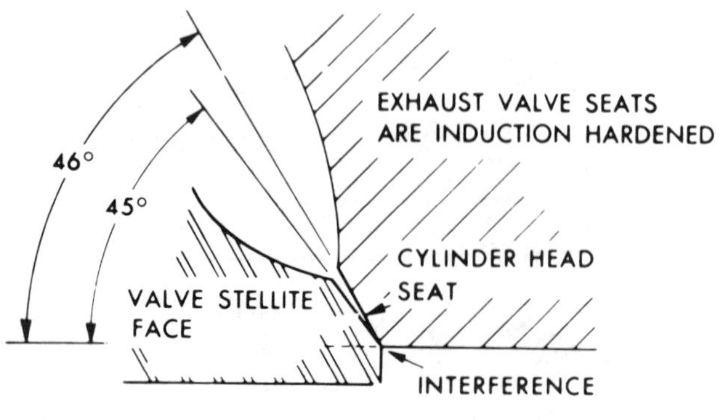

FIG 7-50 Integral valve seat.
(Chevrolet Motor Division, GMC)

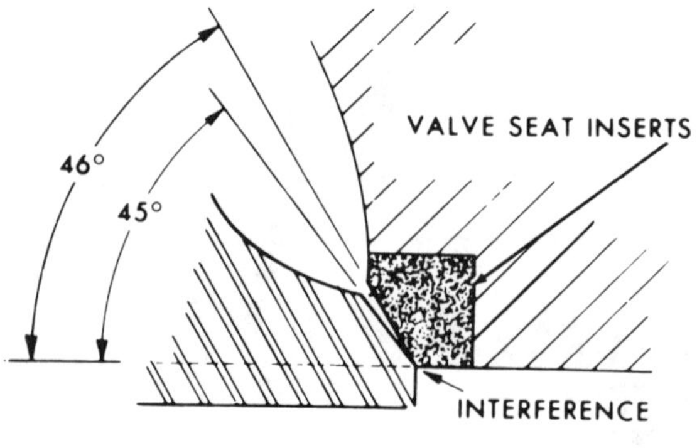

FIG 7-51 Insert valve seats.
(Chevrolet Motor Division, GMC)

of operation, there is a very thin line of contact with a high surface pressure. This is said to assist in the sealing of the gases.

The second design is called the noninterference valve seat (Fig. 7-53). This has both the valve face and seat ground to the same angle. It is recommended by at least one engine manufacturer for engines which burn unleaded fuel.

Valve seat grinding equipment is manufactured by many different firms, each of which has slightly different machine designs and methods of grinding valve seats. It is important to follow the instructions for the set you are using.

OBJECTIVE 7-B4 (manipulative)

Working as a team with one other student and with the aid of Job Report Sheet 7-B4, you will grind the valve seats of a cylinder head with a minimum of four combustion chambers from an instructor-assigned vehicle engine.

Time Recommended:

Twice the flat rate time allowed in *Motors, Chilton,* or *National Service Data* manuals.

Preparation:

1. Review pages 185 through 205.
2. Study pages 205 through 208.
3. Study manufacturer's or general repair manuals for the automobile being serviced.
4. Attend lectures and demonstrations on valve service.

Tools and Materials Needed:

1. Valve seat grinding set
2. Valve seat narrowing tool (optional)
3. Refaced valves

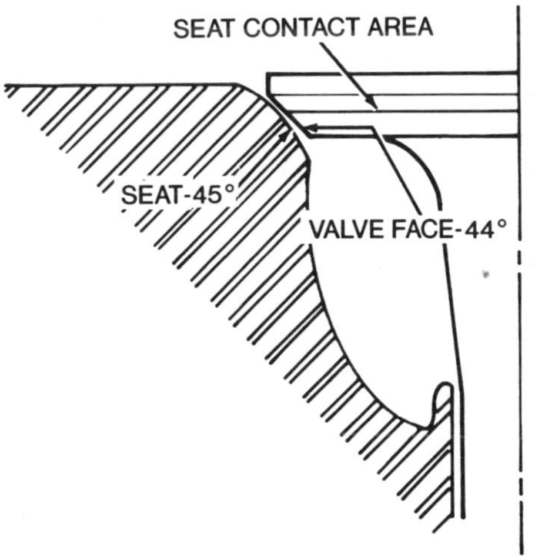

FIG 7-52 Interference valve seat angle. (Dana Corporation)

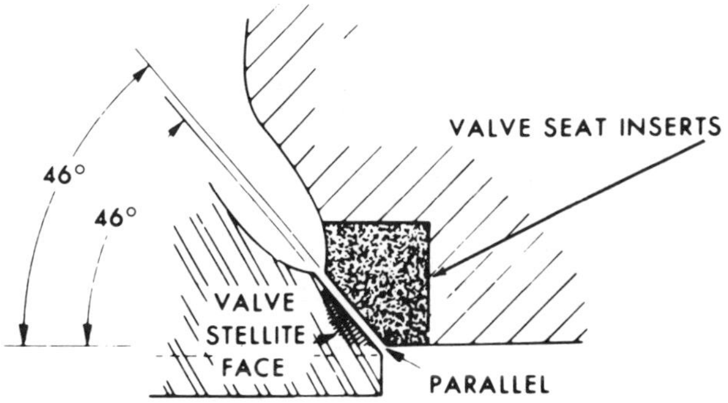

FIG 7-53 Noninterference seat with the same face and seat angle. (Chevrolet Motor Division, GMC)

Procedure:

1. Prepare the cylinder head by cleaning it thoroughly according to processes described earlier in this unit and in Unit 5. Any foreign matter attached to the head will interfere with the grinding operations.
2. Select the correct pilot for the valve guide and insert it into the guide. Some pilots are tapered, and some expand in the guide (Fig. 7-54). The pilot should not turn during the grinding operation.

3a. Select three stones of the correct diameter that have 15° differences in their grinding surfaces. Example: a 45° seat would require a 45° seating stone, a 30° topping stone, and a 60° throating stone for all grinding operations (Fig. 7-55).

Note: The correct diameter is slightly (⅛ in. or 3.175 mm) larger than the valve head diameter (Fig. 7-56).

3b. An alternate method would be to use a seat narrowing tool (Fig. 7-57). This tool eliminates the need for all stones except the actual seat stone.

4. Attach the stone(s) to the stone holder(s).
5. Place stone holder with the attached stone on the dressing stand. Adjust the angle of the dressing stand to match the approximate angle of the stone.
6. Adjust the diamond nib so that it lightly contacts the entire stone as the handle is moved up and down (Fig. 7-58).

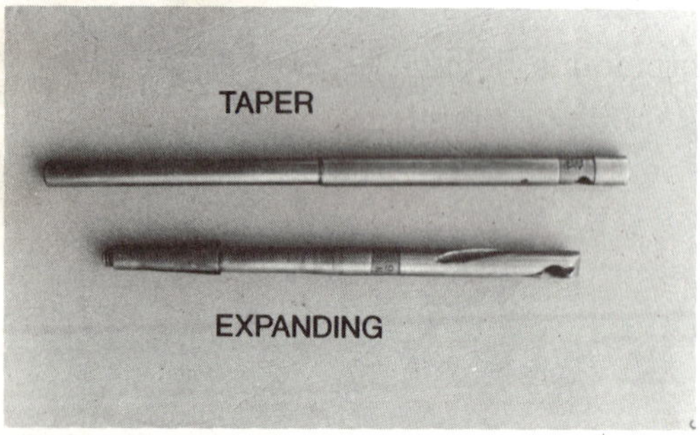

FIG 7-54 Tapered pilot (top); expanding pilot (bottom).

FIG 7-56 Choose the correct diameter seat grinding stone by comparing it to the valve head diameter.

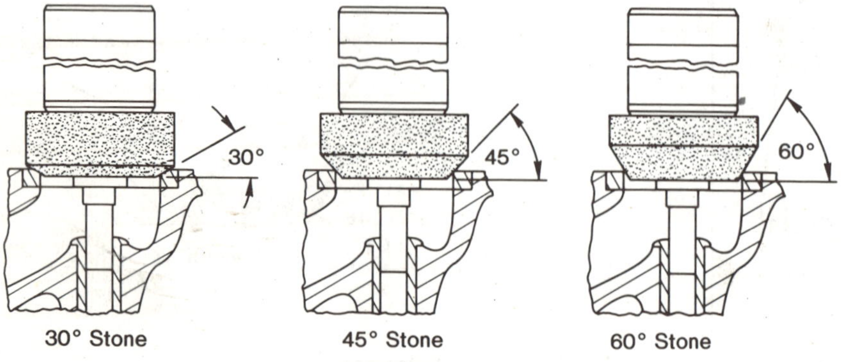

FIG 7-55 Seat grinding and narrowing stones.

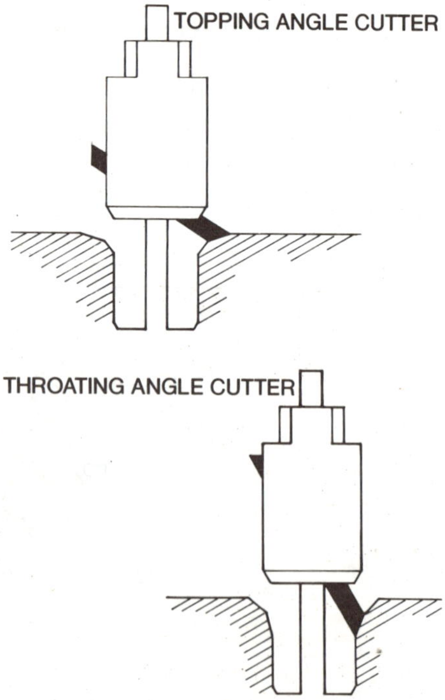

FIG 7-57 Cutting tool used to narrow the valve seat.

(Lisle Automotive Tools)

7. Insert the driver motor and proceed to dress the stone. Turn the diamond nib closer to the stone in small amounts (0.001 in. or 0.0254 mm) after each pass over the stone.
8. Dress the stone only enough to get an even surface (Fig. 7-59). Hold the driver motor as straight as possible in this operation. Unless the valve seats are in very poor condition, the stones should not need dressing before 3–4 seats are ground.
9. Place the stone and its holder on the pilot to be used for the seat grinding.
10. Grind the seat only enough to remove the irregularities on the seat surface. Operate the motor only 1 to 2 seconds at a time. Check the seat by pulling the stone holder up. Notice how the driver motor is being held (Fig. 7-60) to prevent the weight of the motor from applying too much pressure to the grinding stone. The stone should turn as fast as possible to give the best results.
11. With machinists dye or a bright colored felt tip marker, color the freshly ground seat area (Fig. 7-61). This helps distinguish the different grinding or cutting angles: seat, top, and throat.
12. Use a topping stone, usually 15° less than the seat angle, for narrowing the actual width of the seat to meet factory specifications (Fig. 7-62). When specifications are not available, use the following guide:

Intake Seats	Exhaust Seats
0.050–0.070 in. (1.27–1.78 mm)	0.060–0.080 in. (1.52–2.03 mm)

FIG 7-58 Adjust the diamond nib to make slight contact before turning on the driver motor.

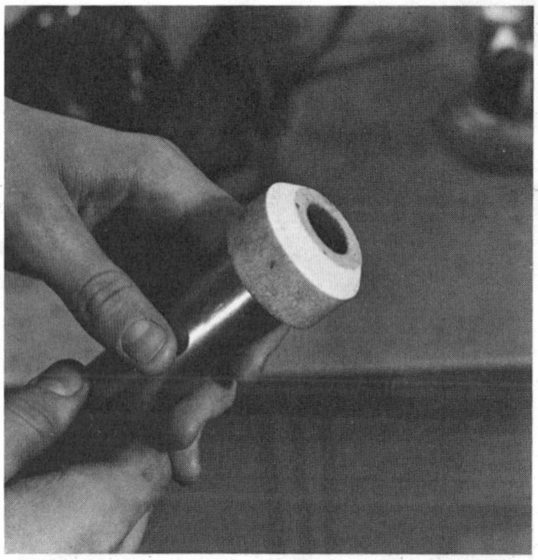

FIG 7-59 Correctly dressed stone.

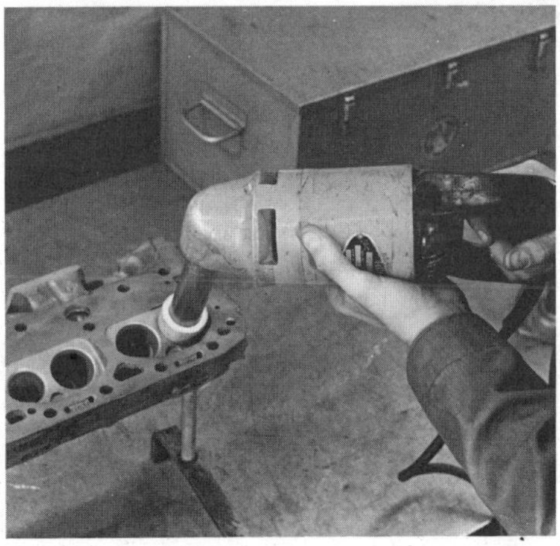

FIG 7-60 Correct motor position for seat grinding operations.

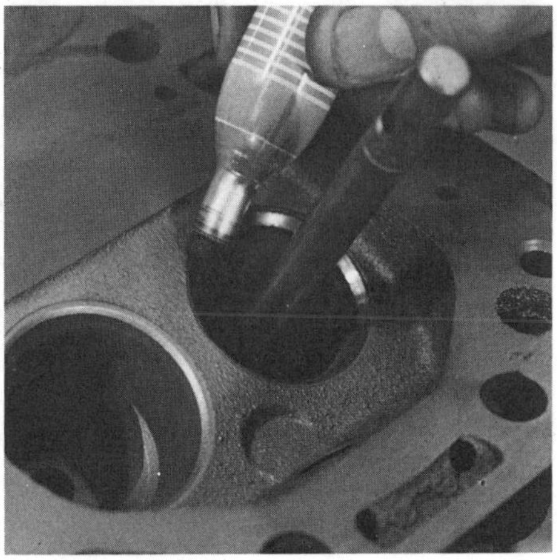

FIG 7-61 Coloring the freshly ground seat.

Note: In most cases, narrowing the seat from the top is adequate to bring it within specifications. If this is not possible, a 60° stone or cutter will have to be used to reduce the width from the bottom (throat).

13. Remove the pilot and locate the valve for the seat just completed.
14. Color the face of the valve with dye or felt marker and insert the valve in its correct guide.
15. Pop it up and down several times and remove it for inspection. There should be a line of contact about 1/4 to 1/3 of the way down the face from the margin. This line is made by the upper edge of the valve seat (Fig. 7-63).
16. A valve seat indicator can be used to check the concentricity of the seat after grinding. Place the indicator over the pilot and adjust the rod to contact the middle of the seat. Rotate the indicator in a full circle while reading any variations on the dial. The maximum allowable variation is 0.0005 in.
17. Grind all the seats in the same manner. Have your instructor check them all before you return the seat grinding equipment.
18. Clean the head thoroughly with a stem cleaner or soap or water. Be absolutely certain the guides are clean.

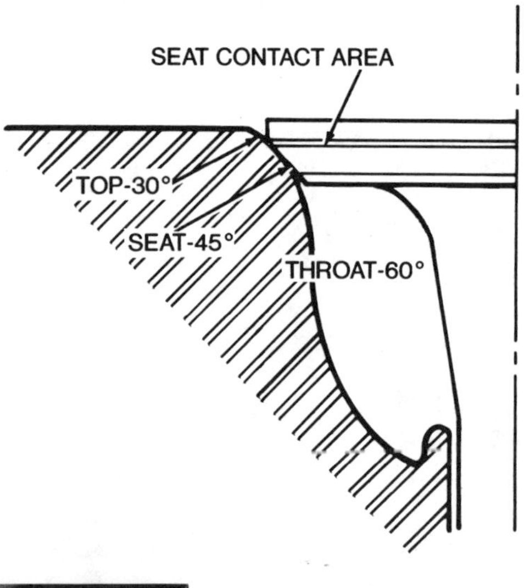

FIG 7-62 Top and throat angles for a 45° seat. (Dana Corporation)

FIG 7-63 After inserting the valve in the guide, check for a contact line on the face.

VALVE SEAT GRINDING

Student Name ______________________________ Date ______________

Make of Car ______________ Model ______________ Year ______________

Time Started ______________ Time Finished ______________ Total Time ______________

Flat Rate Time ______________________________

Specifications:

Valve seat width

Intake ______________ Exhaust ______________

General Specifications:

Intake	Exhaust
0.050–0.070 in. (1.27–1.78 mm)	0.060–0.080 in. (1.52–2.03 mm)

Checkup Questions:

1. Which type of pilot did the seat grinding set have? ______________________________

2. Why must the valves be refaced before the seating is done?

3. Were the valve guides reconditioned prior to the seat grinding? (Yes/No)

4. Did the guide reconditioning affect the seat grinding? (Yes/No) ______________________________

If yes, how? ______________________________

5. List the three grinding or cutting angles used:

Seat	Topping	Throating
°	°	°

6. What process was used to clean the head after all grinding was completed? ____________________

DATE COMPLETED ____________________ INSTRUCTOR'S SIGNATURE ____________________

VALVE RETAINERS AND LOCKS

Every valve must have a spring retainer and some type of lock (keeper) to keep the valve spring in position. The retainer is necessary to keep the spring in line with the valve stem. Most retainers have steps of different diameters on the bottom side (closest to the head) that match the inside diameter of the coils of the spring. If there is a mismatch of springs and retainers, the spring will cause unnecessary valve guide wear.

The inside shape of most retainers used today is a cone. This shape matches the outside shape of the locks. The inside shape of the locks matches the groove or grooves of the valve stem.

Prior to assembly of the valve parts, inspect all the retainers and locks for signs of wear or out-of-shape surfaces.

VALVE SPRINGS

Valve springs play a vital role in the operation of the engine. They close the valve with enough force to seal the combustion gases and to ensure proper contact between the valve face and its seat. This is extremely important, because the valve face is normally much hotter than its seat as the engine runs. When the valve is closed with the correct spring tension, the heat will transfer from the face to the seat. This must be done continuously as the engine operates or the valve face will be damaged and its ability to seal will be lost.

SPRING PROBLEMS

Valve springs with either too much or too little tension can cause problems. Weak springs allow the valve to bounce or float, causing poor performance. Valve failure can occur with weak springs because the combustion heat in the valve head is not transferring to the valve seat when the valve is closed. Springs with too much tension can cause a valve to stretch or neck. When a spring is not straight (the ends are not parallel), the result is poor valve action and uneven valve guide wear.

VALVE SPRING TESTING

The engine technician should test every spring, including new ones, to be certain specified tension and straightness specifications are met. Smooth valve train operation is assured when the springs meet recommended levels of tension and shape.

Engine manufacturers provide specifications for valve spring tension testing. These figures are primarily for checking used springs to see whether they meet tension specifications allowing their further use. For example, one manufacturer allows a 10% loss of tension before replacement of the spring is recommended.

The use of valve spring shims to reduce the installed height will often compensate for minor tension loss. See Installed Height Information, page 215.

OBJECTIVE 7-B5 (manipulative)

Working alone or as a team with one other student and with the aid of Job Report Sheet 7-B5, you will test the straightness and tension of at least 8 valve springs on an instructor-assigned vehicle engine.

Time Recommended:

15 minutes.

Preparation:

1. Review pages 185 through 211.
2. Study page 211.
3. Study manufacturer's or general repair manuals for the engine being serviced.
4. Attend lectures and demonstrations on valve service.

Tools and Materials Needed:

1. Spring tension tester
2. Telescoping gauge
3. 2 in. micrometer

Procedure:

1. Select one spring and place it next to the scale at the right side of the lower platform of the spring tension tester (Fig. 7-64).
2. Rotate the spring through 360° of movement (one complete circle) with some part of the spring coils touching the scale.
3. If either end of the spring moves away from the scale more than 1/16 in. (approximately 1.6 mm), the spring is not straight and should be discarded.
4. If there is less than 1/16 in. gap, move the spring to the center of the lower platform.
5. Bring the upper platform down and compress the spring to the specified height. The height can be checked with a telescoping gauge set to the proper distance (Fig. 7-65).
6. The pointer on the front dial indicates the tension in pounds. Compare this to the specifications.

Note: Some spring testers have an adjustable stop for testing many springs with the same compressed length. Set the stop *only* when the length is correct and there is a spring under tension. This is necessary because the lower platform moves as the spring is being compressed.

FIG 7-64 Check the valve spring for straightness against the tension tester rule.

FIG 7-65 Measuring the compressed height with a telescoping gauge.

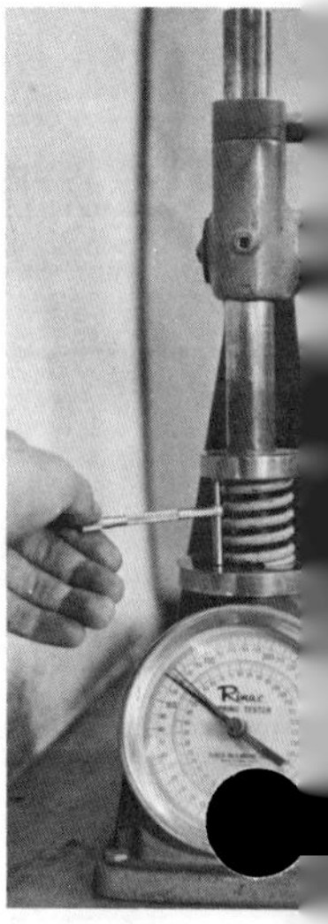

JOB REPORT SHEET 7-B5

VALVE SPRING TESTING

Student Name ______________________ Date ______________

Make of Car ____________ Model ______________ Year ______________

Time Started ____________ Time Finished ____________ Total Time ____________

Flat Rate Time or Recommended Time ______________________

Specifications:

Factory valve spring specifications:

OUTER

Free Length ____________ in or mm.

Valve Closed ____________ lbs or kg @ ____________ in or mm.

Valve Open ____________ lbs or kg @ ____________ in or mm.

INNER (If Used)

Free Length ____________ in or mm.

Valve Closed ____________ lbs or kg @ ____________ in or mm.

Valve Open ____________ lbs or kg @ ____________ in or mm.

Straightness Limit 1/16 in. (1.59 mm) maximum

Checkup Questions:

1. What type (design) of valve spring was used? ______________________

____

2. How many springs were distorted beyond limits? ______________________

3. Can original springs be used as is? ______________________

   With shims? ______________________

4. If new springs are required, will they be tested to check their straightness and tension? ____________

DATE COMPLETED ______________ INSTRUCTOR'S SIGNATURE ______________

VALVE SPRING INSTALLED HEIGHT

Grinding the valve seat and resurfacing the valve face permits the valve to sink deeper into the head or block. When this occurs, the distance the valve spring is compressed (valve closed) will lengthen, thereby reducing its tension force. The compressed spring distance with the valve closed is referred to as *valve spring installed height.*

Working as a team with one other student and with the aid of Job Report Sheet 7-B6, you will measure and adjust to specifications the installed height of all valve springs on an instructor-assigned vehicle engine.

OBJECTIVE 7-B6 (manipulative)

Time Recommended:

Twice the flat rate time allowed in *Motors, Chilton,* or *National Service Data* manuals; approximately 15 minutes.

Preparation:

1. Review pages 185 through 215.
2. Study pages 215 through 216.
3. Study manufacturer's or general repair manuals for the engine being serviced.
4. Attend lectures and demonstrations on valve service.

Tools and Materials Needed:

1. Telescope gauge and micrometer or depth gauge rule
2. Teardown and Assembly Sheet, 3-A1

Procedure:

1. Assemble the refaced valve, spring retainer, and locks in the valve closed position.
2. Measure the distance from the bottom side of the retainer to the spring pocket. This distance is the valve spring installed height (Fig. 7-66).

FIG 7-66 Measuring the valve spring installed height.

3. Compare the measured distance with the manufacturer's specifications (Teardown and Assembly Sheet, 3-A1).
4. When the installed height is too great, select only *one* of three thicknesses of valve spring shims (0.015 in., 0.030 in., or 0.060 in.) for each spring (Fig. 7-67). This should reduce the distance between the retainer and spring pocket to meet specifications.

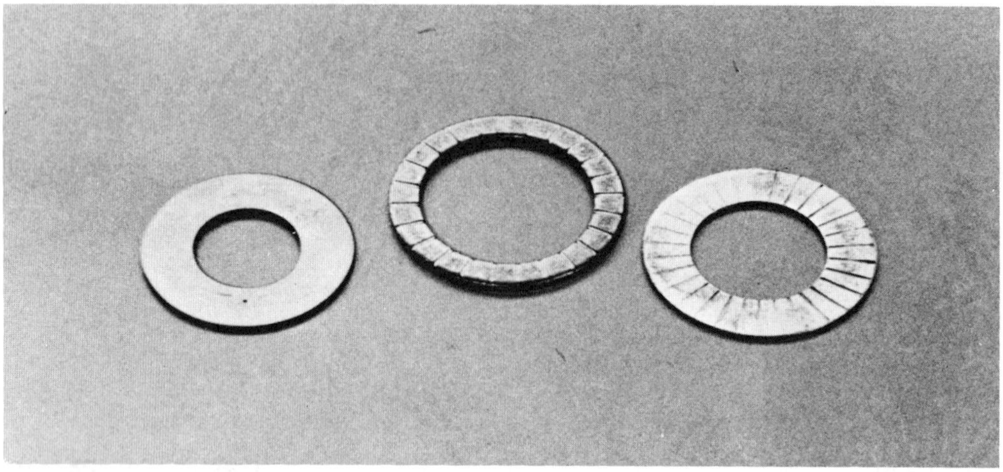

FIG 7-67 Valve spring shims.

JOB REPORT SHEET 7-B6

VALVE SPRING INSTALLED HEIGHT MEASUREMENT

Student Name ______________________________ Date ______________

Make of Car ______________ Model ______________ Year ______________

Time Started ______________ Time Finished ______________ Total Time ______________

Flat Rate Time or Recommended Time ______________________________

Specifications:

Factory valve spring installed height specifications:

Intake ______________ in or mm

Exhaust ______________ in or mm

Checkup Questions:

1. Can the installed height be corrected by using valve spring shims?

   ____

2. What is the maximum thickness of shim recommended? ______________________

DATE COMPLETED ______________________ INSTRUCTOR'S SIGNATURE ______________________

CYLINDER HEAD ASSEMBLY

When no further valve service operations are to be completed, the cylinder head can be assembled. If new parts are to be installed, check their fit very carefully. New valves are usually ready for installation. If a different face angle was used, it will need refacing.

Note: Do not reface valves with plated (aluminum or nickel) head surfaces.

All parts must be absolutely clean. Any materials left, especially from the grinding processes, can cause excessive engine wear in a short period of time.

OBJECTIVE 7-B7 (manipulative)

Working as a team with one other student and with the aid of Job Report Sheet 7-B7, you will assemble a cylinder head following all required service operations from an instructor-assigned vehicle engine.

Time Recommended:

Twice the flat rate time allowed in *Motors, Chilton,* or *National Service Data* manuals.

Preparation:

1. Review pages 185 through 219.
2. Study pages 219 through 220.
3. Study manufacturer's or general repair manuals for the automobile being serviced.
4. Attend lectures and demonstrations on valve service.

FIG 7-68 Installing positive valve stem seal over a plastic sleeve on the valve stem. (Dana Corporation)

Tools and Materials Needed:

1. Valve spring compressor
2. Valve stem seals
3. Small amount of grease
4. Pump oil can

Procedure:

1. Place the cylinder head on the bench with the combustion chambers closest to you.
2. Lightly oil the guides and insert all the valves in their respective guides.
3. If the head was machined for positive valve seals, they should be installed now. Use the plastic sleeve on the valve stem prior to pushing the seals into position (Fig. 7-68).
4. If umbrella seals are to be installed, slip them over the valve stems with the cup side down (toward the guide).
5. Assemble the spring and retainer and place them over the valve stem.

Note: Variable rate springs (those with coils closer together at one end than the other) should be installed with the more closely coiled end nearest the head (Fig. 7-69). Constant rate springs, even those with dampers (flat inner springs), can be assembled with either end nearest the head.

6. Compress the valve spring far enough to install the locks.
7. If the valve stems are designed to have "O" ring seals, install them now.
8. Lightly cover the valve stem end with stiff grease and assemble the locks on the stem (Fig. 7-70). The grease should hold the locks in place while the spring compressor is released.
9. Release the valve spring compressor.
10. Assemble the remaining valves.
11. Cover the head with a plastic bag until it is ready for installation on the block.

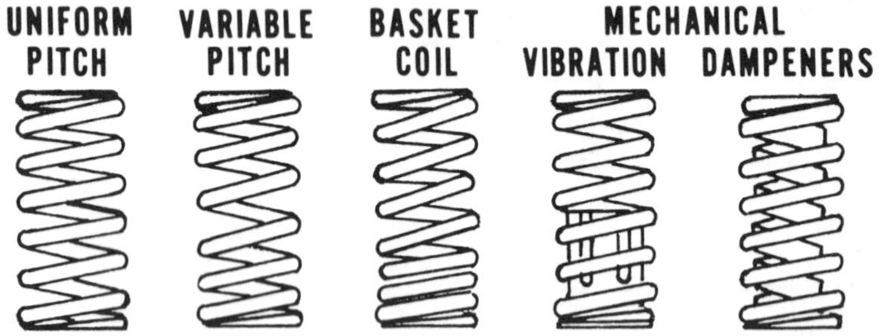

FIG 7-69 Various valve springs and their installation position.
(Dana Corporation)

FIG 7-70 With the spring compressed, install the seal if so equipped. Apply grease to hold locks while releasing the spring compressor.

CYLINDER HEAD ASSEMBLY

Student Name ______________________________ Date ______________

Make of Car ____________ Model ______________ Year ______________

Time Started ____________ Time Finished ____________ Total Time ____________

Flat Rate Time ______________________________

Checkup Questions:

1. What type of valve stem seals was originally installed? ______________________________

____

2. Was there any change in the type of valve stem seals? (Yes/No) ____________

3. What type of valve springs was used? ______________________________

4. Is the head properly prepared for storage (plastic bag)? (Yes/No)____________

DATE COMPLETED ______________ INSTRUCTOR'S SIGNATURE ______________

Name ______________________________ Date ______________

DIRECTIONS:

Circle the best answer to each question.

1. The operation which squares the stem end is called
   I. tipping.
   II. refacing.
   A. I only B. II only C. Both I and II D. Neither I nor II
2. After the valves have been refaced and tipped, they should be cleaned with
   I. a wire wheel.
   II. glass beads.
   A. I only B. II only C. Both I and II D. Neither I nor II
3. The margin of a valve
   I. is the thickness of the head between the face and the top head surface.
   II. should be no less than 1/32″ (0.787 mm) after refacing.
   A. I only B. II only C. Both I and II D. Neither I nor II
4. Dressing the refacing grinding wheel
   I. should be done before any refacing is begun.
   II. can be done any time the face surface is not to be ground smooth.
   A. I only B. II only C. Both I and II D. Neither I nor II
5. Refacing the valves
   I. should remove as little metal as possible.
   II. requires holding the valve in the chuck by its stem on the normal valve guide wear area.
   A. I only B. II only C. Both I and II D. Neither I nor II
6. When refacing a valve,
   I. move the stone away from the valve face when enough grinding has been done.
   II. move the valve back and forth over the face of the stone.
   A. I only B. II only C. Both I and II D. Neither I nor II
7. Grinding excessive metal from the tip (beyond 0.030 in. or 0.762 mm)
   I. can save heavily pitted stems.
   II. reduces the chances of mushrooming when the valve is later assembled and run.
   A. I only B. II only C. Both I and II D. Neither I nor II
8. Valve guide service
   I. is performed to restore the original clearance between the stem and guide.
   II. can be knurling, reaming, or replacement of the guide itself.
   A. I only B. II only C. Both I and II D. Neither I nor II
9. Knurling should be performed
   I. prior to hot tank cleaning.
   II. is recommended for guide wear up to 0.020 in. (0.508 mm).
   A. I only B. II only C. Both I and II D. Neither I nor II

10. Enlarging valve guides by reaming
   I. requires valves with oversize stems.
   II. requires that the guide first be knurled.
   A. I only B. II only C. Both I and II D. Neither I nor II

11. Bronze wall valve guide reconditioning requires
   I. a reversible drill motor.
   II. a special tap.
   A. I only B. II only C. Both I and II D. Neither I nor II

12. Valve seals reduce
   I. the amount of exhaust gases reaching the combustion chamber.
   II. the oil loss past the valve covers.
   A. I only B. II only C. Both I and II D. Neither I nor II

13. Valve seats are of two designs:
   I. integral and nonmetallic.
   II. insert and integral.
   A. I only B. II only C. Both I and II D. Neither I nor II

14. The interference valve seat
   I. has a valve face angle greater than the seat angle.
   II. is recommended for most engines.
   A. I only B. II only C. Both I and II D. Neither I nor II

15. Valve seat grinding stones should be selected from a set according to
   I. their diameter and grinding angle.
   II. color of the contact surface.
   A. I only B. II only C. Both I and II D. Neither I nor II

16. Topping and throating are terms used to describe
   I. the openings of the valve guide.
   II. the two surfaces of the valve face that are 15° less than the original surface.
   A. I only B. II only C. Both I and II D. Neither I nor II

17. Valve seats must
   I. be within specified limits.
   II. contact the valve face within its surface.
   A. I only B. II only C. Both I and II D. Neither I nor II

18. Valve springs
   I. for reuse must be tested for tension and squareness.
   II. that are new do not need to be tested.
   A. I only B. II only C. Both I and II D. Neither I nor II

19. Valve spring installed height
   I. refers to the position of the valve stem.
   II. is measured from the spring pocket to the valve tip.
   A. I only B. II only C. Both I and II D. Neither I nor II

20. Valve stem seals
   I. of the square ring ("O" ring) design should be installed after the spring is compressed.
   II. of the positive type should be installed over a plastic sleeve before the spring is placed in position.
   A. I only B. II only C. Both I and II D. Neither I nor II

DATE COMPLETED ______________ INSTRUCTOR'S SIGNATURE ______________________

unit 8

Camshafts and Valve Train Service

This unit gives the basic information and techniques needed to service the common valve train components according to current engine reconditioning practices. Additional information is included on complete inspection procedures for all valve train parts.

OBJECTIVE 8-A (written test)

After completing the preparation assignment, you will answer a minimum of 14 out of 20 multiple choice questions in a written test, without reference material, on the following topics:

1. Camshafts and valve train operation
2. Timing gears, sprockets, and chains
3. Lifters, pushrods, and rocker arms
4. Inspection and measurements of the valve train

Preparation:

Study pages 225 through 241.

VALVE TRAIN

The valve train consists of those parts of the engine which control the entry of the air-fuel mixture into the cylinder and the exit of the exhaust gases. Valve train components usually include: timing gears or sprockets with a chain or belt, camshaft, valve lifters, pushrods (if so equipped), rocker arms, and valves (Figs. 8-1 and 8-2).

The valve action in a typical OHV engine begins with the crankshaft rotating the camshaft. This, in turn, causes the lifters to move. The pushrod transmits this motion to the rocker arm, which causes the valve to open. As the cam lobe moves past its high point (nose), the valve spring applies enough force to move the rocker arm, pushrod, and lifter to the original (valve closed) position.

Cam goes at ½ the crankshaft speed

The camshaft is driven by the crankshaft at one-half the crankshaft's speed. This is why the camshaft gear is twice the size of the crankshaft gear. All four-stroke cycle engines have this camshaft-crankshaft speed relationship. To properly synchronize the crankshaft and camshaft movements, these parts are indexed together by what are called valve timing marks (Fig. 8-3). These marks are applied to gears or sprockets and to chains to assist the engine in the assembly operation. The movements of the valves are timed with the piston movements to give an engine the best performance possible.

CAMSHAFT

The camshaft controls when and how much the valves are opened by changing rotary motion from the crankshaft into reciprocating (up and down) motion at the lifter. This change in motion takes place because the cam lobe is off center from the camshaft.

CAMSHAFT DESIGN

The first consideration in camshaft design is making it fit the block or head of the engine. Very often the camshaft must drive the distributor, oil pump, and fuel pump (Fig. 8-4).

The cam lobe profile is the single greatest influence on engine performance because of its direct relationship with valve openings. For this reason, cam lobe designs are very carefully considered. The names used in cam lobe design are shown in Figure 8-5. The heel or base diameter is the surface where the lifter is sliding while the valve is closed. The nose is the point of maximum lobe lift. The lift is the distance, in inches or millimetres, the lifter is raised. Each side of the lobe has a flank or ramp called an opening or closing flank, depending on the direction of cam rotation. The timing points are the locations where duration measurements are made, duration being the number of degrees of *crankshaft* rotation that each valve is open.

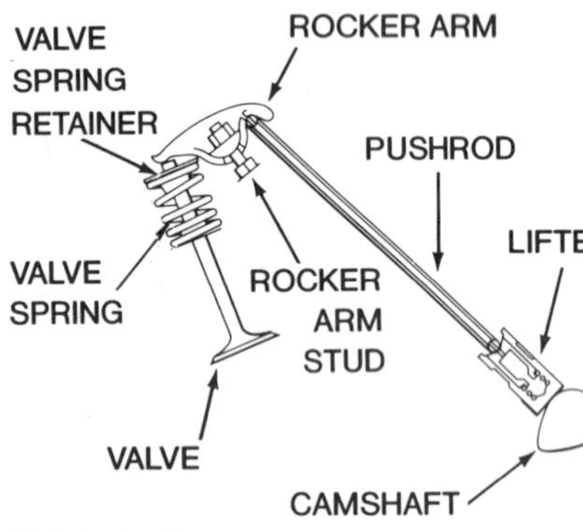

FIG 8-1 Valve train parts. (Buick Motor Division, GMC)

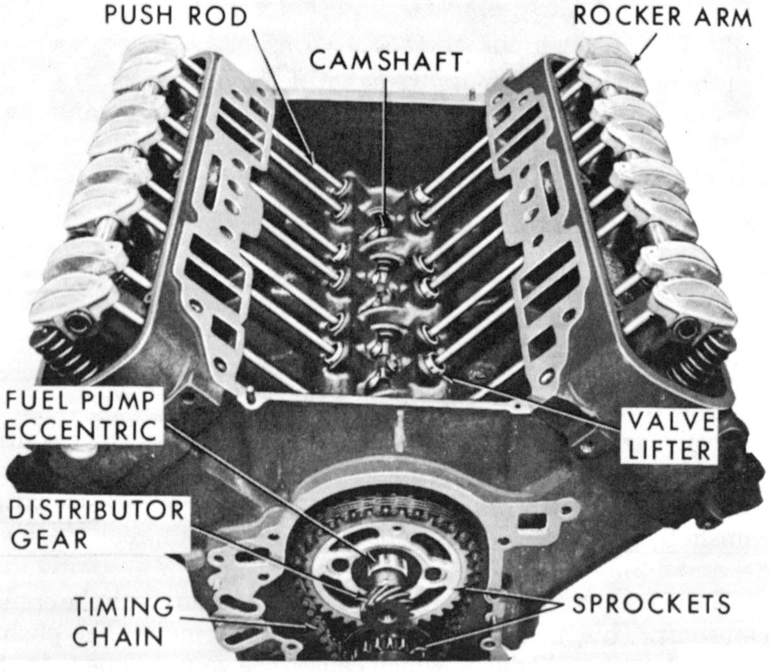

FIG 8-2 Valve train parts assembled in an engine. (Buick Motor Division, GMC)

FIG 8-3 Typical valve timing marks. (Chevrolet Motor Division, GMC)

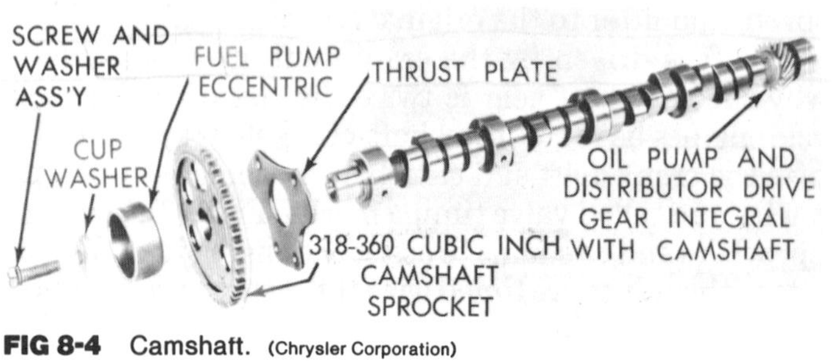

FIG 8-4 Camshaft. (Chrysler Corporation)

CAMSHAFT MANUFACTURING AND MATERIALS

Most camshafts are cast from high quality grey iron, then ground to precise dimensions according to the projected use of the engine. A few camshafts are machine cut from a solid piece of steel called a billet, then ground to specifications. Billet camshafts are used frequently in high performance engines. A cast camshaft will have a pebbly surface between the lobes. A billet camshaft will have machine marks (light grooves) between the lobes.

CAMSHAFT INSPECTION AND MEASUREMENT

When an engine continues to run poorly after all normal tune-up work has been completed, the cause may be a worn cam lobe or lobes. When this occurs, the valve will not open as much as it should, which does not allow that cylinder to "breathe."

When visually inspecting a camshaft for wear, the technician should check for two common cam lobe problems. First, the lobes often experience pitting, edge loading or galling of the surface where the lifter makes contact (Fig. 8-6). Second, the lobes wear until there is full coverage of lifter contact across the top of the lobes. Since the camshaft is a single piece, it must be replaced when one or more lobes are worn.

To measure the lobe lift, several methods can be used. If the camshaft is removed from the engine, a dial indicator or a micrometer can be used. When the camshaft is in the engine, a dial indicator is the best tool to use.

The camshaft journals usually wear only slightly, but they should be checked before new cam bearings are installed.

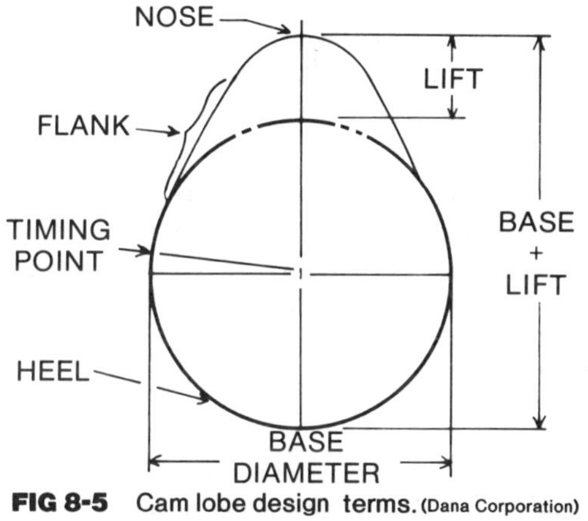

FIG 8-5 Cam lobe design terms. (Dana Corporation)

FIG 8-6 Cam lobe wear. (Dana Corporation)

FIG 8-7 Using a micrometer to measure the cam lobe height.

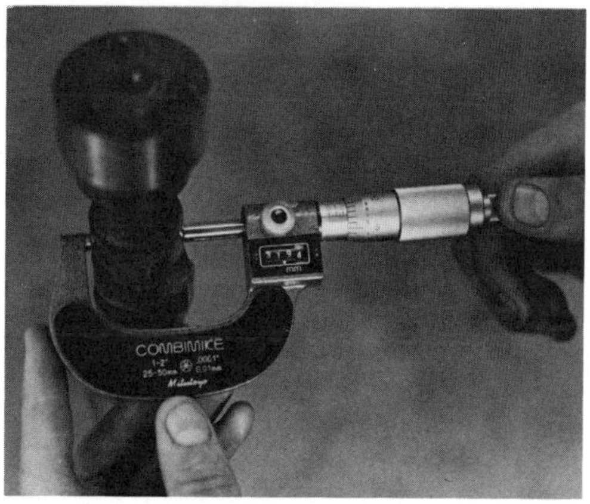

FIG 8-8 Measuring the base circle.

OBJECTIVE 8-A1 (manipulative)

Working as a team with one other student and with the aid of Job Report Sheet 3-A1 and 8-A1, you will visually inspect for wear and measure all journals and cam lobes for lift on a camshaft from an instructor-assigned vehicle engine.

Time Recommended:

Twice the flat rate time allowed in *Motors, Chilton,* or *National Service Data* manuals.

Preparation:

1. Review pages 225 through 227.
2. Study page 228.
3. Study manufacturer's or general repair manuals for the engine being serviced.
4. Attend lectures and demonstrations on valve train service.

Tools and Materials Needed:

1. Camshaft and lifters
2. Micrometer

Procedure:

1. Clean the camshaft, *removed from the engine,* of excessive oil and dust by wiping or washing in solvent. Do not clean camshafts in chlorinated hydrocarbons (carburetor cleaner) or a caustic solution (hot tank). This will remove the protective lube coating needed if the camshaft is to be used again.
2. Visually inspect cam lobes for pitting and excessive wear. If there is any doubt about the lobe wear, inspect each lifter together with its respective cam lobe. The wear pattern on the lifter face can indicate the need to replace a camshaft. If the lobes do not appear to be worn excessively, continue with Step 3. If the lobes are worn excessively, show the camshaft to your instructor or supervisor, who can give you a final decision on whether to discard or reuse the camshaft.

Note: Very few camshafts are reused by engine mechanics. If the camshaft is to be discarded, use the replacement for your measurement exercise to complete this objective.

3. Determine the lobe lift by measuring the maximum distance (height) (Fig. 8-7). Record this information on Job Report Sheet 3-A1.
4. Measure and record the diameter of the base circle (Fig. 8-8).
5. Subtract the base circle figure from the lobe height to find the lift.
6. If the lobes measure within specification, measure and record all the camshaft journal diameters.

JOB REPORT SHEET 8-A1

CAMSHAFT INSPECTION AND MEASUREMENT

Student Name ______________________ Date ______________

Make of Car ____________ Model ______________ Year ______________

Time Started ____________ Time Finished ______________ Total Time ____________

Flat Rate Time or Recommended Time ______________________

Specifications:

Refer to Job Report Sheet 3-A1

Camshaft Life ______________________

Camshaft Journal ______________________

Checkup Questions:

1. What is the maximum lobe wear (the lobe difference of the base and lift and the base diameter)?

   ____

2. What is the cause of lobe wear? ______________________

   ____

   ____

3. What is the maximum journal wear? ______________________

   Which journal (number)? ______________________

4. Were any other defects noted? ______________________

   ____

   ____

DATE COMPLETED ______________ INSTRUCTOR'S SIGNATURE ______________

TIMING GEARS, SPROCKETS, AND CHAINS

To be certain the camshaft opens the valves at the proper instant, a system of gears or sprockets and chain is used (Fig. 8-9). The crankshaft, the driving member of the system, is turned at ½ the crankshaft speed, so the camshaft gear is twice the size of the crankshaft gear.

The term *gear* is applied to a drive system of two gears touching one another. The term *sprocket* describes the driving and driven members between a chain. These parts of the valve train are the sprocket and chain design. Some overhead camshaft engines use gears and a notched belt (Fig. 8-10).

The camshaft drive components wear during normal operation, so they are often replaced during an overhaul. The chain tends to stretch and the sprockets develop notches. The nylon-faced aluminum sprocket is used in many engines today to reduce cam drive noise. Wear or failure often occurs when the nylon breaks away from the aluminum (Fig. 8-11).

When any of these valve train parts wear, the effect is retardation of the camshaft. This means the valves open and close later than they should in relation to the piston movement. When this occurs, the level of performance is reduced.

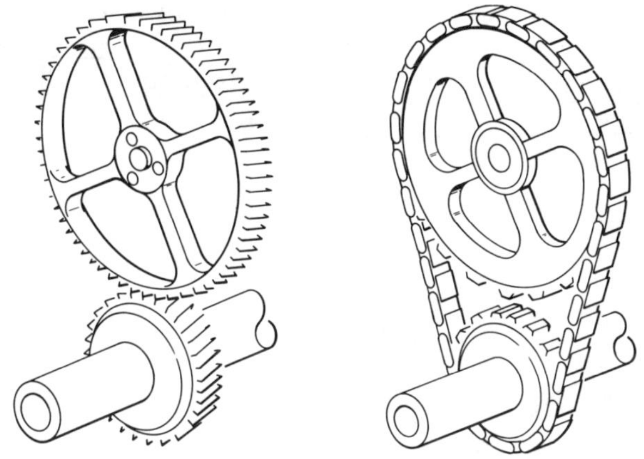

FIG 8-9 Gear drive (left) and chain, and sprocket drive (right). (Ford Motor Company)

FIG 8-10 A notched rubber belt often drives overhead camshafts. (Chevrolet Motor Division, GMC)

FIG 8-11 Carefully inspect nylon-faced timing gears.

LIFTERS OR TAPPETS

The parts of the valve train which move the pushrods or rocker arms are called lifters or tappets. The term *lifter* is more common today as a result of the almost total application of hydraulic lifters to stock domestic engines. However, the word *tappet* is often used to describe this same valve train part, especially the mechanical or adjustable type.

Three lifter designs are used in current engines: the hydraulic; the mechanical or solid; and the roller lifter, used only in high performance engines.

Hydraulic lifters are very quiet running because of their ability to sense the operating clearance necessary between the rocker arm and the valve stem tip. Therefore, there is no clearance adjustment screw, as on a mechanical lifter, and little or no adjustments are required once hydraulic lifters are installed. These features make the hydraulic lifter the most popular choice of engine designers (Fig. 8-12).

The operation of a hydraulic lifter is shown in Figure 8-13. While the lifter is on the base circle of the camshaft, the valve is closed. When the camshaft rotates, the lifter pushes against the oil between the body and the plunger until the disc valve closes and the oil is trapped in the chamber. This causes the lifter to move up as an assembly (Fig. 8-14), which in turn moves the pushrod to open the valve. When the camshaft rotates further, the lifter moves back on the base circle of the cam. The return spring pushes the plunger back up and allows pressured oil from the oil gallery to refill the high pressure area at the bottom of the plunger (Fig. 8-15).

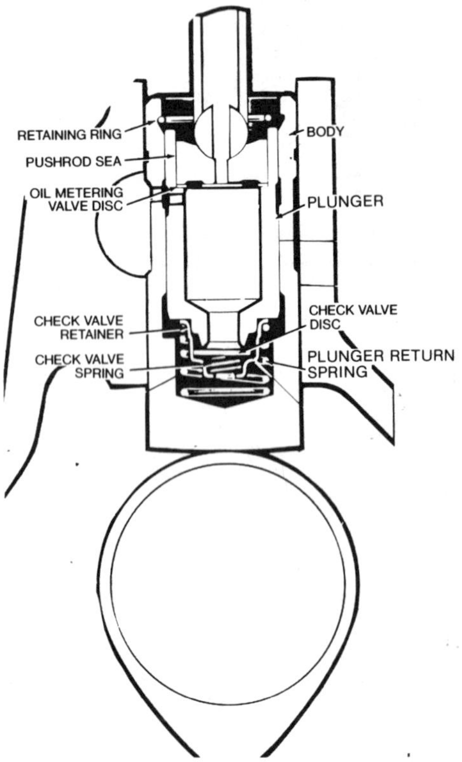

FIG 8-12 Hydraulic lifter.
(Dana Corporation)

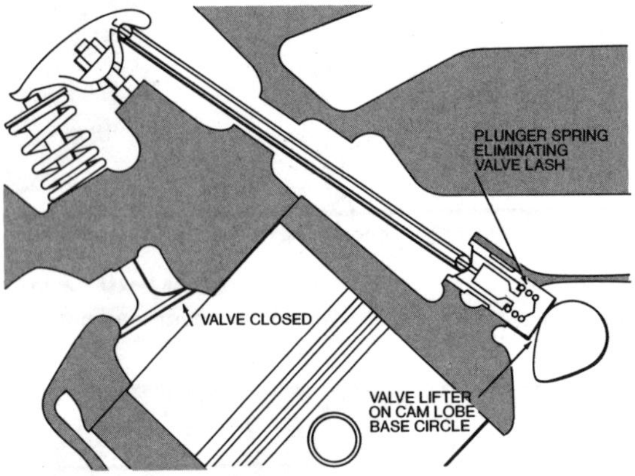

FIG 8-13 Hydraulic lifter on the base circle of the cam lobe.
(Dana Corporation)

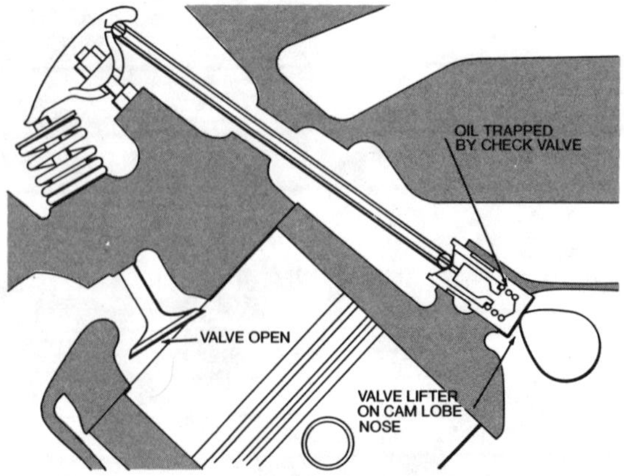

FIG 8-14 Hydraulic lifter in the valve open position. (Dana Corporation)

Hydraulic lifters also meter the oil flow to the upper valve area of many engines.

Mechanical lifters are used in some import and other high performance engines for off-road use. The timing and amount of valve lift are much more predictable with mechanical than with hydraulic lifters. Their major drawbacks are the need for frequent adjustment and noisy valve operation.

Lifter inspection, which is part of any major engine work, should be directed at the camface (cam lobe contact area). This surface should have a slightly convex (rounded) surface (Fig. 8-16). Check all camfaces for pits, grooves, or other unusual marks. Should any of these problems be found, the lifter should be discarded, since the camface cannot be resurfaced.

It is foolish to attempt to disassemble, clean, and assemble any hydraulic lifter. Most often it is less costly to replace both lifters and camshaft than to clean the lifters and take a chance of reusing the camshaft.

PUSHRODS

The pushrods are positioned between the lifter and the rocker arm. They are solid or tubular rods with hardened, spherical ends, either concave or convex (Fig. 8-17). Quite often the tubular pushrods assist in bringing oil to the upper valve areas. As noted in the disassembly unit, pushrods removed from engines with no means of valve clearance adjustment should be kept in order and returned to their original positions during the assembly operation.

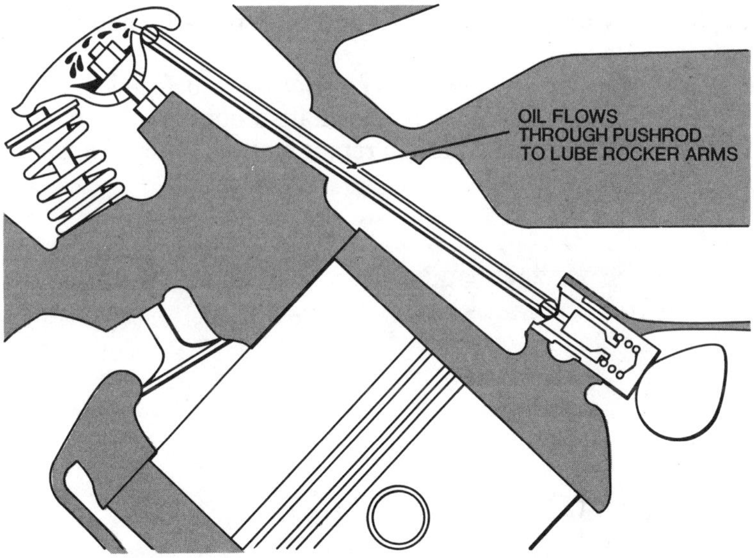

FIG 8-15 Lifter metering oil to the rocker arm. (Dana Corporation)

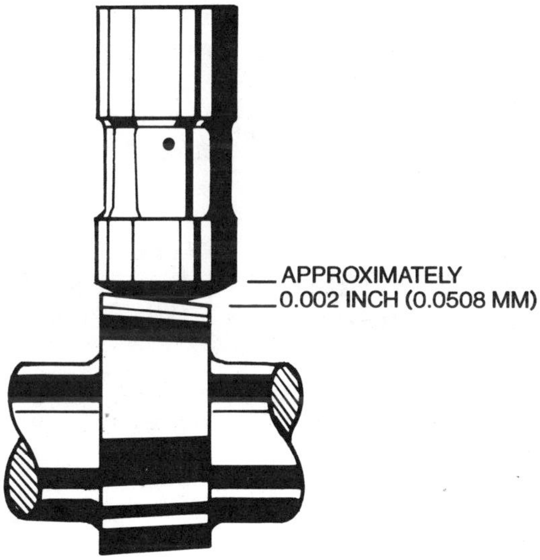

FIG 8-16 Lifter placement on the cam lobe. (Dana Corporation)

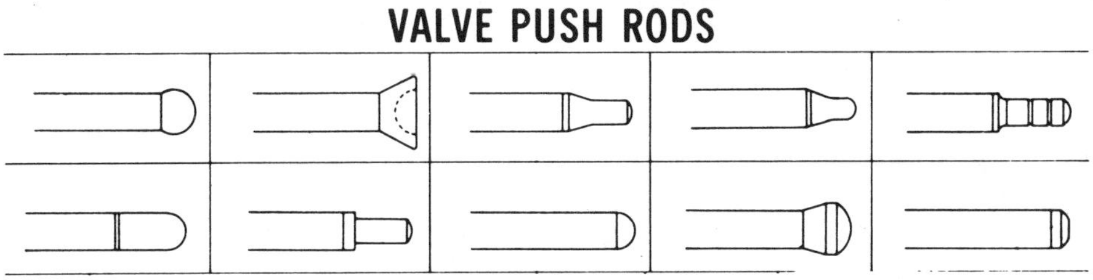

FIG 8-17 Common pushrod end shapes. (Dana Corporation)

Pushrods must be inspected for wear or breakage at the ends, especially the rocker arm end. During their operation, pushrods will flex or deflect slightly when opening the valve. Should a permanent bend develop, the part should be discarded. Pushrods can be checked between centers using a dial indicator or rolled on a perfectly flat surface and visually inspected for deflection. Allowable runout on an indicator is approximately 0.015 in. (0.370 mm); a feeler gauge of more than 0.008 in. (0.203 mm) should not be able to pass between the pushrod and flat surface.

ROCKER ARMS

Rocker arms are levers which transmit the motion created by the camshaft to open the valves. Most rocker arms are designed to multiply the movement of the pushrod and lifter so as to produce greater movement at the valve (Fig. 8-18). A typical camshaft will move a lifter and pushrod 0.250 in. (6.35 mm). The mechanical advantage of a rocker arm with a 1.5:1 ratio will move the valve 0.375 in. (0.53 mm).

The two common methods of mounting or holding rocker arms to the cylinder head are by a shaft or a stud and pivot.

Rocker arms are manufactured either by casting or by stamping (Fig. 8-19). The cast arms are usually made from iron or aluminum; stamped arms are made from steel.

Wear analysis can be made on rocker arms by a complete visual inspection. The stamped types should be examined at the pivot points and at the valve stem and pushrod areas. The shaft type should also be inspected for wear at the shaft opening. The shaft itself experiences some wear and should be replaced when heavy scoring is found.

Stamped rocker arms and some cast rocker arms *cannot* be resurfaced. If their condition is bad, they must be replaced. Most of the shaft-mounted, cast rocker arms should have the stem end resurfaced. Check a repair manual for each engine.

OBJECTIVE 8-A2 (manipulative)

Working as a team with one other student and with the aid of Job Report Sheet 8-A2, you will inspect and measure for wear the valve train components—including timing gears, sprockets, chains, lifters, and pushrods—from an instructor-assigned vehicle engine.

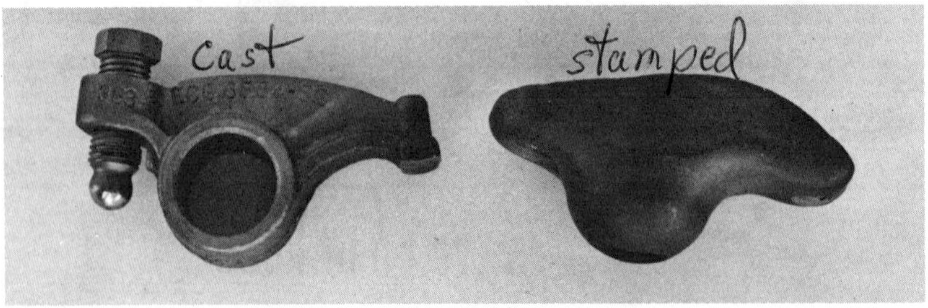

FIG 8-19 Cast (left) and stamped (right) rocker arms.

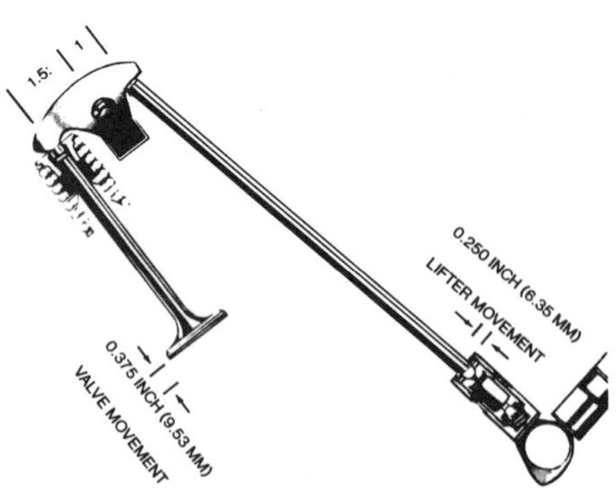

FIG 8-18 Rocker arm ratio.
(Cadillac Motor Car Division, GMC)

Time Recommended:

Twice the flat rate time allowed in *Motors, Chilton,* or *National Service Data* manuals; approximately 30 minutes.

Preparation:

1. Review pages 225 through 234.
2. Study pages 234 through 236.
3. Study manufacturer's or general repair manuals for the engine being serviced.
4. Attend lectures and demonstrations on valve train service.

Tools and Materials Needed:

1. Flashlight
2. Flat surface

Procedure:

Note: Often the valve train parts other than the pushrods are routinely replaced during a major engine repair. Procedures listed below are general in nature. Check a repair manual if in doubt.

1. Clean the valve train components as stated below. As indicated previously, lifters and some pushrods should be kept in order. Check with your instructor.
   a. For timing gears, chains, and sprockets, use solvent carburetor cleaner or caustic solution if all parts are of ferrous metal.
   b. Lifter—Wipe lifters clean.
   c. Wipe pushrods clean if order is to be maintained. Carburetor cleaner, solvents, or caustic can be used otherwise.
2. Visually inspect gears and sprockets for excessive damage. Look at the contact pattern of wear for steps or grooves more than 0.002 in. (0.0508 mm) deep. Nylon-faced camshaft sprockets usually are replaced because of cracking. Check closely.
3. Check the timing chain by holding one end and measuring the amount of sag at the other. More than ½ in. (12.7 mm) sag indicates a need for replacement. Some manufacturers recommend checking chain and sprocket wear when assembled on the engine. Ask for your instructor's help.
4. Lifters can be checked by placing the edge of one against the face of another (Fig. 8-20). When new, the lifter has a slightly convex surface, but wear limit is acceptable if no light will pass through the center area of the face.
5. If one or more lifters show unacceptable wear, the camshaft will probably also be worn. This means the camshaft and all lifters will need replacement (Fig. 8-21).
6. Inspect the pushrods by looking closely at each end for chips, cracks, or loose ends.
7. Roll each pushrod on a very flat surface with a flashlight behind. Look for intermittent light between the pushrod and the surface.
8. If a feeler gauge of more than 0.008 in. (0.203 mm) will pass between the pushrod and the flat surface, replace the pushrod.

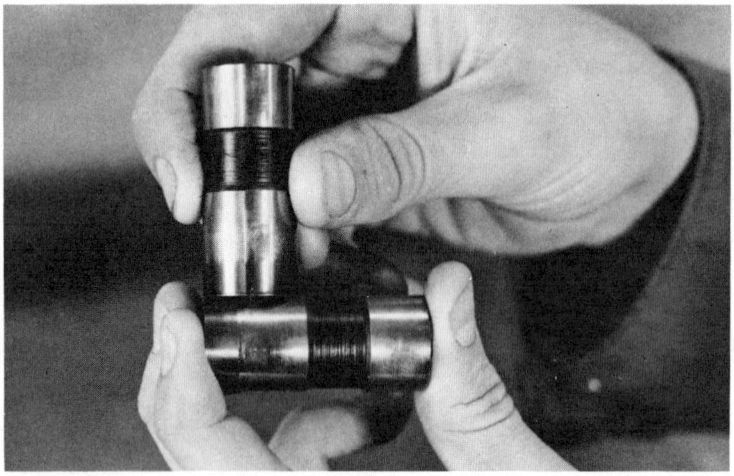

FIG 8-20 Checking a lifter face.

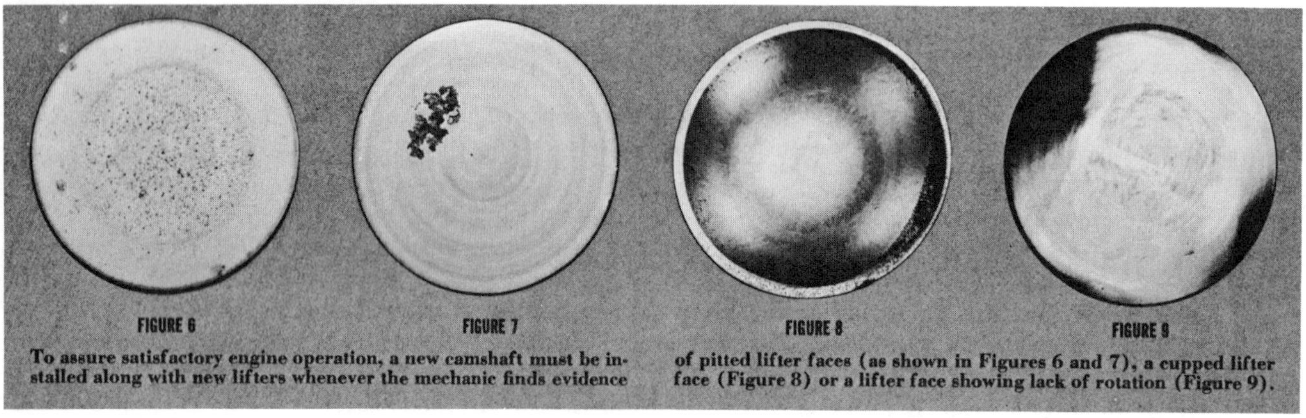

FIG 8-21 Common wear patterns on lifter faces. (Dana Corporation)

JOB REPORT SHEET 8-A2

VALVE TRAIN INSPECTION AND MEASUREMENT

Student Name ______________________ Date ______________

Make of Car ____________ Model ______________ Year ______________

Time Started ____________ Time Finished ____________ Total Time ____________

Flat Rate Time or Recommended Time ______________________

Specifications:

Timing chain wear limit ______________________

Checkup Questions:

1. Did the gears or sprockets show enough wear to warrant replacement? (Yes/No) ____________
2. How much sag did the timing chain have? ______________________
3. Why do lifters and camshafts need to be replaced as a group?

   ____

   ____

   ____
4. How many pushrods need replacing because of bending? ______________________

   because of damaged ends? ______________________

DATE COMPLETED ______________________ INSTRUCTOR'S SIGNATURE ______________________

OBJECTIVE 8-A3 (manipulative)

Working as a team with one other student and with the aid of Job Report Sheet 8-A3, you will disassemble, clean, and resurface the valve stem contact area of all rocker arms on an instructor-assigned vehicle engine.

Time Recommended:

Twice the flat rate time allowed in *Motors, Chilton,* or *National Service Data* manuals. Approximately 5 minutes per rocker arm.

Preparation:

1. Review pages 225 through 236.
2. Study page 241.
3. Study manufacturer's or general repair manuals for the engine being serviced.
4. Attend lectures and demonstrations on valve train service.

Tools and Materials Needed:

Valve refacing machine with rocker arm attachments

Procedure:

1. Rocker arms should be disassembled from the rocker arm shaft, if necessary, and thoroughly cleaned.

   Note: During disassembly, maintain the original order of placement on the shaft.

2. Mount the rocker arms on the machine with the correct fixture (Fig. 8-22).
3. Turn on the machine and resurface the rocker arm by rotating the fixture and rocker arm to maintain a curved surface, as originally machined (Fig. 8-23).
4. Resurface all rocker arms in like manner.
5. Thoroughly clean the shaft, especially the interior surfaces, if it is reusable.
6. Assemble the rocker arms on the shaft in their original order.
7. Store in a dust-free container (plastic bag) until assembled on the cylinder head.

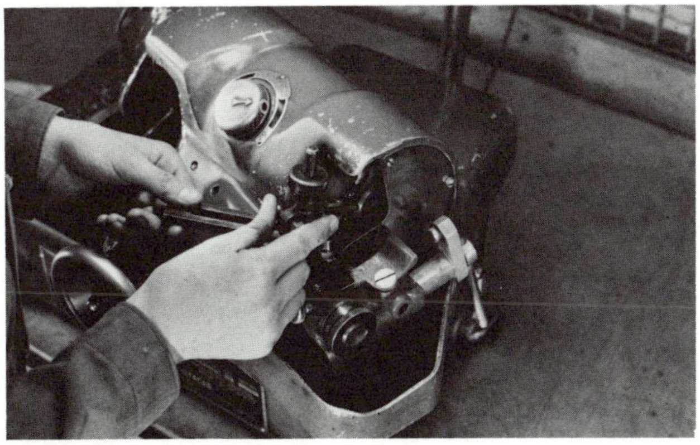

FIG 8-22 Rocker arm being resurfaced. (Dana Corporation)

FIG 8-23 Refinished rocker arm. (Dana Corporation)

ROCKER ARM SERVICE

Student Name ______________________________ Date ______________

Make of Car ______________ Model ______________ Year ______________

Time Started ______________ Time Finished ______________ Total Time ______________

Flat Rate Time or Recommended Time ______________________________

Specifications:

None

Checkup Questions:

1. What cleaning process was used to degrease the rocker arms?

 ____

2. How was the interior of the shaft cleaned? ______________________________

 ____

3. Where were the oil holes located in the rocker arm *and* the shaft?

 ____

 ____

DATE COMPLETED ______________ INSTRUCTOR'S SIGNATURE ______________

Name ______________________________ Date ________________

DIRECTIONS:

Circle the best answer to each question.

1. The valve train consists of
   I. camshaft, pushrods, lifters, and valves.
   II. camshaft, pushrods, rocker arms, and valve springs.
   A. I only   B. II only   C. Both I and II   D. Neither I nor II
2. Camshaft speed is
   I. twice the crankshaft speed.   1/2 the crankshaft speed
   II. same as the crankshaft speed.
   A. I only   B. II only   C. Both I and II   D. Neither I nor II
3. The camshaft is driven by
   I. chains and gears from the crankshaft.
   II. belts from the crankshaft.
   A. I only   B. II only   C. Both I and II   D. Neither I nor II
4. Cam lobe lift is the
   I. distance the valve moves as it opens and closes. NO
   II. difference in diameter between the base circle and the highest point of the lobe. ?
   A. I only   B. II only   C. Both I and II   D. Neither I nor II
5. The camshaft movement to open the valves
   I. is determined by design according to projected engine use.
   II. is not the same for all engines.
   A. I only   B. II only   C. Both I and II   D. Neither I nor II
6. Two terms used to describe camshaft action are
   I. lobe profile and ramp angle. ← not an action terms
   II. duration and lift.
   A. I only   B. II only   C. Both I and II   D. Neither I nor II
7. When only one or two lobes of a camshaft are pitted, the camshaft
   I. can be used if new lifters are installed.
   II. should be replaced.
   A. I only   B. II only   C. Both I and II   D. Neither I nor II
8. As timing gears or sprockets and chains wear, the valve action   what does lag mean ?
   I. improves for high speed operation. - NO
   II. reduces the lag of air and fuel entering the cylinder.
   A. I only   B. II only   C. Both I and II   D. Neither I nor II
9. The camshaft for a V-8 engine has
   I. 8 lobes for the intake valves.
   II. 16 lobes.
   A. I only   B. II only   C. Both I and II   D. Neither I nor II

10. The valve clearance normally maintained by hydraulic lifters is
I. ¼ turn
II. zero
A. I only B. II only C. Both I and II D. Neither I nor II

11. When the hydraulic lifter is on the base circle of the cam
I. the return spring takes up any clearance.
II. the area below the plunger is refilled with oil.
A. I only B. II only C. Both I and II D. Neither I nor II

12. On some engines, the pushrods are designed to carry oil
I. to the lifters.
II. only when the lifter is on the base circle of the cam.
A. I only B. II only C. Both I and II D. Neither I nor II

13. The major drawback of mechanical lifters is that
I. their action is not predictable.
II. they are more costly than hydraulic lifters.
A. I only B. II only C. Both I and II D. Neither I nor II

14. Lifters that are to be reused
I. should be disassembled and cleaned.
II. should have a slight concave surface.
A. I only B. II only C. Both I and II D. Neither I nor II

15. The type of rocker arms that can often be resurfaced is
I. cast.
II. stamped.
A. I only B. II only C. Both I and II D. Neither I nor II

16. Rocker arm ratio refers to
I. the number of rockers per cylinder.
II. the amount of valve spring compression.
A. I only B. II only C. Both I and II D. Neither I nor II

17. Camshaft sprockets should be inspected for wear
I. along with the timing chain.
II. by checking with a straightedge.
A. I only B. II only C. Both I and II D. Neither I nor II

18. Pushrods should be inspected
I. for bends or broken ends.
II. by rolling each one on a flat surface to check for bends.
A. I only B. II only C. Both I and II D. Neither I nor II

19. Lifters require replacement when
I. the camshaft is pitted or worn.
II. the lifter face is pitted or worn.
A. I only B. II only C. Both I and II D. Neither I nor II

20. Rocker arms that can be reconditioned are most often resurfaced
I. on a valve facing machine.
II. by grinding the stem end to a smooth rounded surface.
A. I only B. II only C. Both I and II D. Neither I nor II

DATE COMPLETED ______________ INSTRUCTOR'S SIGNATURE ______________

unit 9

Cylinder Block Preparation

Cylinder block service procedures are discussed in this unit, with special emphasis on inspection practices, cylinder wall preparation, and cleaning operations.

After completing the preparation assignment, you will correctly answer a minimum of 7 out of 10 multiple choice questions in a written test, without using reference material, on the following topics:

1. Cylinder block design and materials
2. Cylinder block checks and measurements
3. Cylinder wall preparation

OBJECTIVE 9-A (written test)

Preparation:

Study pages 245 through 263.

CYLINDER BLOCK MATERIALS AND DESIGN

The engine block or cylinder block is the basic structure of the engine. Almost all other parts are attached to the block in some manner.

Two types of blocks are used in today's passenger vehicles: gray cast iron and die cast aluminum alloy. The iron block is more popular than aluminum for several reasons:

1. It is less expensive to produce.
2. Iron makes a stronger casting.
3. Heat does not distort cast iron.
4. Iron has good corrosion resistance.

Aluminum blocks have the advantage of being lighter and are able to dissipate heat more quickly than iron. Most aluminum block engines have cast iron cylinder walls called liners or sleeves because aluminum cylinder walls are not compatible with aluminum pistons and iron piston rings for long-term engine operation.

Cylinder blocks designed for liquid cooling systems have open areas, called water jackets, within the casting where a water-base coolant can circulate, taking heat away from the block and releasing it into the air passing through the radiator.

On the outer surfaces of most blocks are round openings that are filled with heavy sheet metal plugs (Fig. 9-1). There are many different names used to describe these plugs: core, core hole, freeze, expansion, Welsh, water jacket, and soft. In this text, they will be referred to as core plugs. These plugs are used to retain the coolant as it circulates within the block. The core is a mold formed in the shape of the water passages and placed within the mold of the outside shape of the block. When the molten metal is poured into these two molds, the core keeps the metal from flowing into the areas needed for water jackets. When the metal hardens, the core mold is washed away, leaving rough holes on the outside surface of the block, which are then machined to allow core plugs to be installed.

Core plugs are manufactured from three common metals: steel (plated), brass, and copper. Plated steel is the most common core plug material. These plugs are designed in several styles (Fig. 9-2). The cup plug has different depths of flanges; the expansion plug has one variation with a flange and several without flanges.

In most engines, the cylinders are integral (cast all together) with the block. The cylinders must be true round surfaces, and they must be perpendicular to the center line of the crankshaft for peak engine efficiency. A few engines have liners or sleeves which are replaceable during an overhaul.

Within the large cylinder block casting are openings machined to hold the camshaft and crankshaft. Various engine designs have different amounts of these supporting features to strengthen the block. The openings holding the camshaft are usually called housing bores or bores; the crankshaft supports are termed saddles or webs. When the main bearing caps are attached to the saddles, the openings are called housing bores.

The lower engine area which contains the crankshaft, lubricating oil, and oil pump is usually referred to as the crankcase. The oil pan serves as the container for the crankcase.

The service information following is meant for a cylinder block cleaned in a hot tank. If this has not been accomplished, do it now. Refer to Unit 5 for details.

A thorough visual inspection should be made to detect major problems, such as extremely worn or irregular surfaces and evidence of fractures (cracks). If a crack can be seen, notify your instructor for further directions. Crack repair is discussed in Unit 7, pages 175 to 176. Review that material and read the following information on epoxy repair before beginning any service work.

EPOXY REPAIR

Minor cracks and sand holes in the lower areas of the block which cause leakage may be repaired with epoxy-metallic adhesive. Repair with this material should be confined to those areas of a casting whose inner wall surface is not exposed to engine combustion, coolant, or oil pressure (Fig. 9-3). These surfaces should be repaired with taper plug.

Follow the directions on the container if they conflict with the following steps.

1. Clean the surface to be repaired by grinding or rotary filling followed by lacquer thinner or acetone.
2. Mix the epoxy until it appears uniform in color and texture.
3. Apply the epoxy to the repair area with a clean tool (gasket scraper, putty knife), forcing the mixture into the opening.
4. Allow the mixture to harden. A heat lamp can be used to speed the curing process. Ten to fourteen hours is required for air drying.
5. Grind or file the surface smooth when adhesive is fully dry. Repaint the surface to blend with original color.

FIG 9-1 Cylinder block core (freeze) plugs.

FIG 9-2 Core plug designs. (Dana Corporation)

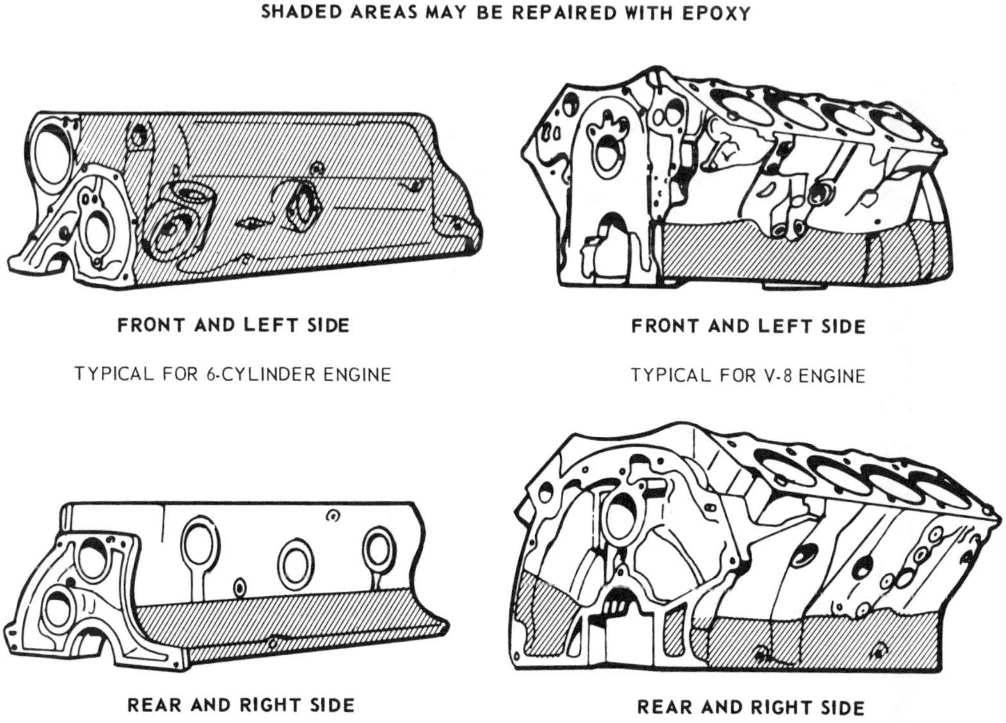

FIG 9-3 Epoxy repair areas. (Ford Motor Company)

CYLINDER BLOCK INSPECTION AND MEASUREMENTS

All of the cylinder walls of an engine to be repaired should be thoroughly inspected and measured to determine what service procedures are necessary to restore the engine to its original operating efficiency. The information presented here is general; however, most piston ring and engine manufacturers recommend approximately the same procedures and guidelines.

During the lower engine disassembly process, it is important to remove the ring ridge before removing the pistons. The ring ridge is an unworn area above the highest point where the top piston ring contacts the cylinder wall (Fig. 9-4), the difference in bore diameter between the ridge area and the area just below it is an indication of the amount of wear the cylinder wall has experienced.

CYLINDER CHECKS

Under new operating conditions, a cylinder does not wear the same from top to bottom. The top will always wear more than the bottom, resulting in a cylinder with a taper. Excessive cylinder bore taper (+0.010 in. or 0.254 mm) is probably the biggest reason why an engine needs to be rebored.

To calculate the taper, subtract the BDC bore diameter measurement from the TDC bore diameter measurement, for example:

TDC	4.016 in.
BDC	4.002 in.
	0.014 in. taper

The total amount of cylinder bore wear can be calculated by subtracting the original bore specification from the maximum bore diameter measured just below the ridge at 90° from the piston pin, for example:

Bore measurement (worn)	4.016 in.
Original bore	4.000 in.
	0.016 in. wear

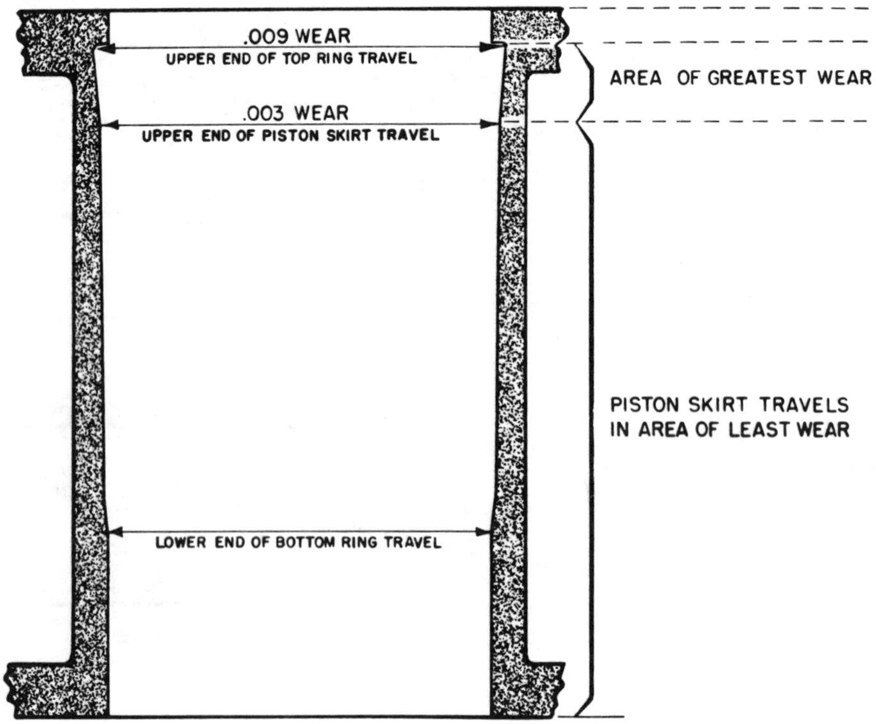

FIG 9-4 Typical cylinder wall wear.

Out-of-round cylinder bores can greatly reduce the effectiveness of piston rings. When the bore is not round, the piston rings will not conform to the irregular cylinder shape. This produces lost combustion gases during the power stroke and less compression pressure on the compression stroke.

To determine the out-of-round in a cylinder, measure the diameter of the bore crosswise of the block. Next, take a measurement lengthwise of the block. The difference between the readings is the out-of-round wear. Example:

A measurement	4.016	
B measurement	4.015	
	0.001	out-of-round

Some cylinder walls will show wavy lines, which are caused by ring flutter (rings being too loose in their grooves) or excessive ring pressure (rings contacting too tight in the bore).

The terms *scuffing* and *scoring* are often used to describe problems relating to cylinder walls, piston rings, and pistons. Both of these irregularities are the result of the actual melting and welding together of parts of the ring face, piston, and cylinder wall (Fig. 9-5). This occurs when the cylinder lubrication is lost, usually as the result of excessive heat.

Scuffing starts as a small surface disturbance. It often affects adjacent areas, which causes further disturbance. Additional friction heat develops and finally a noticeable void and scratch pattern appears (Fig. 9-6). Scoring can be considered scuffing advanced to the stage that it is readily apparent during a visual inspection. Scuffed cylinder walls must be deglazed or honed until the marks disappear.

Before proceeding further, review or complete the cylinder measurements for taper, out-of-round, and wear, using Job Report Sheet 3-A1 on pages 63–72.

Nearly everyone agrees that a rebored engine will be superior in performance and will outlast one which had only its piston rings replaced. The majority of engines disassembled today need more than a rering overhaul; they need reboring.

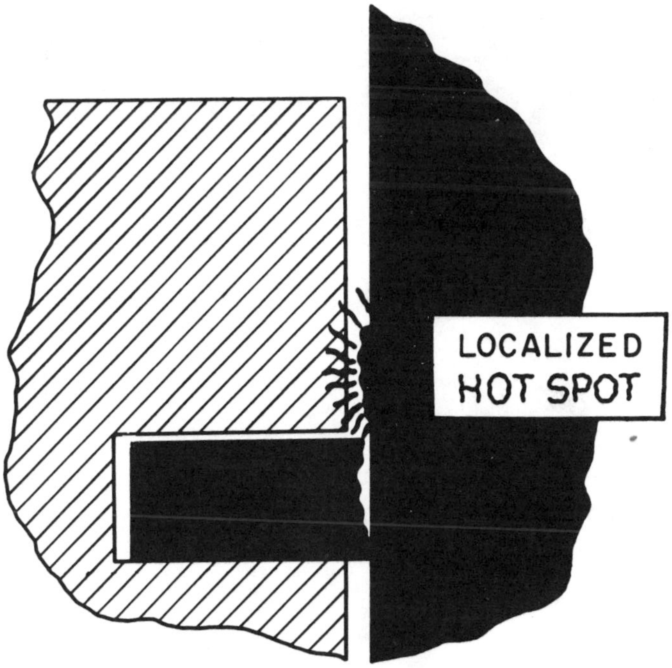

FIG 9-5 Closeup view of scuffing / scoring action. (Dana Corporation)

FIG 9-6 Scoring damage. (Dana Corporation)

With the following guidelines and the help of your instructor, decide whether to rering or rebore your engine. For a successful rering overhaul,

1. Cylinder taper should be less than 0.010 in. (0.254 mm).
2. Cylinder out-of-roundness should be less than 0.001 in. (0.025 mm).
3. Scuff or score marks (grooves in the cylinder wall) should be less than 0.001 in. (0.025 mm) deep.
4. There should be no cracks or pits and little or no wavy lines.

Rebore (oversize bore) is necessary when *any* of the following are observed:

1. Cylinder taper is more than 0.010 in. (0.254 mm).
2. Cylinder out-of-roundness is more than 0.001 in. (0.025 mm).
3. Scuff or score marks are more than 0.001 in. (0.025 mm) deep.
4. Cracks are detected (consult your instructor).
5. Pits in the ring travel area are larger than half the ring width.
6. There are excessive wavy lines of 0.0005 in. (0.0127 mm) maximum depth.

ALIGNMENT CHECKS

All engine blocks sustain certain amounts of distortion (warpage, twist, etc.) during their machining and operating periods. A check of the amount of block distortion should be made in the earliest stages of repair. If there is excessive warpage, it may be more economical to replace the block with a used or new one rather than to machine it.

The surfaces which must be checked on all blocks are the main bearing housing bores and the deck surfaces. The main bearing housing bores are the openings in the lower block holding the crankshaft. The deck surfaces are those where the cylinder head is attached to the block. To determine whether this distortion is excessive, use a straightedge and a set of feeler gauges.

OBJECTIVE 9-A1 (manipulative)

Working as a team with one other student and with the aid of Job Report Sheet 9-A1, you will check the alignment and straightness of the deck and main bearing bore surfaces on an instructor-assigned vehicle engine.

Time Recommended:

Twice the flat rate time allowed in *Motors, Chilton,* or *National Service Data* manuals; approximately 20 minutes.

Preparation:

1. Review pages 245 through 249.
2. Study pages 250 through 251.
3. Study manufacturer's or general repair manuals for the engine being serviced.
4. Attend lectures and demonstrations on cylinder block preparation.

Tools and Materials Needed:

1. Straightedge
2. Assorted feeler gauges

Procedure:

1. Remove any foreign matter or slight nicks from the deck surface by draw filing.
2. Lay the straightedge front to rear on the deck surface and attempt to place a 0.006 in. (0.152 mm) feeler gauge anywhere between the deck and straightedge (Fig. 9-7).
3. Lay the straightedge diagonally on the deck surface and attempt to place a 0.003 in. (0.076 mm) feeler gauge anywhere between the two surfaces.
4. Lay the straightedge side to side in various positions from the front to the rear. Use a 0.003 in. (0.076 mm) gauge in the same manner as before (Fig. 9-8). If the feeler gauge slips between the straightedge and the deck surface, the block needs to be milled.
5. Clean any foreign matter from the main bearing bore surfaces (Fig 9-9).
6. Lay the straightedge in the deepest part of the bearing bore.
7. With a feeler gauge that is half the maximum specified oil clearance (0.001 in. or 0.0254 mm), try to slide the feeler under the straightedge at each bearing (Fig. 9-10). If the feeler gauge slips under the straightedge at any one of the bearing bores, the block needs align boring or align honing.

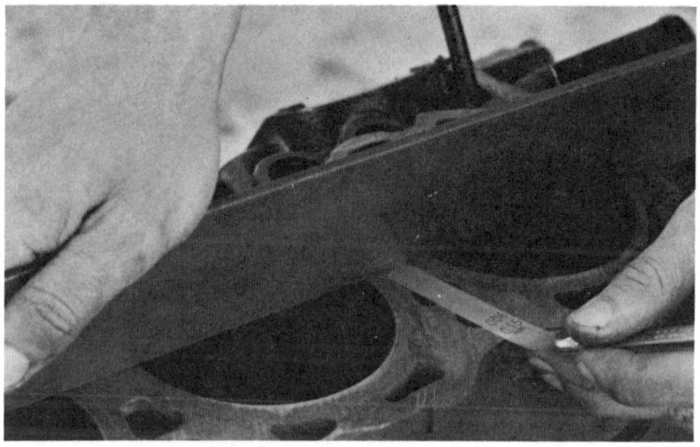

FIG 9-7 Using a straightedge and feeler gauge to check lengthwise deck surface alignment.

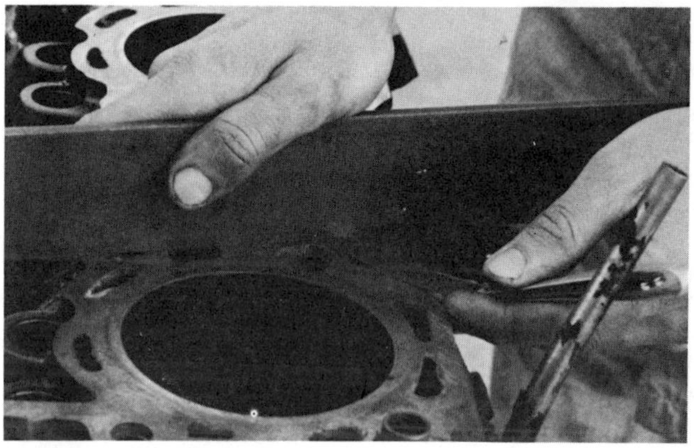

FIG 9-8 Crosswise deck alignment check.

FIG 9-9 Using lacquer thinner to clean main bearing housing bore.

FIG 9-10 Checking bore alignment.

CYLINDER BLOCK ALIGNMENT CHECKS

Student Name ______________________ Date ______________

Make of Car ______________ Model ______________ Year ______________

Time Started ______________ Time Finished ______________ Total Time ______________

Flat Rate Time ______________________

Specifications:

Deck Surface Flatness

0.003 in. (0.076 mm) in any 6 in. (152 mm)

0.006 in. (0.152 mm) overall

Main Bearing Bore Straightness

No more than half the maximum specified oil clearance (approximately 0.001 in. or 0.0254 mm). Maximum oil clearance ______________________

Checkup Questions:

1. What was the maximum thickness of feeler gauge that would slide under the straightedge? __________
2. What effect would a warped deck surface have on the head gasket?

   ____

3. What was used to clean the main bearing bores before they were checked? ______________

   ____

4. What is the minimum oil clearance for the main bearing on the engine being checked? __________

DATE COMPLETED ______________ INSTRUCTOR'S SIGNATURE ______________

MAIN BEARING HOUSING BORE MEASUREMENT

During the initial machining operation at the factory, the main bearing housing bore size and alignment is determined. The block is cast without main bearing caps; however, the surface where the cap is eventually attached is milled to a smooth finish. Main bearing caps are cast and machined to a smooth surface and then bolted to the block. An operation called align or line boring is then performed in which a straight opening of consistent size that will hold the crankshaft in its proper position is made.

After the engine has been in operation, especially if it has been lugged often, the housing bores may not be round. Every main bearing housing bore should be measured for its diameter and out-of-roundness.

Working as a team with one other student and with the aid of Job Report Sheets 3-A1 and 9-A2, you will measure all the cylinder block main bearing bores for size and out-of-roundness on an instructor-assigned vehicle engine.

OBJECTIVE 9-A2 (manipulative)

Time Recommended:

Twice the flat rate time allowed in *Motors, Chilton,* or *National Service Data* manuals. Approximately 8 minutes per cylinder.

Preparation:

1. Review pages 245 through 251.
2. Study pages 255 through 256.
3. Study manufacturer's or general repair manuals for the engine being serviced.
4. Attend lectures and demonstrations on cylinder block preparation.

Tools and Materials Needed:

1. Torque wrench and socket set
2. Telescoping gauge set and micrometer *or* dial bore gauge and setting fixture

Procedure:

1. Clean any foreign matter from the bore and parting surfaces of the main bearings (Fig. 9-11).
2. Torque the main bearing caps to specifications with a torque wrench (Fig. 9-12).
3. In a vertical (top to bottom) position in main bearing #1, expand the telescoping gauge to its maximum diameter and lock it.
4. With a micrometer, measure the telescoping gauge and record on Job Sheet 3-A1, Teardown and Assembly. Compare the reading with specifications. Reading should be within 0.001 in. (0.0254 mm) of specifications.
5. Rotate the telescoping gauge 90°, expand it in a horizontal position, and lock it. Watch for irregular settings due to the locking grooves.
6. Mike the telescoping gauge and record on Job Report Sheet 3-A1. Subtract to find the difference readings in steps 4 and 6. This will be the out-of-roundness.

7. Using specifications, set the dial bore indicator to original manufacturer's specifications and place it in a vertical position in main bearing #1. Record this reading at "A" on Job Report Sheet 3-A1, Teardown and Assembly. Reading should be within 0.001 in. (0.0254 mm) of O.E.M. specifications (Fig. 9-13).
8. Rotate the indicator approximately 90° for a horizontal reading (position "B"). Watch for irregular readings when passing over bearing lock groove or dowels. The difference between this reading and step 7 is the out-of-roundness. Record the reading at "B" on Job Report Sheet 3-A1.
9. Measure all main bearing bores in like manner. If the readings are not within specification, notify your instructor.

FIG 9-11 A file, used lightly, will clean main bearing parting surfaces.

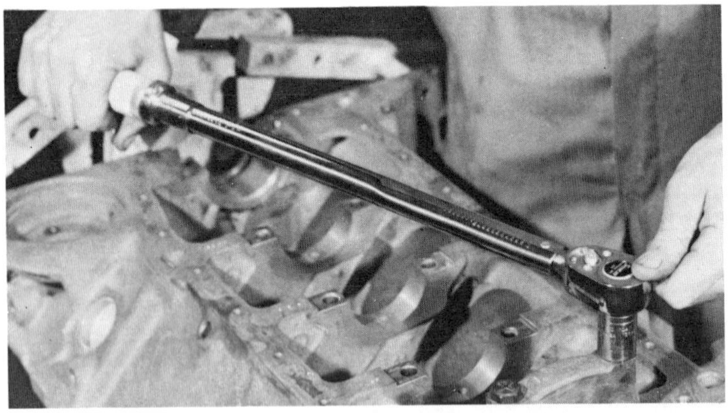

FIG 9-12 Torque main caps to specifications.

FIG 9-13 A dial bore gauge can be used to check main bearing housing bores.

MAIN BEARING BORE MEASUREMENT

Student Name ______________________________ Date ______________

Make of Car ______________ Model ______________ Year ______________

Time Started ______________ Time Finished ______________ Total Time ______________

Flat Rate Time or Recommended Time ______________________________

Specifications:

Main bearing bore specifications (see Job Report Sheet 3-A1)

Checkup Questions:

1. List the causes of main bearing bore out-of-roundness.

____

____

____

2. Did any of the main bearing caps show evidence of cracking or damage? ______________

3. List position and amount of irregularities in measurements:

#1 ______________ #2 ______________ #3 ______________

#4 ______________ #5 ______________ #6 ______________ #7 ______________

DATE COMPLETED ______________ INSTRUCTOR'S SIGNATURE ______________

ALIGN BORING

If, during the alignment checks and measurements of the main bearing housing bores, the readings were *not* within specifications, align or line boring must be completed. Most align boring is done with a cutting tool to remove the desired amount of metal, but another method uses a long stone hone (Fig. 9-14). Both methods require accurate equipment and precision operation. If you are asked to align bore a block, follow the directions very carefully.

The following information is general because of differences in operation and availability of this equipment.

Whichever method is used to renew the centerline of the crankshaft in the block (align boring), the following basic steps apply:

1. Remove the main bearing caps and remove approximately 0.010 in. (0.254 mm) from the mating surfaces.
2. Clean the block mating surfaces and chase the threads of the bolt holes (Fig. 9-15).
3. Tighten the caps to specified torque.
4. Set the boring bar in position and adjust the cutting tool so that it removes enough material to enlarge the bore to its original specified diameter.
5. With the bar centered, test the amount of stock removed from each bearing bore. It is important to remove as little as possible from the crown (center) of the block side of the bore, because this influences the distance to the deck surface.
6. Before the actual cutting begins, measure the distance between the crankshaft and camshaft. This distance is critical because the timing gears and/or chains are not usually adjustable.
7. Cut all bores to original diameter and lightly chamfer the edges and oil holes before cleaning.

FIG 9-14 Align honing machine.
(Sunnen Products Company)

DECKING

The term *decking* is used to describe the machining process used to even and align the block surface that mates with the cylinder head. Two types of decking machines in use today are the rotating cutter and the wet grinder (Fig. 9-16). The objective of the decking process is to produce a block with a flat deck surface that is parallel with the crankshaft center line and perpendicular to the cylinder bores.

Because of vast differences in decking equipment, no procedures will be given. Carefully follow the directions for whatever machine you use.

CYLINDER WALL PREPARATION

Deglazing and honing are often confused when discussing cylinder wall reconditioning. Deglazing is a *light* abrasive treatment given to cylinder walls to break the shininess caused by friction and petroleum vanish. It can also remove very slight scores or wavy lines. There are many deglazing tools in use today (Fig. 9-17). Those tools with stones use only light spring pressure to hold the abrasive against the cylinder wall.

Honing removes more metal than deglazing. This is usually accomplished by using an abrasive stone tool with a positive tensioning device to apply pressure to the cylinder walls (Fig. 9-18).

Although there are exceptions, deglazing is most often done during a rering overhaul, and honing is done after a rebore or when heavy stock removal is required.

All cylinders must be honed after they have been bored, for the cutting tool of the boring bar leaves a rough surface which is best removed with a hone. After the rough surface is removed, the operator should apply a cross-hatch pattern with the hone (Fig. 9-19). This will provide the ideal surface for ring seating (initial wear in) and retention of cylinder wall lubricant.

OBJECTIVE 9-A3 (manipulative)

Working as a team with one other student and with the aid of Job Report Sheet 9-A3, you will deglaze at least 4 cylinder walls on an instructor-assigned vehicle engine.

Time Recommended:

Twice the flat rate time allowed in *Motors, Chilton,* or *National Service Data* manuals; approximately 6 minutes per cylinder.

Preparation:

1. Review pages 245 through 259.
2. Study pages 260 through 262.
3. Study manufacturer's or general repair manuals for the engine being serviced.
4. Attend lectures and demonstrations on cylinder block preparation.

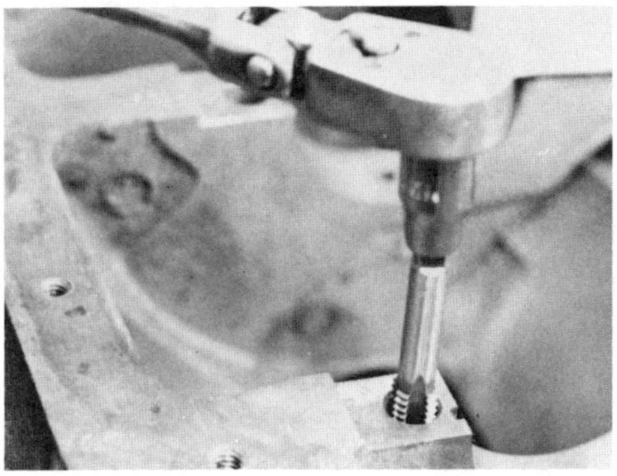

FIG 9-15 Chasing the threads in the main bearing bolt holes.

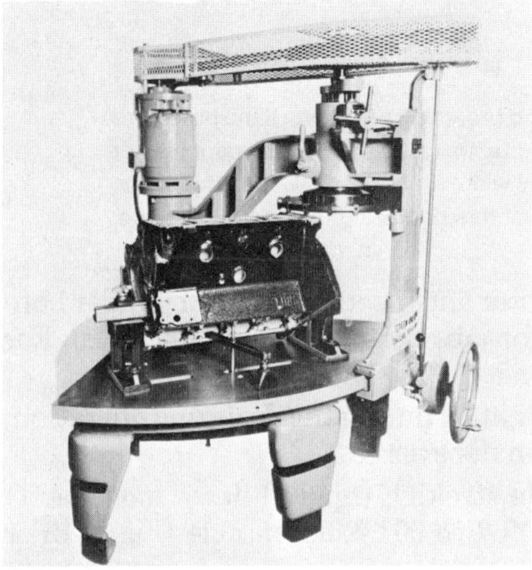

FIG 9-16A Rotary cutter block and head surfacing machine.
(Storm-Vulcan, Inc.)

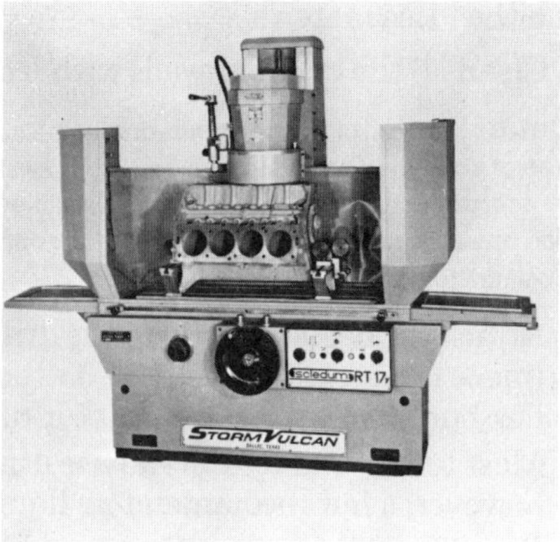

FIG 9-16B Wet grinder block and head surfacing machine. (Storm-Vulcan, Inc.)

FIG 9-17 Popular deglazing tools.

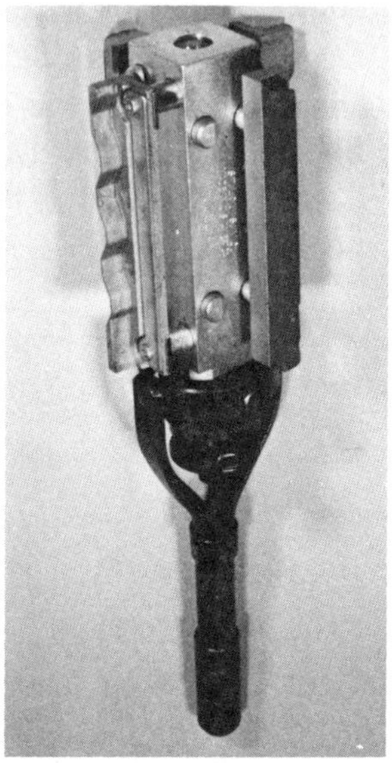

FIG 9-18 Positive stone hone.

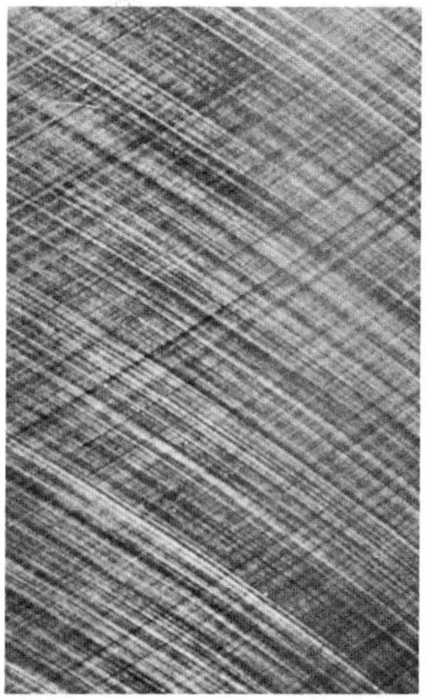

FIG 9-19 Abrasive pattern (cross-hatch) resulting from proper deglazing or honing.
(Dana Corporation)

Tools and Materials Needed:

1. Deglazing tool
2. Variable speed drill motor
3. Lacquer thinner or acetone
4. Clean wiping rags

Procedure:

Since deglazing is performed only for rering purposes, it is often done while the crankshaft is in the block and the block is in the vehicle. It is extremely important to protect the internal engine area (crankcase) from the abrasives and metal that break loose during this operation. Place clean rags below the cylinders to catch the loose materials.

1. Clean all the cylinder walls with lacquer thinner or acetone.

   Caution: Avoid prolonged contact with the skin and breathing of vapors. If the bores are free of scratches and waviness, repeated swabbings with a rag soaked in lacquer thinner or acetone can deglaze cylinder walls satisfactorily without abrasives. Check the piston ring manufacturer's instruction sheet for approval of this deglazing method. Otherwise, continue to step 2.

2. Attach the deglazing tool to the drill motor and insert it into the cylinder bore (place clean rags in the bore bottoms for jobs performed in the vehicle). For flex hones, insert with the motor running.
3. Most tool manufacturers advise using light oil during the deglazing operation; however, a few recommend no liquid whatsoever.
4. Run the deglazer at a slow speed while stroking (up and down movement) 15–20 times. A cross-hatch pattern of 45° to 60° is acceptable (Fig. 9-19 or 9-17).
5. With a flat stone deglazer, stop the deglazer motor and compress the stones together before removing from the bore. With a flex hone, remove the deglazer from the bore while the motor is running.
6. Wipe the cylinder clean and make a close inspection to see if the deglazing is complete. Check with your instructor.
7. Deglaze all remaining cylinders in a like manner.
8. Clean all cylinders as described in Objective 9-B3, steps 7–11, using soap and hot water followed by repeated oil swabbings.
9. Clean the deglazing tools and equipment and return them to their storage area.

JOB REPORT SHEET 9-A3

DEGLAZE CYLINDER WALLS

Student Name ______________________ Date ______________

Make of Car ______________ Model ______________ Year ______________

Time Started ______________ Time Finished ______________ Total Time ______________

Flat Rate Time ______________________

Specifications:

Cross-hatch pattern 45°–60°

Checkup Questions:

1. What type of deglazing tool was used? ______________________

2. Why is a cross-hatch pattern needed? ______________________

3. Was a lubricant used during the deglazing? ______________________

   If yes, what type? ______________________

4. Were all cleaning procedures followed exactly? ______________________

DATE COMPLETED ______________ INSTRUCTOR'S SIGNATURE ______________

Name ______________________________ Date ______________

DIRECTIONS:

Select the best answer to each question and circle the letter at the left.

1. The heavy sheet metal plugs on the sides of most blocks are called
   I. stop or water plugs.
   II. core or freeze plugs.
   A. I only
   B. II only
   C. Both I and II
   D. Neither I nor II
2. Epoxy crack repair works well on
   I. cylinder walls.
   II. main bearing housings.
   A. I only
   B. II only
   C. Both I and II
   D. Neither I nor II
3. Cylinder ring ridge
   I. is the raised portion of the cylinder wall designed to stop the piston movement at TDC.
   II. should be removed after the piston is removed.
   A. I only
   B. II only
   C. Both I and II
   D. Neither I nor II
4. Cylinder taper
   I. is the difference in cylinder size at TDC when two points are measured at 90° from one another. no
   II. is the normal wear a cylinder experiences measured at TCD and BDC 90° from the piston pin position. ?
   A. I only
   B. II only
   C. Both I and II
   D. Neither I nor II
5. A successful rering overhaul requires
   I. less than 0.010 (0.254 mm) taper. yes
   II. score marks less than 0.001 in. (0.0254 mm) deep. no ?
   A. I only
   B. II only
   C. Both I and II
   D. Neither I nor II

ask Roger

6. Alignment checks should be made on
   I. all flat surfaces.
   II. main bearing, deck surfaces, and cylinder walls.
      A. I only
      B. II only
      C. Both I and II
      D. Neither I nor II

7. The main bearing housing bore measurements
   I. are necessary only if the bearings are severely worn.
   II. should be made with the caps off.
      A. I only
      B. II only
      C. Both I and II
      D. Neither I nor II

8. Align boring
   I. renews the surface where the heads are attached.
   II. should be performed during each major engine repair.
      A. I only
      B. II only
      C. Both I and II
      D. Neither I nor II

9. Deglazing is performed on cylinder walls
   I. when heavy stock removal is required.
   II. with a positive tension stone device.
      A. I only
      B. II only
      C. Both I and II
      D. Neither I nor II

10. The cross-hatch pattern on cylinder walls
   I. assists in holding oil.
   II. improves the seating of the piston rings.
      A. I only
      B. II only
      C. Both I and II
      D. Neither I nor II

DATE COMPLETED ______________ INSTRUCTOR'S SIGNATURE ______________________

After completing the preparation assignment, you will correctly answer a minimum of 14 out of 20 multiple choice questions in a written test, without reference material, on the following topics:

1. Cylinder boring
2. Honing
3. Extracting broken fasteners and thread replacement
4. Final cleaning
5. Core and oil gallery plugs
6. Camshaft bearing

OBJECTIVE 9-B (written test)

Preparation:

Study pages 267 through 295.

CYLINDER BORING

New pistons must be purchased before the boring operation begins. This is necessary because piston manufacturers finish their pistons in one of two ways. One method is to finish the piston to the exact bore size stated so that the operating clearance must be figured in the boring and honing. The other finishing method is to remove the operating clearance from the piston during its manufacture to fit the exact size cylinder.

When the decision has been made to bore the cylinder, a determination must be made of how much bore size must be increased so that oversize pistons can be purchased. This estimate can be made by adding 0.010 in. to the amount of wear (0.016 in. in the last example) and moving to the next common size. In this case, 0.030 in. oversize pistons would be satisfactory. Oversize pistons are commonly sold in +0.020 in., +0.030 in., +0.040 in., and +0.060 in. Most engine manufacturers and rebuilders discourage the use of more than 0.030 in. oversize bore.

The boring operation employs a machine called a boring bar that uses a rotating cutter to enlarge the cylinder opening. The boring bar is attached to the block at the deck (head) surface (Fig. 9-20) or is attached to a large fixture which holds the block at the main bearing saddles (Fig. 9-21). The actual cutting operation is done in two or three passes after the bar is centered exactly over the cylinder. The cylinder is enlarged to a point approximately 0.0015 in. (0.038 mm) *less than* the final bore. Fine honing will bring the cylinder to the exact finished size.

Note: If the cutting tool of the boring bar is not perfectly sharp, allow up to 0.003 in. (0.076 mm) for honing. A dull cutting tool plows through the cylinder metal gouging out portions below the surface. A sharp one cuts a chip that is even with the desired surface.

An alternative to the traditional cutting boring bar is the Sunnen CK-10 Cylinder Resizing Machine (Fig. 9-22). This machine resizes cylinder bores by rough honing to size and fine honing to correct clearance with accurate, automatic operation. It can also be used to fine hone engines after boring and is recommended by many high performance engine builders.

After boring or resizing process is complete, the top edge of the bore at the deck surface will be quite sharp. This sharp edge can cause abnormal combustion problems (pre-ignition) if not removed. Some boring bars have cutting tools designed to chamfer this edge. A chamfering cone, used with a drill motor, may also be used. The chamfer should be approximately 1/16 in. (1.58 mm) wide at a 45° angle.

OBJECTIVE 9-B1 (manipulative)

Working as a team with one other student and with the aid of Job Report Sheet 9-B1, you will bore the cylinders to recommended size on an instructor-assigned vehicle engine.

Time Recommended:

Twice the flat rate time allowed in *Motors, Chilton,* or *National Service Data* manuals; approximately 40 minutes per cylinder.

Preparation:

1. Review pages 267 through 268.
2. Study pages 268 through 270.
3. Study manufacturer's or general repair manuals for the engine being serviced.
4. Attend lectures and demonstrations on cylinder block preparation.

Tools and Materials Needed:

1. Boring bar with attachments
2. Assorted wrenches and handles
3. Torque wrench
4. Oversize pistons
5. Telescoping gauge set
6. Micrometer—bore size
7. Chamfering cone and drill motor

Procedure:

The general information in this objective is not intended to replace specific instructions for the boring bar to be used. Your instructor will assist you in this operation. Fill in all specifications on Job Report Sheet 9- B1 before beginning.

1. Place the block on a bench or fixture so its deck surface is parallel to the floor.
2. Remove any foreign matter from the deck surface by draw filing and clean the cylinder wall with a solvent. Also, clean the areas where the main bearing caps attach.
3. Torque the main bearing caps to specifications.
4. Position the boring bar approximately over the cylinder to be bored by attaching it lightly to its holding device.
5. Crank the cutting head into the cylinder about 2 in. (0.51 mm) and expand the centering jaws.
6. When the bar is located in the center of the bore, tighten the hold-down clamp and withdraw the centering jaws.

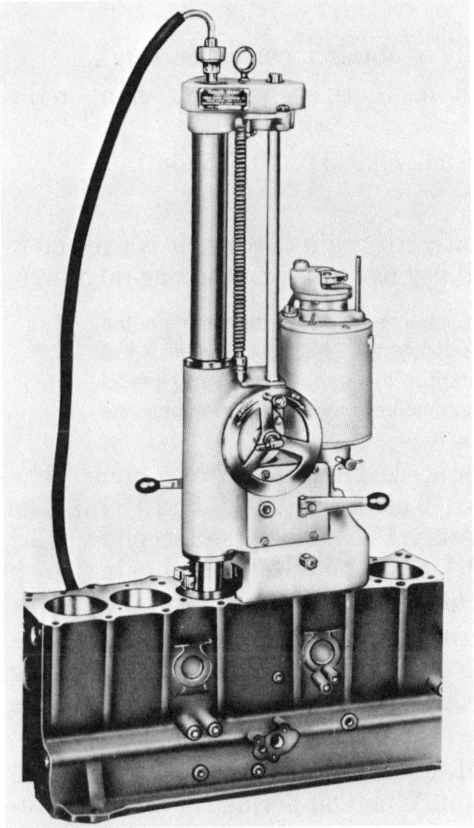

FIG 9-20 Boring bar attached to the block surface.

(Kwik-Way Manufacturing Co.)

FIG 9-21 Boring machine.

(Rottler Boring Bar Company)

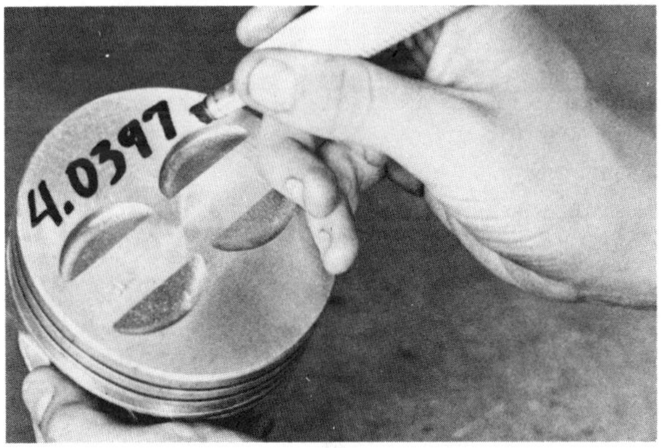

FIG 9-23 Measure and mark the top of each piston.

FIG 9-22 Automatic honing machine.

(Sunnen Products Company)

7. Return the cutter head to the "up" position and locate the cutting tool holder assembly and the sharpening fixture.
8. Follow the sharpening directions exactly to obtain a perfect cutting edge.
9. With a micrometer, measure all the pistons. Mark each measurement on the piston head (top) (Fig. 9-23).
10. Set the cutting tools for the first cut, usually 0.025 in. (0.635 mm).

Note: *Measure Twice—Cut Once!*

11. Turn on the motor and engage the clutch to begin cutting. It is extremely important to watch the cutting tool as it begins to cut into the ring ridge area.

Note: If there is any doubt about the centering position of the bar *or* the setting of the cutting tool, disengage the clutch or gear and stop the motor. This must be done within the first ¼ in. below the deck surface. Visually inspect (with a good light source) the cut and, even better, use a telescoping gauge and micrometer to measure it. Compare the micrometer reading with that used to set the cutting tool.

12. Make additional cuts until the cylinder is sized and ready to be honed. The last cut should be a small one (0.003 in.–0.005 in. or 0.076–0.127 mm). The sharper the tool, the less honing is necessary. If available, use a second cutting tool and set it for the finish cut. Enough stock should be left for honing which will allow the removal of the cutting marks from the bore.
13. Some models of boring bars have a cutting tool to chamfer the top of the bore. If this is not available, use a chamfering cone with drill motor after all boring is completed (Fig. 9-24). The chamfer should be 1/16 in. (1.58 mm) wide at a 45° angle (Fig. 9-25).
14. Move the bar to the next cylinder and follow steps 7–11. This includes the sharpening step which is very important. When all boring is completed, return the equipment after it has been cleaned to its point of origin.

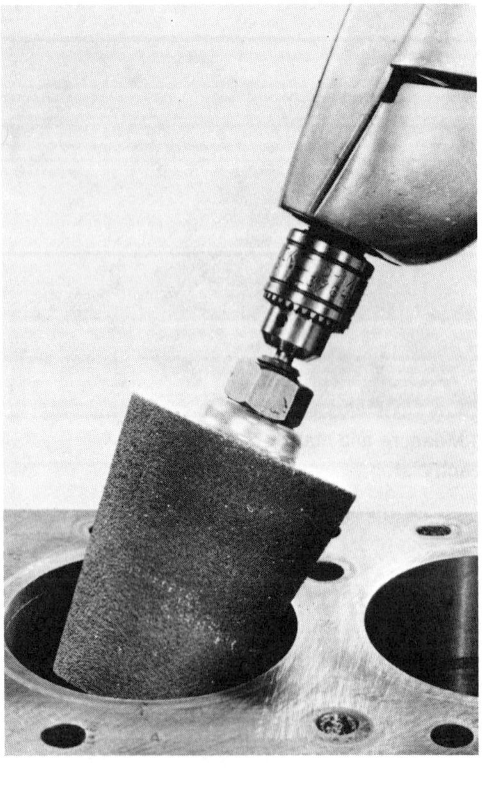

FIG 9-24 Chamfering cone and drill motor being used to cut the sharp edge at the top of the bore.

(Rottler Boring Bar Company)

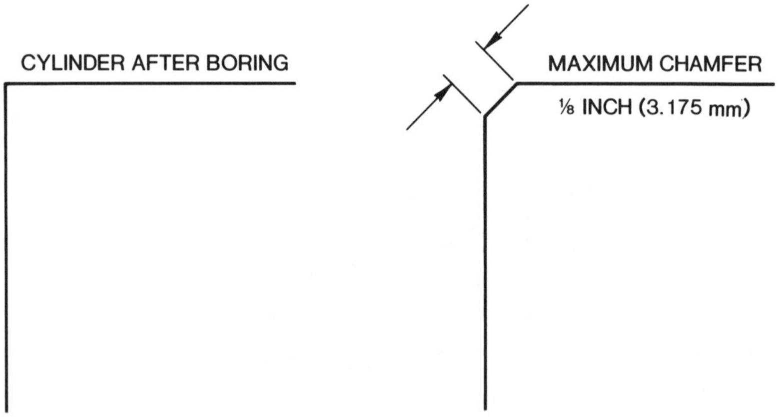

FIG 9-25 Bore chamfer dimensions.

JOB REPORT SHEET 9-B1

CYLINDER BORING

Student Name ______________________ Date ______________

Make of Car ____________ Model ______________ Year ______________

Time Started ____________ Time Finished ____________ Total Time ____________

Flat Rate Time ______________________________

Specifications:

Present cylinder size (unworn area)

Note: Make certain all cylinders are approximately the same size.

______________ in. or mm

New piston sizes (measured on lower skirt, opposite the piston pin):

1. ____________ 2. ____________ 3. ____________ 4. ____________

5. ____________ 6. ____________ 7. ____________ 8. ____________

Recommended piston clearance:

______________ – ______________

Average piston size + clearance = finished bore diameter

______________ + ____________ = ____________

Bore diameter after:

1st Cut ____________ 2nd Cut ____________ 3rd Cut ____________

Finish Cut (Allow for increase in bore size for honing to finish bore diameter.) ____________

Checkup Questions:

1. Why is it necessary to torque the main bearing caps? ______________________________

____

2. How many angles are used to sharpen the cutting tool? ______________________

3. What is the width of the actual cutting edge? ______________________

4. Why is it important to stop the cutting action in the first ¼ inch (6.35 mm) of the bore if problems are suspected? ____

____

____

5. How much stock was left for honing to finish bore size? ______________________

____

6. How was the block chamfered? ____

____

DATE COMPLETED ______________________ INSTRUCTOR'S SIGNATURE ______________________

DRY CYLINDER SLEEVE INSTALLATION

Dry sleeves are centrifugally cast sleeves of high quality cast iron. They are made with ⅛ in. (3.175 mm) or 3/32 in. (2.36 mm) wall thicknesses, preferably ⅛ in. They have a smoothly ground outside finish to assist in installation. These sleeves are used to repair cylinder bores with deep scores, cracks, or other irregular surfaces. Installation is as follows:

1. Obtain a sleeve 0.250 in. (6.35 mm) larger than the standard bore and longer than the bore by at least ⅛ in. (3.17 mm).
2. Measure the outside diameter of the sleeve and record this figure on the sleeve.
3. Bore the cylinder(s) to be repaired with sleeves first. The finished bore diameter should be 0.0025 in. (0.0635 mm) *smaller* than the sleeve. This provides the interference fit necessary to dissipate the cylinder heat and hold the sleeve firmly in place.
4. With a hacksaw, cut off the end of the sleeve which does not have the chamfer so it will protrude above the deck surface ⅛ in. (3.17 mm) when installed. The sleeve length should be cut to the length of the block bore plus ⅛ in. if it is being driven into the bore. Sleeves being pressed into position can be cut to exact length.
5. Assemble the tools necessary to drive or press the sleeve into the block. A hydraulic press works well with in-line engines; a front end coil spring compressor or a wood block and sledgehammer will work on V-type blocks.
6. Warm the block slowly and evenly with an acetylene torch to expand the sleeve opening (bore). Cool the sleeve with ice to shrink its diameter.
7. Apply sealant (green loctite, red or white lead) to the lower area of the bore and the upper area of the sleeve. This will prevent any leakage between the sleeve and block.
8. Drive or press the sleeve (chamfered end first) into position *as quickly as possible.* When the block and the sleeve reach the same temperature, it becomes extremely difficult to move the sleeve.
9. With the sleeve in position in the block, *carefully* file the sleeve until it is even and smooth with the deck surface.
10. The cylinder can now be bored to fit the piston.

HONING

After the block has been bored, the cylinder walls should be refinished. This involves making the clearance correct and making the finish texture suitable for piston ring seating.

The amount of clearance or fit between a piston and the cylinder wall will vary because of two major factors: the piston design (refer to Unit 10) and the type of operation the engine will be asked to perform. When an engine is to be operated consistently at higher than normal (5000) rpm, clearance must be increased.

The majority of original and replacement pistons take into account a wide variety of operational requirements. For original equipment manufacturers (O.E.M.), pistons use the factory specification to determine the piston clearance. For replacement pistons, always follow the piston maker's clearance recommendations.

During the actual honing process, several safety precautions should be observed to prevent possible injury:

1. Do not allow the hone to go too deep into the cylinder, for it may catch on the main bearing support. Check the maximum depth and don't exceed that amount.
2. Do not remove the hone while it is still turning. Keep it within the cylinder, never allowing more than 1 in. (25.4 mm) to be exposed at the deck surface.

OBJECTIVE 9-B2 (manipulative)

Working as a team with one other student and with the aid of Job Report Sheet 9-B2, you will hone the cylinder walls of an instructor-assigned engine using a rigid or flex-type hone.

Time Recommended:

Twice the flat rate time allowed in *Motors, Chilton,* or *National Service Data* manuals; approximately 15 minutes per cylinder.

Preparation:

1. Review pages 245 through 273.
2. Study pages 274 through 275.
3. Study manufacturer's or general repair manuals for the engine being serviced.
4. Attend lectures and demonstrations on cylinder block preparation.

Tools and Materials Needed:

1. Cylinder hone with assorted grit stones
2. Variable speed ½ in. drill motor
3. Coolant or honing oil
4. Pistons
5. Piston pull-gauge with long feeler strip

Procedure:

1. Just as boring requires the installation and torquing of the main bearing caps, so does honing.
2. Place the block in a position where the cylinders to be honed are parallel to the floor. This will keep the coolant or oil from draining out of the cylinders too quickly.
3. Select the honing stones according to the amount of stock to be removed and install them in the hone (Fig. 9-26).
4. Place the hone in the cylinder and expand it until there is a slight drag.
5. Attach the motor and start the hone at a slow speed.
6. Begin stroking (up-and-down motion) as the coolant is being applied. If hon-honing oil is not available, use a high concentrate detergent solution.

Note: Do not go too deep into the cylinder or remove the hone while it is turning. Injury and/or damage to the hone will result.

7. Depending on the stone grit used and the amount of stock to be removed, stop the hone in the bore after approximately 15 seconds of operation.

Loosen the pressure of the stones and withdraw the hone. Failure to loosen the pressure will leave "stripes" down the cylinder walls as the hone is removed.

8. Wipe the cylinder clean and attempt to set the piston in the cylinder. If it does not go in or is dragging heavily, continue honing as before until the fit appears to be close to correct.
9. Change the stones to a fine grit (280) when there is 0.0005 in. (0.0127 mm) left to hone.
10. Hone till the desired cylinder size and cross-hatch pattern are achieved (Fig. 9-27). The size can be checked with a piston and piston pull gauge (Fig. 9-28) or a dial bore gauge set to the correct diameter.
11. Finish the rest of the cylinders and clean with soap and hot water followed by an oil swab. Refer to steps 4–11 of Objective 9-B3, page 282.
12. Clean up all tools and equipment used in this objective.

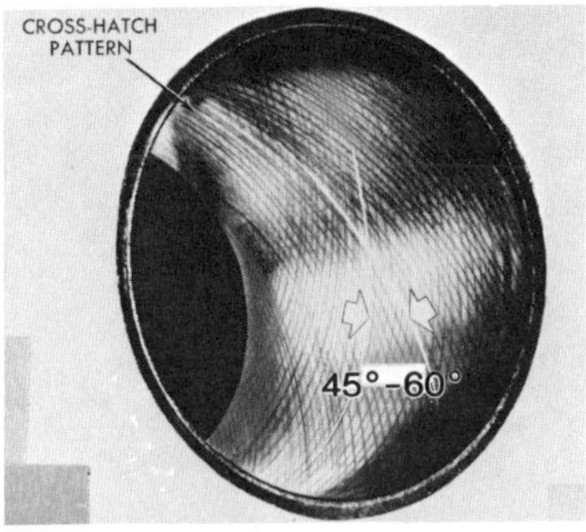

FIG 9-27 Desired cross-hatch cylinder wall pattern.
(Chrysler Corporation)

ROUGHING	70 GRIT
COARSE FINISHING	150 GRIT
MEDIUM FINISHING	220 GRIT
POLISHING	280–300 GRIT

FIG 9-26 Hone stone set.

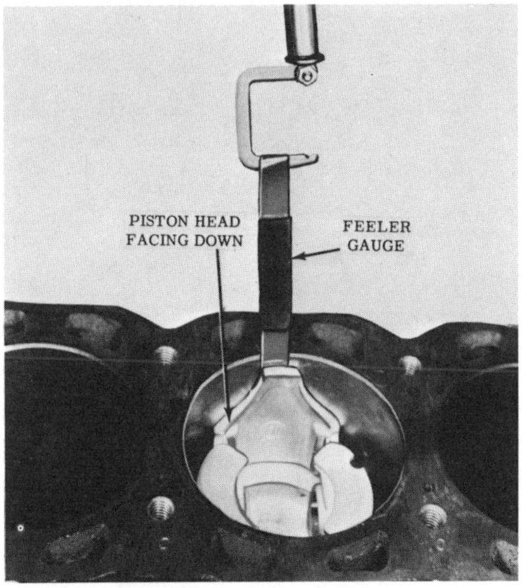

FIG 9-28 Using a piston pull gauge with a long feeler strip.
(Oldsmobile Division, GMC)

CYLINDER WALL HONING

Student Name ____________________ Date ____________

Make of Car ____________ Model ____________ Year ____________

Time Started ____________ Time Finished ____________ Total Time ____________

Flat Rate Time ____________________

Specifications:

Cylinder Size after Boring

____________ in. or mm

Finished Bore Size (Piston size + piston clearance)

____________ in. or mm

Checkup Questions:

1. List the stone grit(s) used: ____________________
2. What type of coolant/lubricant was used in this operation? ____________________

   ____
3. What thickness is the feeler gauge strip on the piston pull gauge?

   ____
4. Why is soap and hot water needed to clean the cylinders after honing? ____________________

   ____

DATE COMPLETED ____________ INSTRUCTOR'S SIGNATURE ____________

EXTRACTING BROKEN FASTENERS

During disassembly or after the cleaning operations have been completed, the technician should closely examine all parts of the engine for broken fasteners and remove them as part of the machining operations. This is necessary since their removal may create metal chips through drilling or loosen other materials not desirable in assembled, finished engine components. The common reasons for fastener breakage are

a. Overtightening—stretched apart
b. Damaged or mismatched threads
c. Bottoming out
d. Physical damage—a striking blow usually to an extended portion

The removal process should extract the broken fastener in the easiest possible manner. If the remaining portion of the fastener is above the surface, attempt to remove it with vise grips, a stud puller, or even a pipe wrench. Grip it as close to the surface as possible.

When the fastener is *at* or slightly below the surface level, first attempt to remove the broken piece with a sharp prick punch or small center punch. Hold it at the angle and location shown in Figure 9-29. As the punch is hammered in a counterclockwise direction for right-hand threads, the broken piece may begin to unscrew. If this occurs, continue hammering to chase the punch around the stub until it can be unscrewed by hand.

When the broken stub is at surface level or below and cannot be moved by the punch-and-hammer method, a screw extractor must be used. The type and size range of the set shown in Figure 9-30 will be adequate for most engine fasteners. Follow the directions on the inside of the cover.

Note: By far the most important step in using any type of screw extractor is center punching and then drilling the hole(s) *exactly in the center* of the broken piece.

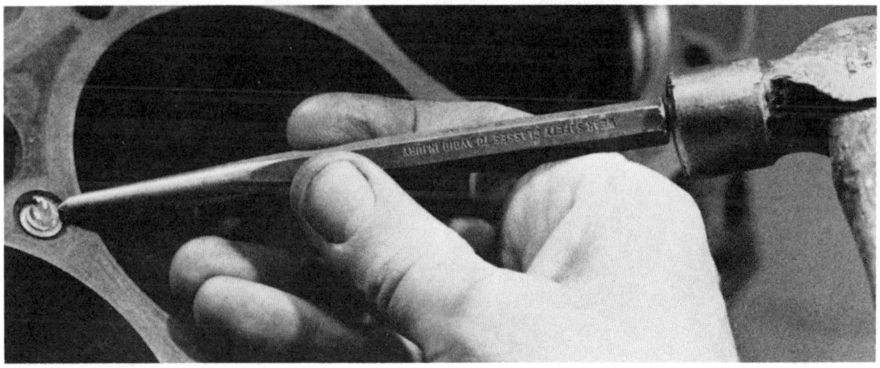

FIG 9-29 A center punch and hammer being used to remove a broken bolt.

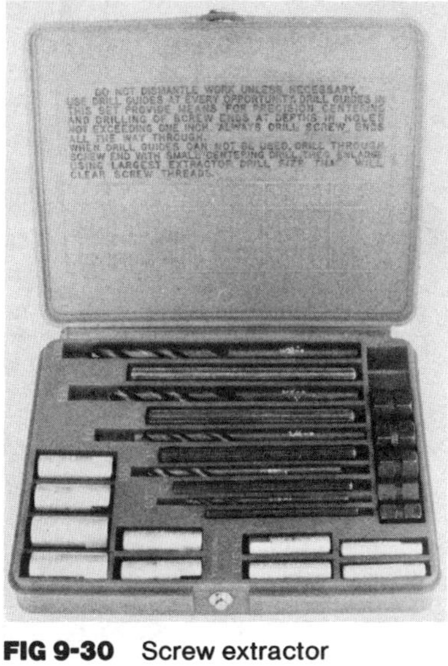

FIG 9-30 Screw extractor set.

THREAD RECONDITIONING AND REPLACEMENT

When threads become damaged so that they are not strong enough to hold a fastener or a fastener cannot be turned into the hole, reconditioning or replacement must be done.

Slightly damaged threads may be reconditioned satisfactorily by chasing them with a tap or die. Chasing involves using a tap (preferably a bottoming type) to clean and reform the threads in a hole or a die to recondition the threads of a bolt or stud.

Heavily damaged threads in a hole will have to be enlarged to the next common size *or* have a thread replacement or insert installed. To enlarge a threaded hole, simply drill the hole to the correct tap-drill size and rethread it. Thread replacements such as Keenserts, Slimserts, or Heli-coils require special drills, taps, and procedures (Fig. 9-31). Follow the advice of your instructor or supervisor to replace any threads in this manner.

Bolt or stud threads that are heavily damaged must be replaced.

Anything foreign that remains in a threaded hole or on the threads of a fastener will cause resistance to movement during tightening. This results in improper tension between mated parts and can cause distortion and gasket leakage.

All bolt or stud holes should have a slight chamfer or counterbore to prevent the top thread of the hole from lifting during the assembly operation. This could cause poor gasket sealing and/or interference with the seating and alignment of the assembled pieces.

FINAL CLEANING

When all machining operations have been completed, the block should be thoroughly cleaned. This should be the last time a complete cleaning is necessary. Between all assembly operations, the block should be covered (plastic bags) and stored in an area that is as dust free as possible.

FIG 9-31 A. Keensert thread inserts.

(Irontite Product Co., Inc.)

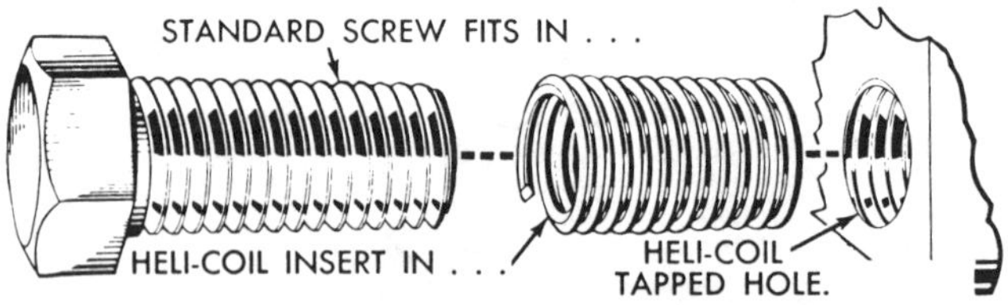

FIG 9-31 B. Heli-Coil insert.

(Chrysler Corporation)

Working as a team with one other student and with the aid of Job Report Sheet 9-B3, you will recondition (chase) all threads, replace damaged threads, and thoroughly clean for the final time all surfaces of an instructor-assigned vehicle engine.

OBJECTIVE 9-B3 (manipulative)

Time Recommended:

Twice the flat rate time allowed in *Motors, Chilton,* or *National Service Data* manuals.

Preparation:

1. Review pages 245 through 280.
2. Study pages 281 through 283.
3. Study manufacturer's or general repair manuals for the engine being serviced.
4. Attend lectures and demonstrations on engine block preparation.

Tools and Materials Needed:

1. Tap and die set
2. Stiff bristle brush
3. Soap
4. Oil can
5. Plastic bag

Procedure:

1. Select the tap that matches the hole to be chased. (Most holes have National Course threads.) When in doubt about the proper tap, locate the bolt that fits the hole and compare it to each tap. The correct tap will "fit" the bolt threads evenly (Fig. 9-32). For holes that held threaded plugs, select the correct pipe tap.

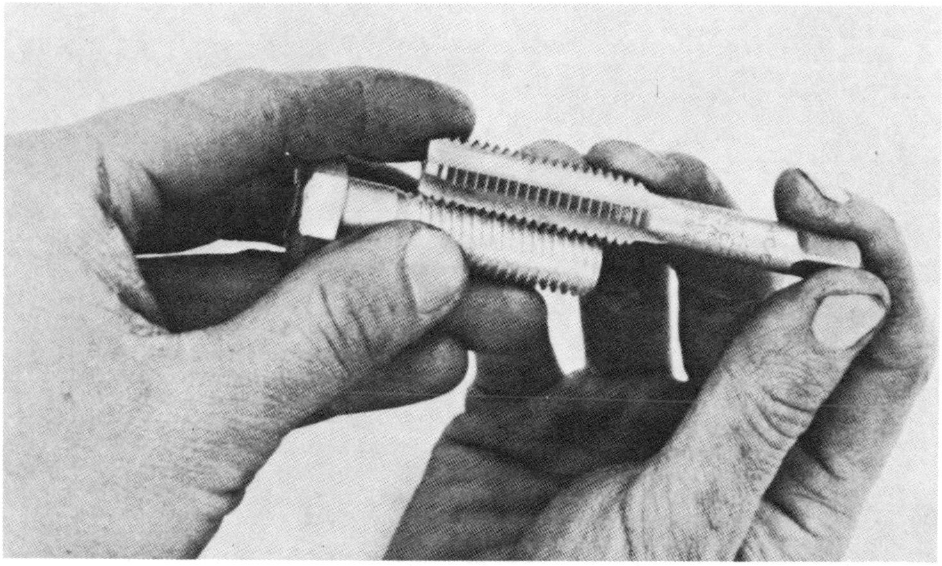

FIG 9-32 Compare a tap and a bolt to assist in tap selection.

2. Turn the tap until strong resistance is felt; then remove it (Fig. 9-33).

Note: No metal should be cut during this operation unless there is slight thread damage—the threads may appear shiny. If thread replacement seems necessary, consult with your instructor or supervisor for instruction on the type of replacement materials available.

3. Chamfer any head bolt holes necessary, (Fig. 9-34) then blow out all holes with compressed air.

Caution: Use eye protection

4. Depending on the equipment available, thoroughly clean the entire engine block. If a hot tank or steam cleaner is available, start your final cleaning using one of these machines.
5. Use soap (detergent) and water applied with a brush on the cylinder walls and lower engine areas. Look for the light gray material left from the deglazing or honing jobs.
6. Use a rifle bore brush with soap and water to clean all the oil galleries (Fig. 9-35).
7. Use soap and hot water on all surfaces until the bubbles are pure white; then rinse with water (Fig. 9-36).
8. Blow off all water and swab the cylinder wall surfaces with light oil (SAE 10) (Fig. 9-37).
9. With a clean rag or towel, wipe all cylinder wall surfaces. Repeat this step as necessary to remove grit and metal (Fig. 9-38).
10. As a final check for cleanliness, take a clean white rag or towel dampened with some light oil and rub it across the cylinder walls. If there is no color change in the rag, the block is clean.
11. At all times when work is not being done on the block, keep it covered with a plastic bag (clothing bags or trash bags work well).

FIG 9-33 Chasing the threads, using an 8-point socket, to turn the tap.

FIG 9-34 Chamfering the bolt holes.

FIG 9-35 Cleaning oil galleries with a rifle brush.

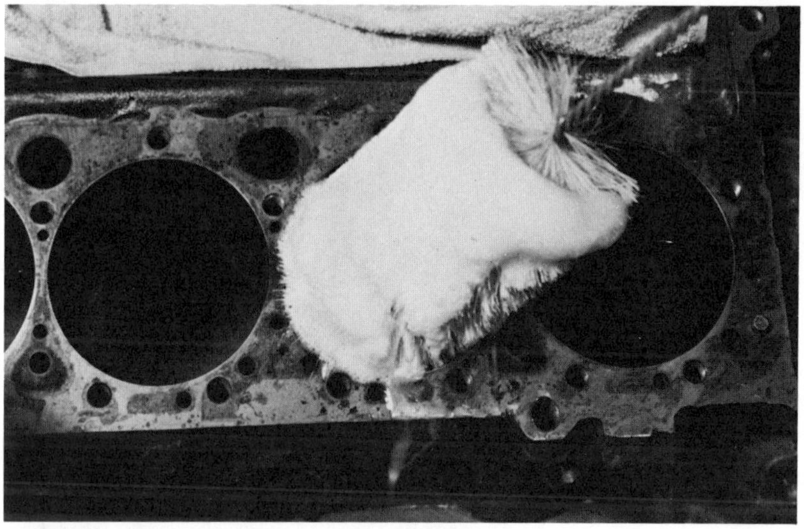

FIG 9-36 Using soap and warm water to clean cylinder bores.
(Chevrolet Motor Division, GMC)

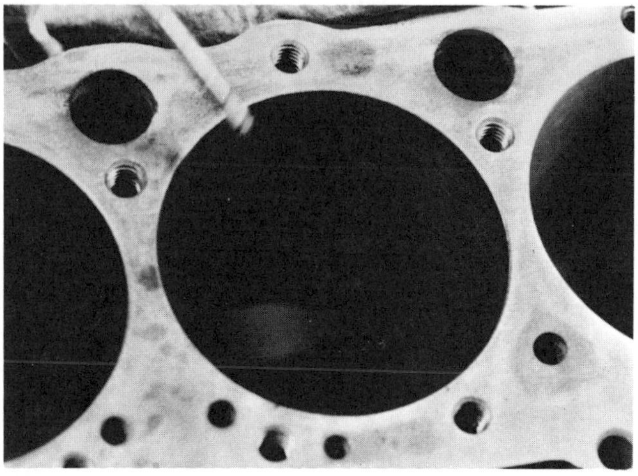

FIG 9-37 Apply oil to bore surfaces.

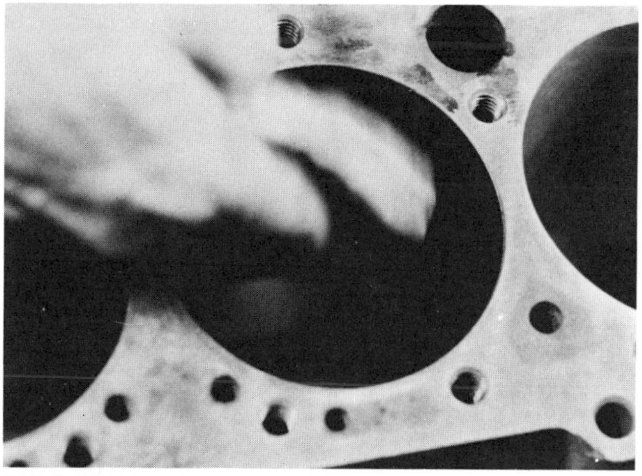

FIG 9-38 Wipe the cylinder walls with oil to remove the last of the honing materials.
(Oldsmobile Division, GMC)

THREAD RECONDITIONING, CHASING, AND FINAL CLEANING

Student Name ______________________________ Date ________________

Make of Car ____________ Model ________________ Year ________________

Time Started ____________ Time Finished ____________ Total Time ____________

Flat Rate Time ____

Specifications:

Checkup Questions:

1. Why does a bottoming tap work best for thread chasing? ________________________

   ____

   ____

2. What effect would an uncleaned threaded hole have on a bolt being tightened with a torque wrench?

   ____

   ____

3. The final cleaning operation should take place only when all ________________ operations have been completed.

DATE COMPLETED ________________ INSTRUCTOR'S SIGNATURE ________________

CORE PLUG INSTALLATION

Before installing new core plugs, be certain the holes in the block are free of foreign matter or rough surfaces. It is a good idea to make a preliminary check by placing the new plug next to the hole to see if the correct size was purchased.

A sealant (RTV Silicone or Permatex) should be applied to the sealings edges of the plugs before they are driven into place.

OIL GALLERY PLUG INSTALLATION

All oil gallery plugs should be returned to their places in the block. Most engines use plugs with pipe threads; however, some plugs are of the cup expansion design (Fig. 9-39). Whichever plug is used, first clean the opening in the block. Next, apply a sealing compound (RTV, Loctite, or Permatex) to the threads. Screw the plug firmly into position or drive into place as directed in the core plug installation instructions.

OBJECTIVE 9-B4 (manipulative)

Working as a team with one other student and with the aid of Job Report Sheet 9-B4, you will replace all core and oil gallery plugs on an instructor-assigned vehicle engine.

Time Recommended:

Twice the flat rate time allowed in *Motors, Chilton,* or *National Service Data* manuals; approximately 4 minutes per plug.

Preparation:

1. Review pages 245 through 283.
2. Study pages 287 through 288.
3. Study manufacturer's or general repair manuals for the engine being serviced.
4. Attend lectures and demonstrations on cylinder block preparation.

Tools and Materials Needed:

1. Core plugs
2. Driving tool
3. Hammer
4. Oil gallery plug wrenches
5. Sealant (RTV, Loctite, Permatex)

Procedure:

1. Clean all hole openings (core and oil gallery). Use abrasive paper or a wire brush if necessary.
2. Check the new core plug diameter to match that of the hole.
3. Apply sealant to outer edge of the plug and drive into position (Fig. 9-40).

Note: The cup plug should be driven from the bottom of the cup and *not* the flange; the expansion plug should be driven into position by contacting outer diameter of the plug *not* the crown (center).

4. Apply chemical locking material to the threads of the oil gallery plugs; then screw firmly into position (Fig. 9-41).
5. If cup core plugs are being used, install as stated in step 3; then stake the hole edges to prevent the plugs from leaving the holes if they loosen.

FIG 9-39 Different types of oil gallery plugs.
(Oldsmobile Division, GMC)

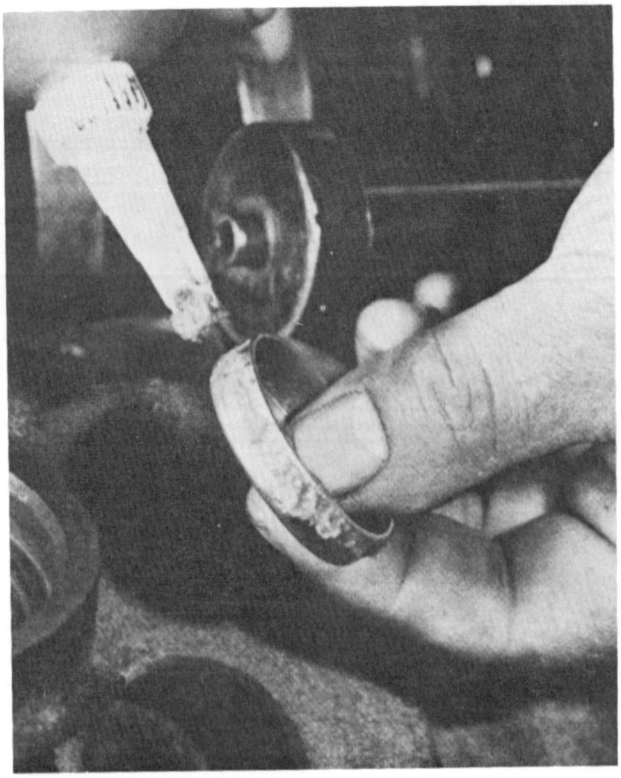

FIG 9-40 Sealing the edge of the core plug prior to installation.

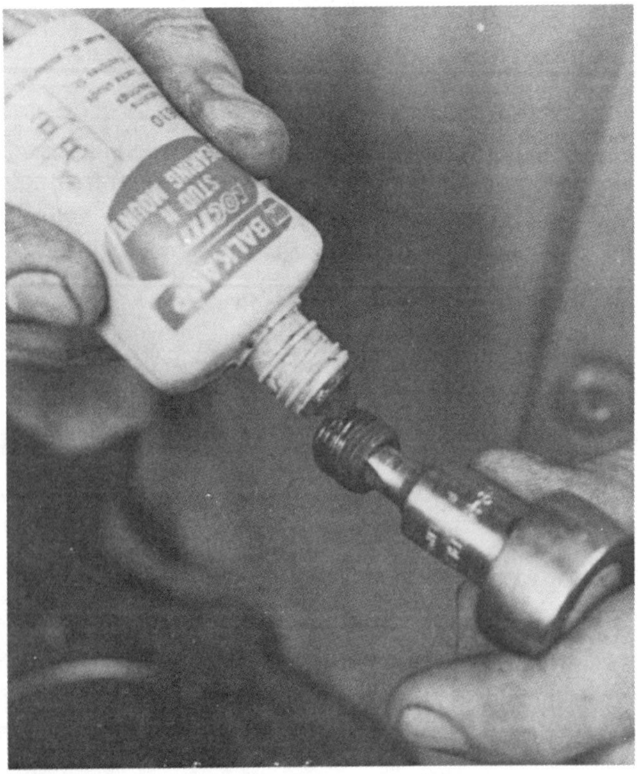

FIG 9-41 Locking chemical being applied to an oil gallery plug.

CORE AND OIL GALLERY PLUG INSTALLATION

Student Name ______________________ Date ____________

Make of Car ____________ Model ______________ Year ____________

Time Started ____________ Time Finished ____________ Total Time ____________

Flat Rate Time or Recommended Time ______________________

Specifications:

Indicate inches or millimetres.

Core Plug Sizes	Location on Block
A. ____________	______________________
B. ____________	______________________
C. ____________	______________________
D. ____________	______________________

Oil Gallery Plug Sizes	Location on Block
A. ____________	______________________
B. ____________	______________________
C. ____________	______________________

Checkup Questions:

1. What type of sealant was used on
   a. core plugs? ______________________
   b. oil gallery plugs? ______________________
2. What design of plug was used for
   a. core plugs? ______________________
   b. Oil gallery plugs? ______________________

DATE COMPLETED ____________ INSTRUCTOR'S SIGNATURE ____________

CAM BEARINGS

Precision-made camshaft bearings are designed to give the camshaft a surface on which it can rotate during all operating conditions. The cam bearing most common today is the full round design. The greatest portion of the cam bearing is a steel strip rolled into a circle with a thin layer of bearing material on the inside surface. This metal is usually babbitt, an alloy of tin, copper, and antimony.

Oil clearance between camshaft and journals and cam bearings is approximately 0.001 in. (0.0254 mm) when originally manufactured. With precise manufacture and careful installation, the cam bearing will return to the original clearance specification.

The outside diameter of a cam bearing is made slightly larger (0.003 in. or 0.0762 mm) so it will have an interference fit. This is necessary to prevent the bearing from turning in its bore in the block and to allow for the greatest amount of heat transfer from the bearing to the block.

Before hot tanking a disassembled cylinder block, the cam bearings should be removed whether or not they are to be reused. Until the engine technician gains considerable experience in cam bearing replacement, all bearings removed should be labeled for position in the engine block by numbering front to rear and indicating oil hole locations. These bearings should be saved as a reference so that when new ones are ready to be installed, their position can be accurately determined.

New cam bearings are sold in sets which contain information on the position of installation. The positions relate to the cam bearing housing bores in the block, numbered from front to rear.

The tools required to remove and replace cam bearings are of three types: the expanding mandrel (universal) type, the solid mandrel type, and the puller type. The first two drive the bearings in and out, and the puller type uses a threaded shaft to draw the bearing.

OBJECTIVE 9-B5 (manipulative)

Working as a team with one other student and with the aid of Job Report Sheet 9-B5, you will remove and replace the camshaft bearings and install the camshaft of an instructor-assigned vehicle engine.

Time Recommended:

Twice the flat rate time allowed in *Motors, Chilton,* or *National Service Data* manuals; approximately 40 minutes.

Preparation:

1. Review pages 245 through 288.
2. Study pages 291 through 293.
3. Study manufacturer's or general repair manuals for the engine being serviced.
4. Attend lectures and demonstrations on cylinder block preparation.

Tools and Materials Needed:

1. Cam bearing removal tool
2. Replacement cam bearings
3. Camshaft

Procedure:

Bearing Removal

1. Select the expanding mandrel which is closest in fit to the cam bearing inside diameter (Fig. 9-42).
2. Assemble the parts of the tool as shown in Figure 9-43.
3. Insert the tool into the cam bearing opening until the flange of the mandrel touches the edge of the bearing.
4. Turn the driving bar to expand the mandrel firmly in the bearing.
5. With a large hammer (40 oz.), drive the bearing from the block (housing bore) (Fig. 9-44).
6. Mark the *number and position* of the bearing for reference when installing new bearing.
7. Repeat steps 1–6 until all bearings are removed.

Bearing Installation

1. Locate worn camshaft bearings, replacement bearing, and the camshaft to be used.
2. Make a preliminary fit check by sliding each new bearing into position on each camshaft journal (Fig. 9-45). There should be a slight amount of play at each bearing and journal, approximately 0.002 in. (0.0508 mm). If necessary, measure the bearing I.D. (inside diameter) and its journal O.D. (outside diameter). I.D. minus O.D. = clearance. The container holding the new bearings or the bearings themselves should have a listing of their installed positions (Fig. 9-46). Should any problems arise as to the fit, notify your instructor.
3. Place the rear bearing (last position) into the block nearest its bore opening so that the chamfered edge of the bearing faces the bore.

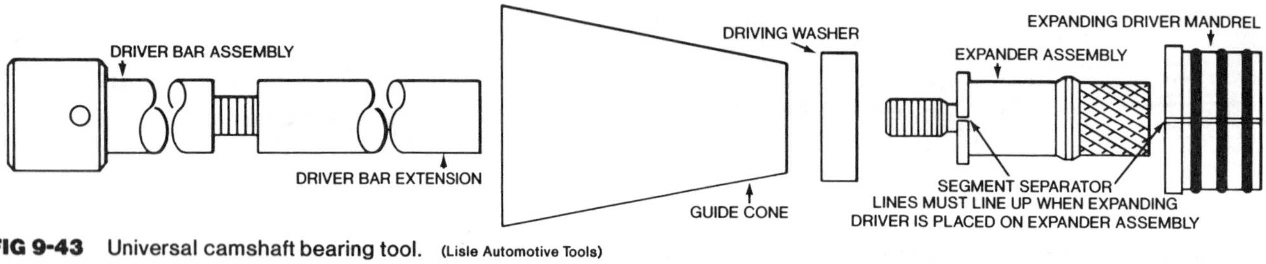

FIG 9-43 Universal camshaft bearing tool. (Lisle Automotive Tools)

FIG 9-42 Selecting expanding mandrel to fit the bearing.

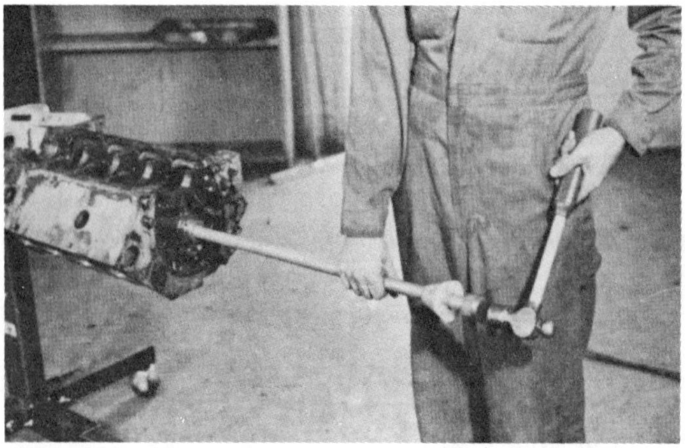

FIG 9-44 Removing the number one cam bearing.

Note: If there is no chamfered edge or there appears to be insufficient chamfer, file the edge until there is an adequate amount.

4. Insert the tool in the block and place the bearing on the expanding mandrel.
5. Turn the driver bar clockwise until the mandrel is snug inside the bearing.
6. Start the bearing into the block by hand pressure. Remove the tool slightly to be certain the oil holes in the bearing and block are lined up (Fig. 9-47).
7. Tap the end of the driver with a large hammer while holding the centering cone (if so equipped).
8. Continue hammering until the bearing edge is flush with the housing bore and the oil holes are aligned.
9. Install all remaining bearings except #1. This bearing is best installed from the rear to the front.
10. Check the bearing fit by carefully installing the camshaft. If the camshaft will not enter any of the bearings, look for rolled-up edges or bright spots. Should either of these be detected, scrape the area lightly until the raised area is removed. *Never* use sandpaper because the grit will become imbedded in the bearing material.
11. Apply assembly lube (molykote or moly grease) to the cam lobes and journals before installing the camshaft (Fig. 9-48). This will help prevent cam lobe failure during the early stages of operation.

Note: If the engine is equipped with mushroom lifters, they must be installed now.

12. Install the camshaft very carefully, trying not to gouge the cam bearings.

FIG 9-45 Checking the cam bearing fit on the camshaft prior to installation in the block.

FIG 9-46 Cam bearing set with location information.

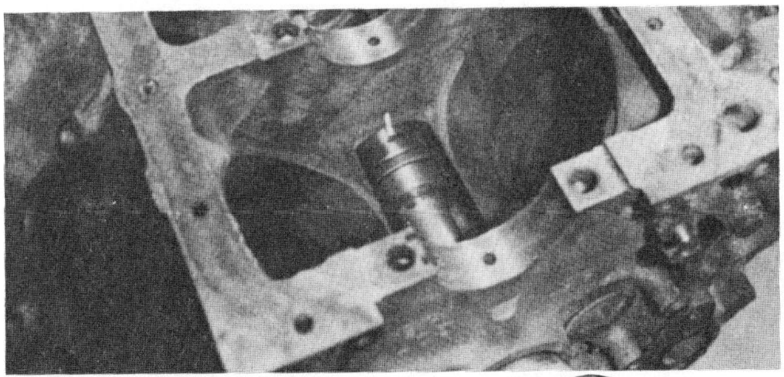

FIG 9-47 Inserting a cam bearing with the oil hole lined up with the oil passage of the block.

important

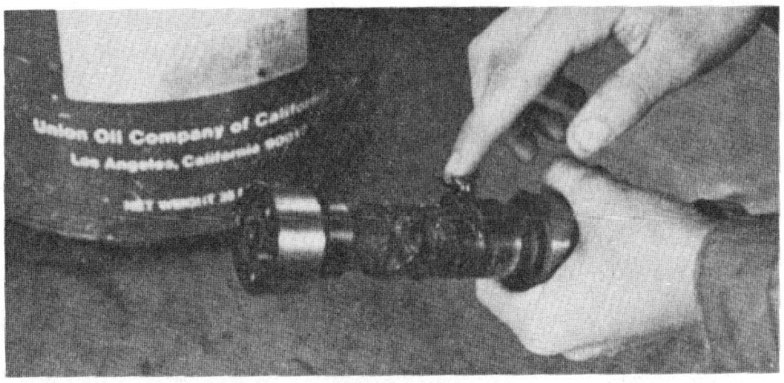

FIG 9-48 Applying moly grease to the cam lobes prior to installation.

JOB REPORT SHEET 9-B5

CAMSHAFT BEARING REMOVAL AND REPLACEMENT

Student Name ____________________ Date ____________

Make of Car ____________ Model ____________ Year ____________

Time Started ____________ Time Finished ____________ Total Time ____________

Flat Rate Time or Recommended Time ____________________

Specifications:

Cam bearing location information.

Checkup Questions:

1. Why should the cam bearings be labeled and saved? ____________________

____

2. Which end of the cam bearing should enter the block first?

____

3. When should you check to be certain the oil holes in the bearing and block will align? ____________

____

4. Did any bearings have to be scraped to install the camshaft? ____________________

Why? ____

____

____

DATE COMPLETED ____________________ INSTRUCTOR'S SIGNATURE ____________________

Name ______________________________ Date ______________

DIRECTIONS:

Circle the best answer to each question.

1. Boring the cylinders to fit oversized pistons
   I. should be done only when the taper exceeds 0.016 in.
   II. can be done before the pistons are purchased.
   A. I only B. II only C. Both I and II D. Neither I nor II
2. Before attaching the boring bar,
   I. center the cutting tool.
   II. chamfer the top edge of the bore.
   A. I only B. II only C. Both I and II D. Neither I nor II
3. If there is any doubt about the centering position of the bar or the amount of cut,
   I. disengage the clutch or gear and stop the motor.
   II. do not cut deeper than ¼ in. (6.35 mm) into the cylinder.
   A. I only B. II only C. Both I and II D. Neither I nor II
4. Dry sleeves
   I. are installed in cylinders to repair the chamfer edge.
   II. require an interference fit in the block.
   A. I only B. II only C. Both I and II D. Neither I nor II
5. Honing
   I. is not necessary after installing dry sleeves.
   II. should not be done to a cylinder bored to fit oversize pistons.
   A. I only B. II only C. Both I and II D. Neither I nor II
6. Honing is required to
   I. refinish the cylinder walls with the correct pattern.
   II. heat treat the cylinder walls.
   A. I only B. II only C. Both I and II D. Neither I nor II
7. The final cleanup of cylinder walls following honing includes
   I. washing with warm soap and water.
   II. swabbing with oil and a clean rag.
   A. I only B. II only C. Both I and II D. Neither I nor II
8. Broken fasteners can often be extracted
   I. with a ring compressor.
   II. by cutting off with a hacksaw.
   A. I only B. II only C. Both I and II D. Neither I nor II
9. The most important step in the use of screw extractors is
   I. cutting off the broken stub so it is level with the surface.
   II. center punching and drilling the hole exactly in the center of the broken piece.
   A. I only B. II only C. Both I and II D. Neither I nor II
10. Thread replacement can be accomplished by
    I. threading the hole to the next smaller size.
    II. coating the damaged threads with gasket sealer.
    A. I only B. II only C. Both I and II D. Neither I nor II

11. Thread chasing means
   I. using a tap to clean a threaded hole of foreign matter.
   II. using a die to clean the threads of a bolt.
   A. I only   B. II only   C. Both I and II   D. Neither I nor II

12. The oil galleries of a block should be cleaned with
   I. fresh oil.
   II. a deglazing tool.
   A. I only   B. II only   C. Both I and II   D. Neither I nor II

13. Core plugs are installed
   I. with a sealant around the edges
   II. by driving them into place using the flange surface.
   A. I only   B. II only   C. Both I and II   D. Neither I nor II

14. Oil gallery plugs
   I. that are threaded should have sealing or locking compound applied during their installation.
   II. of the cup type should not have a sealant applied during their installation.
   A. I only   B. II only   C. Both I and II   D. Neither I nor II

15. When removing cam bearings,
   I. mark the number of each one if it is not already visible.
   II. note the position of the oil holes.
   A. I only   B. II only   C. Both I and II   D. Neither I nor II

16. A preliminary check of the fit of a cam bearing can be made by
   I. installing it half-way and measuring the diameter.
   II. placing each cam bearing on its respective journal.
   A. I only   B. II only   C. Both I and II   D. Neither I nor II

17. Cam bearings
   I. in most blocks can be reused if they have been cleaned in the hot tank.
   II. must be installed with the oil holes facing to the front.
   A. I only   B. II only   C. Both I and II   D. Neither I nor II

18. If the camshaft does not enter any of the bearing easily or at all,
   I. remove the bearing and check its position with the indications on the box or the instruction sheet.
   II. check for rolled edges or bright spots and remove with sandpaper or emery paper.
   A. I only   B. II only   C. Both I and II   D. Neither I nor II

19. Cam bearings are manufactured from
   I. steel.
   II. babbitt.
   A. I only   B. II only   C. Both I and II   D. Neither I nor II

20. Assembly lube (molykote) should be applied to
   I. cam lobes.
   II. lifter faces.
   A. I only   B. II only   C. Both I and II   D. Neither I nor II

DATE COMPLETED ____________ INSTRUCTOR'S SIGNATURE ____________________

unit 10

Crankshaft and Bearing Service

This unit covers the topics normally associated with the lower engine area: the crankshaft, bearings, and lubrication system. Background information as well as service techniques for crankshaft and bearings are presented.

OBJECTIVE 10-A (written test)

After completing the preparation assignment, you will answer a minimum of 14 out of 20 multiple choice questions in a written test, without using reference material, on the following topics:

1. Bearings and lubrication
2. Lubrication system
3. Bearing material and design
4. Crankshafts and their service

Preparation:

Study pages 299 through 323.

BEARINGS AND LUBRICATION

An engine relies on its moving parts to convert heat energy into controlled motion or mechanical energy. The parts rotate (turn) and reciprocate (move up and down) at varying speeds. These movements generate a great deal of resistance between parts, which is called friction. Engine bearings perform the job of reducing friction and providing support under different operating conditions. To reduce the internal friction of an engine to a minimum, lubricating oil is supplied to critical points of bearing materials. Bearings absorb the explosive shocks of the power strokes under various amounts of engine load and speed.

BEARING CLEARANCE AND THE OIL FILM THEORY

Good, clean lubricating oil is essential to efficient engine performance. It must help cool bearing surfaces, absorb some of the shock between engine parts, and lubricate. If oil is to do its job of reducing friction (and, thus, engine wear), it must be supplied between the bearing surfaces in the correct amounts. This is possible because bearing diameter is slightly larger than the piece against which it must turn. The term used to describe this size difference is oil clearance (Fig. 10-1). Most bearings have but a few thousandths of an inch of oil clearance.

Critical load bearings receive a pressured supply of oil at all engine speeds. Other bearings in the engine rely on oil splash or oil vapor for their lubrication. Bearing wear occurs when there is friction caused by metal-to-metal contact. The greater the contact, the more wear takes place. When two parts are at rest there is no oil film. As rotation begins, the oil film builds up in a wedge shape (Fig. 10-2), but some metal contact has already taken place. When an engine is cold, there is more contact. This is why most wear takes place before the engine reaches operating temperature. If bearing clearances are correct, if the proper oil is used, if the temperature is normal, and if there are no excessive loads imposed on the bearings, there will be no metal-to-metal contact. But even under the very best conditions there are occasional periods when the oil film is thin enough to allow slight contact. This is referred to as boundary lubrication.

LUBRICATING SYSTEMS

The type of lubricating system found in modern engines is shown in Figure 10-3. This is referred to as a full pressure system because oil is supplied to the critical rotating surfaces by an oil pump. As the engine runs, there is enough agitation from these rotating surfaces to splash oil onto other surfaces requiring lubrication, such as the cylinder walls and cam lobes.

The lubricating system depends on an oil pan filled with good oil (4–6 quarts). The oil pump pickup assembly is submerged in the oil or mounted outside the block with a pickup tube below the oil level. The oil pump, driven by the camshaft, pushes oil into the passages after it is drawn through a screen attached to the pump. The screen prevents large bits of unwanted foreign matter from entering the system. When the oil leaves the pump, it passes through the filter to the oil galleries to be distributed to the camshaft and lifters, and to the crankshaft and its bearings. Oil is also carried through oil passages to the cylinder head so that the rocker arms and valve stems are lubricated. Oil return holes allow the oil to flow back to the pan at the bottom of the engine. From there, it is picked up again by the oil pump and circulated through the engine.

OIL PUMP RATING

The lubricating system relies on a pump to circulate oil. The pump is usually rated in a measurement of operating pressure. This rating is calculated in pounds per square inch (psi).

The operating oil pressure in modern engines is approximately 40 psi. This is the pressure at which the relief valve in the system begins to bypass the oil back to the oil pan. The amount of oil pressure designed for an engine is a compromise. A high pressure supplies plenty of oil at high speeds and heavy loads, but tends to erode (wash away) the bearing materials. A low pressure does not supply enough oil under high speeds or heavy loading conditions and does not cool the bearings.

If an engine is to maintain good oil pressure during operation, the oil must have the proper viscosity, or resistance to flow (a thin oil has less viscosity than a thick oil), the pump must be in good condition, and all the bearing clearances must be correct. Figure 10-4 illustrates the effects of oil bearing clearance on oil "throw off."

The lubricating system is a group of channels and containers that should be nearly closed at the ends. To have pressure, the pump must work within this nearly closed system. This makes the oil pressure gauge a valuable instrument for determining the performance of the engine lubricating system and, therefore, the condition of the bearings.

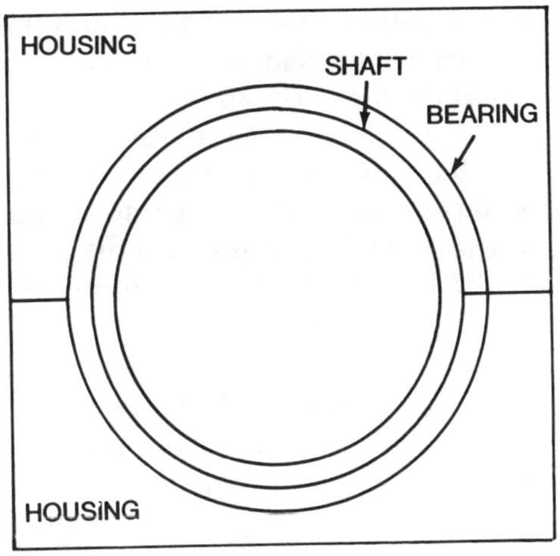

FIG 10-1 The space between the shaft and bearing is called clearance. (Federal-Mogul)

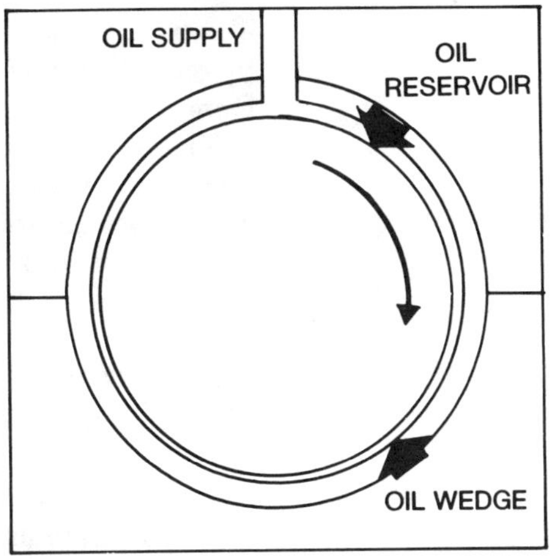

FIG 10-2 Oil wedge buildup during shaft rotation. (Federal-Mogul)

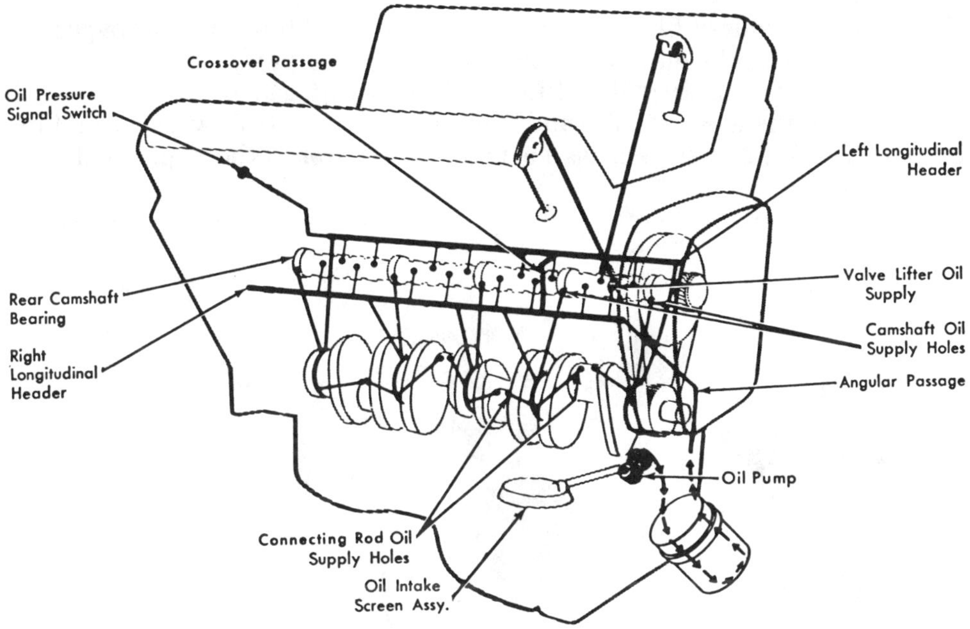

FIG 10-3 Typical lubricating system. (Cadillac Motor Car Division, GMC)

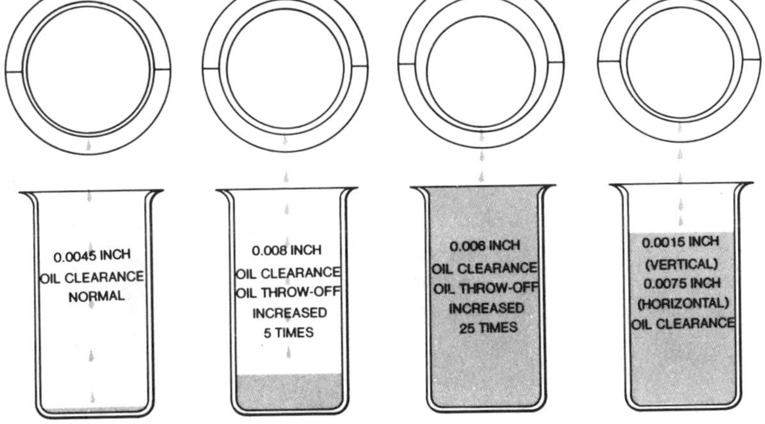

FIG 10-4 Effects of bearing clearance on oil throw-off. (Federal-Mogul)

OIL PUMP DESIGNS

Two designs of oil pumps are used in modern vehicle engines: the rotor pump used by most Chrysler and Ford engines (Fig. 10-5), and the gear pump found in most General Motors, American Motors, and import engines (Fig. 10-6).

The rotor pump has an inner and outer rotor. The four-lobe inner rotor is driven and causes the outer rotor to turn with it. As this happens, the spaces between the rotor lobes become filled with oil. The lobes of the inner rotor move into the spaces of the outer rotor. This squeezes the oil past the gears. The gears push the oil to the pump outlet to provide a pressured flow of oil to all parts of the lubricating system.

OIL PUMP SERVICE

Since oil pumps operate in a constant bath of oil, they should sustain very little wear. Damaging wear occurs when severe bearing failure has taken place or the engine has not received proper maintenance, especially oil changes at recommended intervals.

OIL FILTERS

Oil filters have the job of cleaning the recirculating oil of as many contaminants as possible, including metal, carbon, dust, water, and oil.

There are two kinds of filter systems: the full flow and the partial or bypass. In the full-flow system, the oil filter is located between the pump and the bearings. This means all the oil in circulation through the lubricating system is filtered. This system is by far the most popular and is usually distinguished by a spin-on filter (Fig. 10-7). The full-flow system has a bypass valve to protect the engine against oil

FIG 10-6 Gear oil pump.
(Chevrolet Motor Division, GMC)

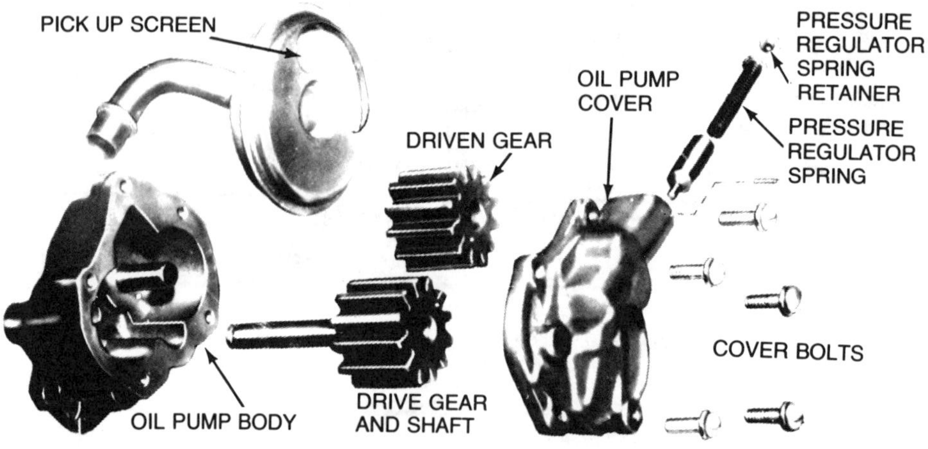

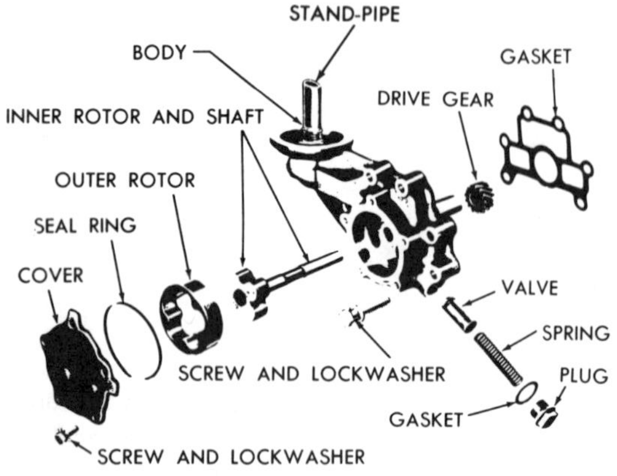

FIG 10-5 Rotor oil pump.
(Chrysler Corporation)

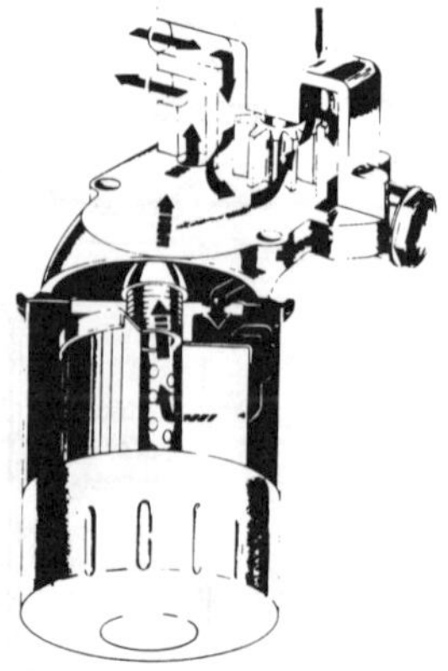

FIG 10-7 Oil path through a full flow filter.
(Buick Motor Division, GMC)

starvation. This means that if the filter becomes plugged the bypass valve is opened by the increased oil pump pressure. When the bypass valve is opened, oil bypasses the filter. Thus, the engine is insured of sufficient oil.

Bypass systems filter only part of the oil from the oil pump. Flexible or rigid tubing brings the oil from any pressure point on the engine. The oil flows through the filter and to the bottom of the housing, where it returns to the crankcase area. When the filter becomes plugged, the oil stops flowing to the filter, but not to the bearings.

Oil filters have different filtering materials and opposing claims are made as to which is best. A good filter will do an adequate job regardless of the filtering material. The two most popular materials used today are paper and fibrous material.

Filter paper is assembled in pleats within the filter housing. The paper traps contaminants as the oil passes through the filter.

Fibrous materials are a combination of cotton, jute, and other materials that provide a trap for heavy contamination.

OBJECTIVE 10-A1 (manipulative)

Working as a team with one other student and with the aid of Job Report Sheet 10-A1, you will disassemble, clean, measure clearances, prime, and assemble the oil pump assembly from an instructor-assigned vehicle engine.

Time Recommended:

Twice the flat rate time allowed in *Motors, Chilton,* or *National Service Data* manuals; approximately 20 minutes.

Preparation:

1. Review pages 299 through 303.
2. Study pages 303 through 304.
3. Study manufacturer's or general repair manuals for the engine being serviced.
4. Attend lectures and demonstrations on lubrication systems.

Tools and Materials Needed:

1. Feeler gauge set
2. Straightedge (small)
3. Chassis or general purpose grease

Procedure:

1. Mark the pump cover before removal to insure its replacement exactly as disassembled.
2. Remove the cover and check to see if a gasket is used.
3. Mark the gears or rotor so they will be replaced in the same manner; then remove them from the housing.
4. Remove the pin, cap, or plug holding the relief spring and withdraw the spring and valve.
5. Degrease all of the oil pump parts (carburetor cleaner works well).
6. Wash and blow dry all pump parts.

7. Inspect all the interior surfaces for excessive wear (scoring) and cracks. The pickup tube and screen should also be checked.
8. Assemble the gears or rotor and relief valve parts. The gears and rotors should turn freely and the relief valve should move within its bore.
9. Measure the side clearance by placing a feeler gauge between the body and gear of the pump (Fig. 10-8). Compare to specification. 0.005 in. or (0.127 mm) is considered maximum.
10. Measure the end clearance by placing a straightedge (6 in. rule) across the end of the housing (Fig. 10-9). Attempt to fit a 0.004 in. (0.102 mm) feeler gauge between the straightedge and either the housing or gears-rotors. If this can be done, replace the pump.
11. Before replacing the pump cover, coat (pack) the gears-rotors with petroleum jelly or chassis grease. This will help prime the pump when the engine first runs.
12. Assemble any remaining pump parts, replacing the gaskets if necessary.

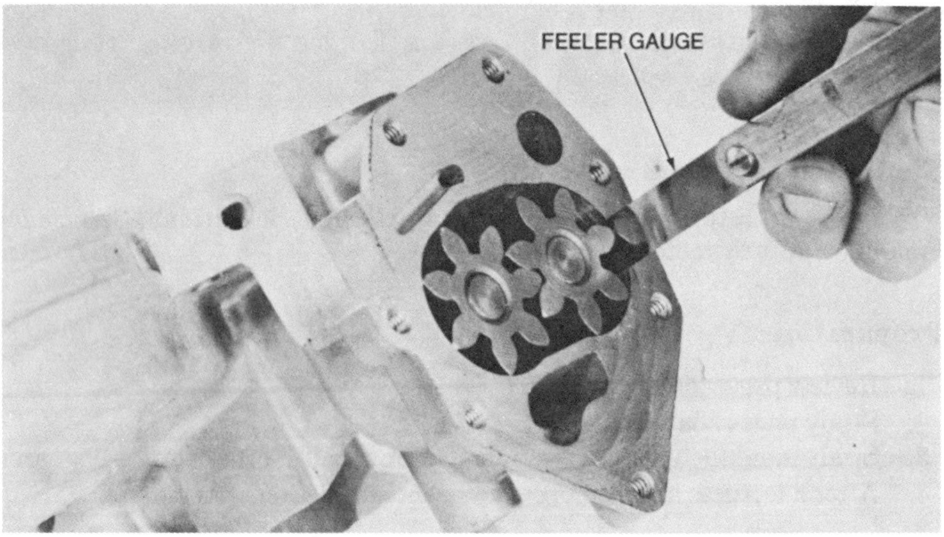

FIG 10-8 Measuring oil pump side clearance.
(American Motors Corporation)

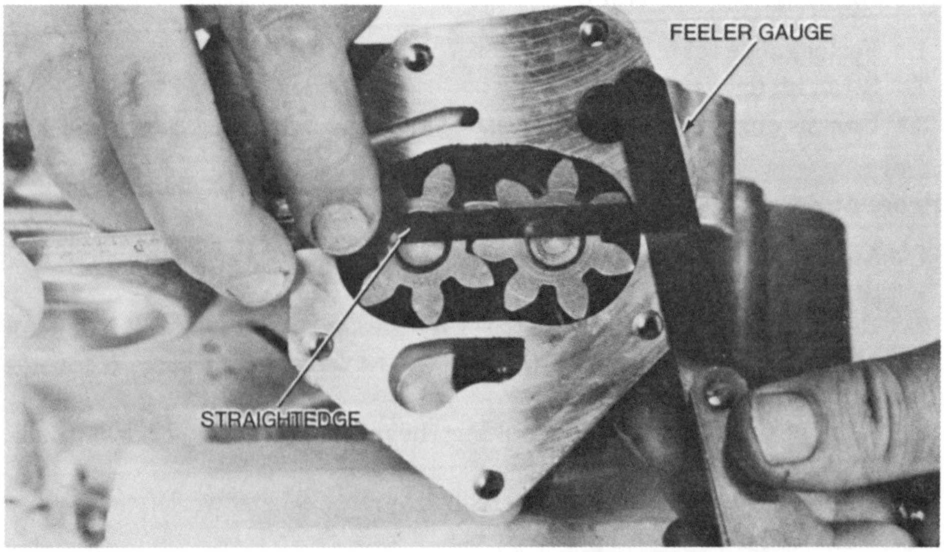

FIG 10-9 Measuring oil pump end clearance.
(American Motors Corporation)

OIL PUMP SERVICE

Student Name ______________________ Date ______________

Make of Car ____________ Model ______________ Year ______________

Time Started ____________ Time Finished ____________ Total Time ____________

Flat Rate Time or Recommended Time ______________________

Specifications:

Clearance specification for the pump being serviced.

End clearance ______________ Side clearance ______________

Others ______________ ______________ ______________

Checkup Questions:

1. Why is the end and side clearance important? ______________________

2. What effect does the spring have on the relief valve? ______________________

3. Why is grease used in assembling the pump? ______________________

DATE COMPLETED ______________ INSTRUCTOR'S SIGNATURE ______________

BEARING MATERIALS AND DESIGN

The bearings used for connecting rods and main bearings consist of two half-circle pieces, usually called a pair. Each piece is a steel backing to which one or more soft metal alloys are applied. This alloy is called the bearing material. Bearing materials are judged on compatibility, load-carrying capacity, conformability, and embedability.

Compatibility is the ability of materials to function, or work, together in harmony. To make a compatible bearing, a metal is normally chosen that is not of the same material as the supported base surface. For example, engineers found that bronze bearing material is more compatible than steel for a steel crankshaft.

Load-carrying capacity is a judgment of different materials according to their strength. Heavy duty bearings usually have a high load-carrying capacity.

Conformability is the ability of a bearing material to adapt, or change shape, in order to conform to journals that are not quite round or smooth.

Embedability is the ability of a bearing material to absorb, become embedded with, minor amounts of foreign matter that circulate in the oil.

The selection of a particular bearing material should be made according to the type of conditions under which the engine will be operated, the condition of the engine parts, and the method of parts reconditioning employed.

Bearing manufacturers make many different alloys for the same engine. Anyone doing major engine repair should be acquainted with the different bearing materials available. This information is usually found in the first pages of a bearing catalog.

The term precision insert is used to describe a type of engine bearings that can be installed without reaming, honing, or grinding (Figs. 10-10 and 10-11). Their

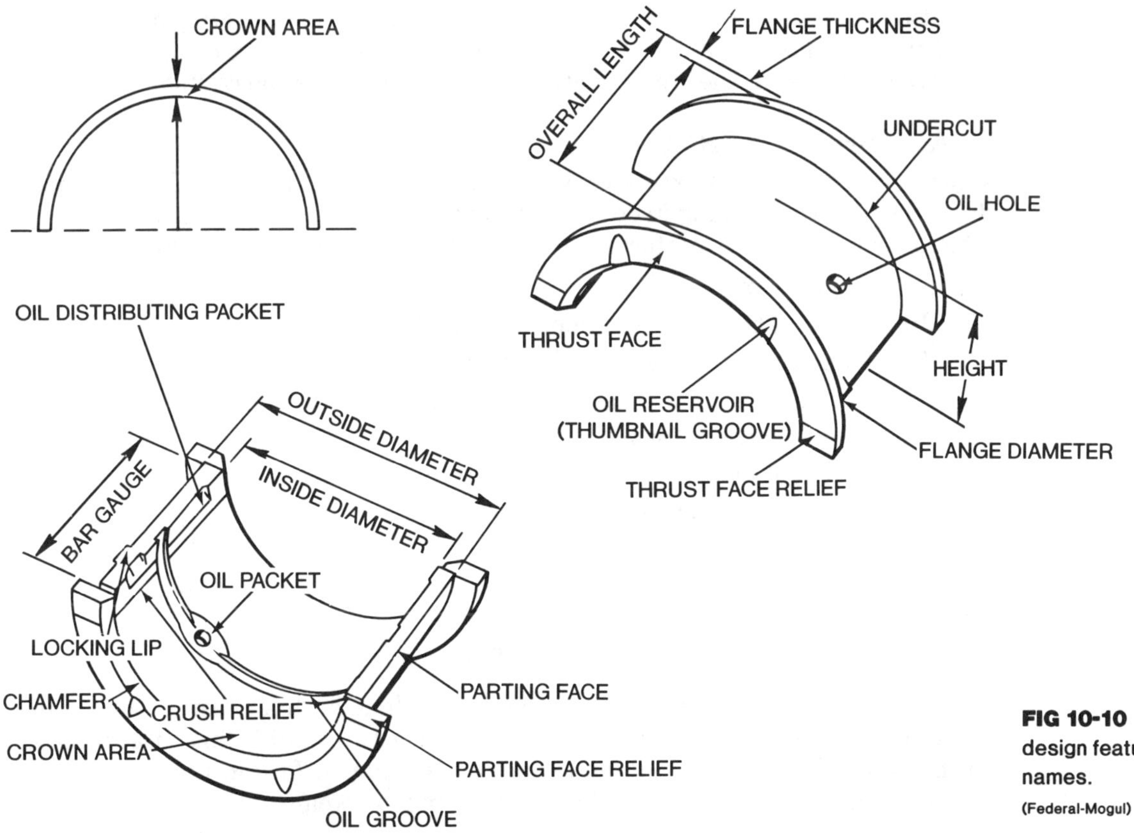

FIG 10-10 Main bearing design features and their names.
(Federal-Mogul)

sizes and materials are manufactured to very precise tolerances so that they exactly fit the engine block and crankshaft. The major portion of a precision insert bearing is a steel piece formed into an arc and then punched or drilled for oil holes. Since the bearing must not rotate in its housing, it is also provided with a locking lip to correspond with a notch in the bore. This locking lip also prevents the bearing from moving laterally during engine operation.

Connecting rod and main bearings are purposely manufactured with *spread.* This is an extra distance across the parting faces of each bearing half (Fig. 10-12) which is greater than the bearing bore diameter and thus requires the bearing to be "snapped" into place. The amount of spread varies between 0.005 in. and 0.030 in. (0.127 mm and 0.762 mm) depending on the thickness, size, and stiffness of the bearing. Bearings have spread to keep them in place and assure a definite fit.

Bearing crush is a term used to describe the additional height of a bearing half (Fig. 10-13). This ensures complete contact of the bearing back when the engine is assembled to the housing bore in the block. It also allows good heat transfer and prevention of bearing rotation. The amount of crush varies, and the engine technician should never modify this extra height.

Other bearing design features include oil grooving (annular) to assist in the distribution of oil in the bearing, chamfering to help the oil and wear particles leave the bearing, and oil holes to meter the oil to cylinder walls or camshaft areas.

CRANKSHAFTS

The crankshaft is the engine part which changes the reciprocating motion of the piston and connecting rod into rotary motion (Fig. 10-14). It is the largest rotating part in the engine.

Crankshafts are designed to rotate at high speeds while subjected to the power impulses from the piston and connecting rod assembly. This rotating mass of metal must turn as smoothly as possible. To do this, crankshafts have counterweights to balance the forces of the cranks so there is no vibration as the crankshaft rotates.

CRANKSHAFT SERVICE

When a crankshaft has excessive wear, it must be reconditioned. This is usually done in shops specializing in crankshaft work or by production engine rebuilders. The following steps are performed to properly recondition a crankshaft:

1. Thorough cleaning
2. Inspection of the front and pilot bearing surfaces
3. Crack detection (magnetic or dye penetrant)
4. Measurement
5. Checking for straightness
6. Rebuilding of journal surfaces if necessary
7. Regrinding of journal and thrust surfaces
8. Polishing
9. Cleaning off grinding materials
10. Coating with protective material (wax)
11. Storing upright (vertical) in a dust free area

The cleaning process (operations 1 and 9) is done with a caustic solution in a device that agitates the solution and protects the crankshaft from contact damage to the journals. The front end and the pilot bearing surface in the center of the flywheel end should be visually checked for irregular wear.

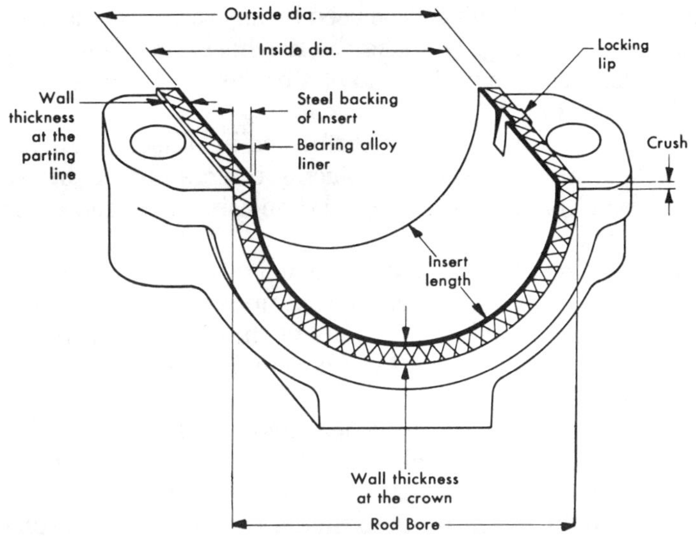

FIG 10-11 Connecting rod bearing design features.
(Sunnen Products)

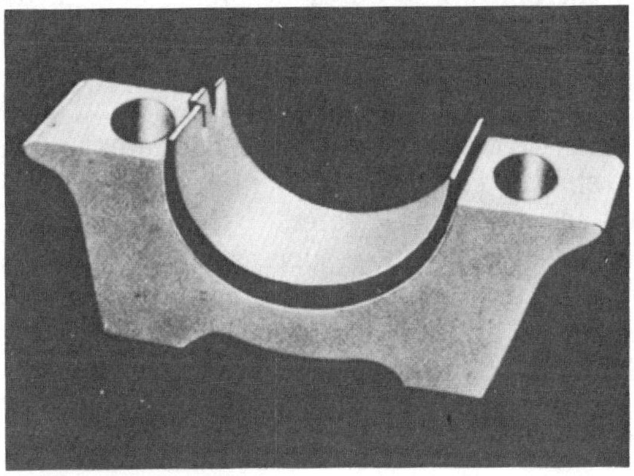

FIG 10-12 Bearing spread.
(Federal-Mogul)

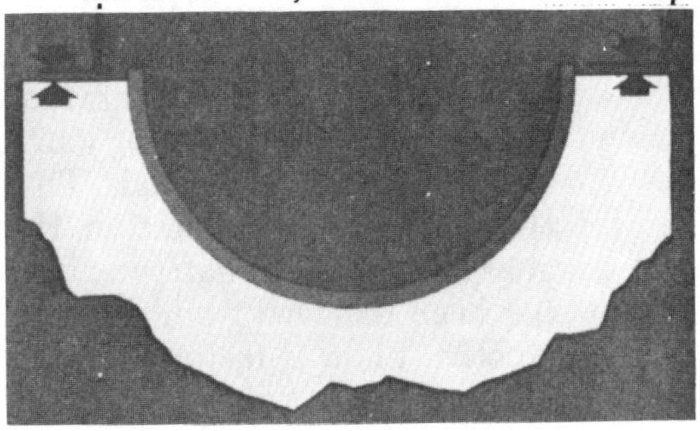

FIG 10-13 Bearing crush.
(Federal-Mogul)

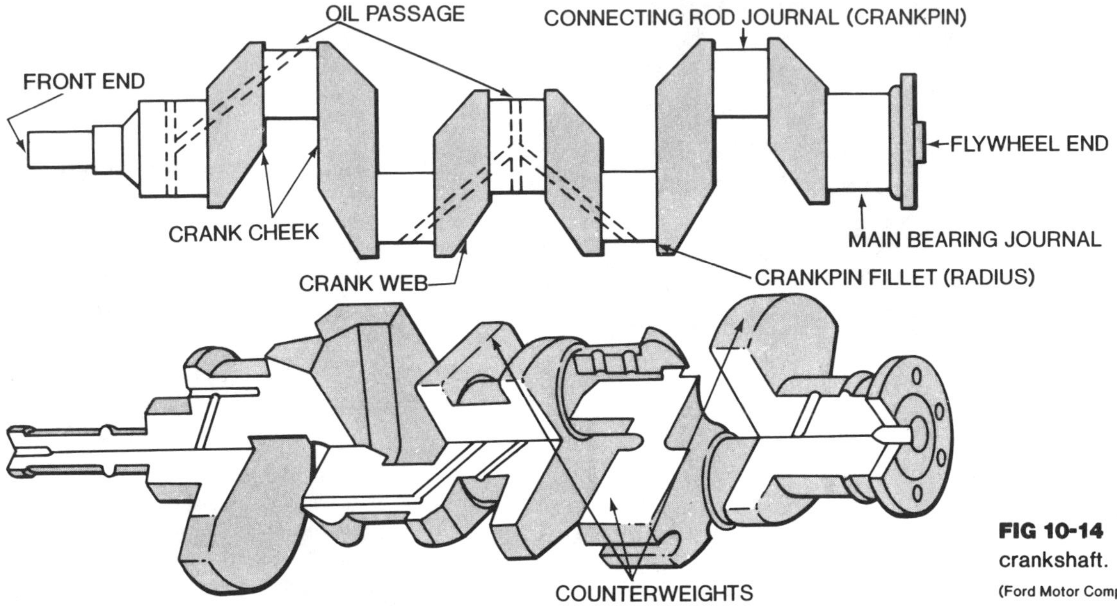

FIG 10-14 Typical crankshaft.
(Ford Motor Company)

Crack detection, the next job, is usually performed on a magnetic machine. A very crude, but often reliable, method that can be used by any technician is to hold the crank up by a piece of wire and lightly strike the flywheel flange with a hard hammer. A cracked crank will have the dull sound of a broken plate, while a good one will produce a clear, ringing tone.

The measurement operation is similar to that performed in Objective 10-A1.

Checking for straightness is accomplished by placing the crankshaft between centers on a lathe or on a V-block on a flat surface and then using a dial indicator to measure the amount and position of deflection.

Rebuilding of the journal surfaces is often accomplished by an electro-weld process. Metal is electrically welded to a distressed journal under inert gas in such a manner that the diameter returns to original size. This welding operation is usually done to only one or two journals, since it is time-consuming and expensive. A replacement shaft can be less costly.

The grinding operations are done on a machine designed specifically for this job (Fig. 10-15). The grinding process removes only enough crankshaft metal to bring the journal(s) to a smooth, even surface. This is done in increments of 0.010 in. for domestic crankshafts and 0.25 mm for imports to match the majority of replacement bearings available. A typical crankshaft will be reground when the journal surface is not within specifications. If the shaft was never before reground, its measurements would be the same as factory specifications. This is referred to as a "standard" shaft, often abbreviated STD.

If any of the journal surfaces, main, or rod bearing have been reground previously, the shaft is considered "undersize." An example of the terminology used by crankshaft reconditioners is a "10-10 shaft." The first number indicates the undersize of the main bearing journals; the second number refers to the undersize of the connecting rod journals. The total amount that can be successfully removed is the subject of some question. However, most crankshaft regrinding shops will not guarantee a 30-30 shaft, one from which 0.030 in. (domestic) or 0.75 mm (import) has been removed from the diameter of both the main and rod bearing journals. Another limiting factor on the amount removed from a shaft is the availability of replacement bearings (Fig. 10-16).

FIG 10-15 Grinding a crankshaft.
(Federal-Mogul)

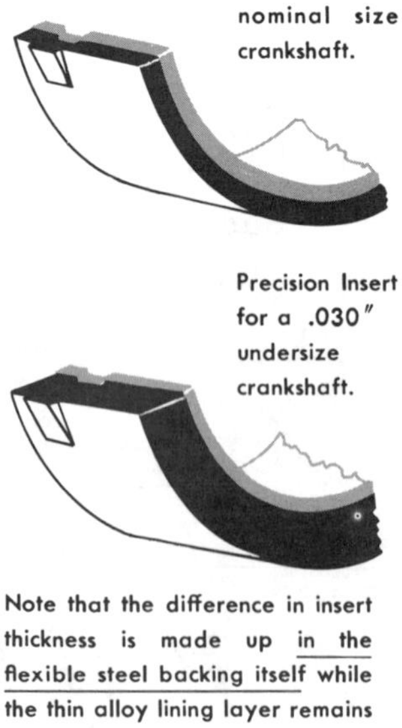

FIG 10-16 Standard and undersize bearing.
(Sunnen Products)

Polishing a crankshaft is done with a long, abrasive motor-driven belt. The shaft should be rotated in the direction it turns within the engine and the polishing belt should be turned so it is pulling the metal away from normal rotation.

All crankshaft reconditioners spray their completed crankshafts with a protective material to prevent rusting. The engine technician should remove this coating just prior to installing the shafts in the block. To remove, wash the crankshaft in clean solvent with a soft brush or rag. Use a soft brush to clean the drilled passages (Fig. 10-17).

When storing a crankshaft, it is recommended that it be set on the flywheel flange in an upright position. This helps maintain the straightness and protects the journal surfaces.

CRANKSHAFT AND BEARING INSTALLATION

Proper crankshaft and bearing installation is necessary if an engine is to have a long, power-producing life. If a crankshaft is finished to correct dimensions with a matching set of replacement bearings, there is usually no problem in bringing about the correct oil clearance.

Two trade-accepted methods of determining bearing oil clearance are given in Objective 10-3A: the Plastigage method and the measurement method.

In the Plastigage method, a soft wax string is compressed between the shaft and the bearing and spreads to a given width. This width is then compared to the clearance dimension stripes on the Plastigage package.

In the measurement method, the bearing insert is installed and the diameter of the opening is measured. The crankshaft is then measured, and this second figure is subtracted from the first. The difference is the bearing's oil clearance.

CROWN WALL THICKNESS

Occasionally, ill-fitting bearings are the result of improper purchase, mishandling by dealers, or incorrectly labeled boxes. This can be avoided if the bearings are measured for wall thickness at the crown before assembling them in the engine.

FIG 10-17 Cleaning the oil holes of a crankshaft.
(Federal-Mogul)

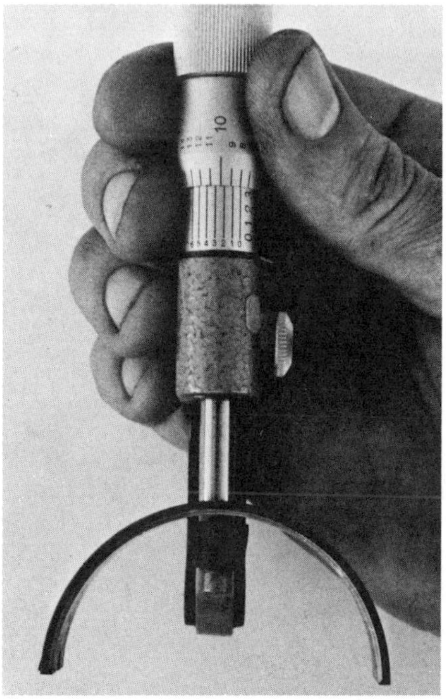

FIG 10-18 Substitute ball anvil micrometer with a steel ball and a piece of hose.

A ball anvil micrometer is necessary to make this measurement because of the arc of the bearing. A substitute ball mike can be made by using a ¼ in. ball (alternator drive-end bearing) and a 3/16 in. (2.40 mm) piece of ⅛ in. hose. Place the ball halfway into the hose and slip the hose over the anvil or spindle of a 1 in. micrometer (Fig. 10-18). When making a reading, be certain to subtract the size of the steel ball bearing from your figures.

To check the bearings, first measure the crown wall thickness. Compare this reading to the figures listed in a bearing manual (Fig. 10-19). Listings in the manual are for "standard" bearings, which fit a crankshaft of original journal diameter. Bearings to fit a 10-10 reground crank would measure 0.005 in. larger than the listed number. Remember, you are measuring only half of the bearing at one time so 2 × 0.005 in. = 0.010 in., the undersize of the crank journal. All bearing halves will measure 0.005 in. more for each 0.010 in. undersize of the crankshaft.

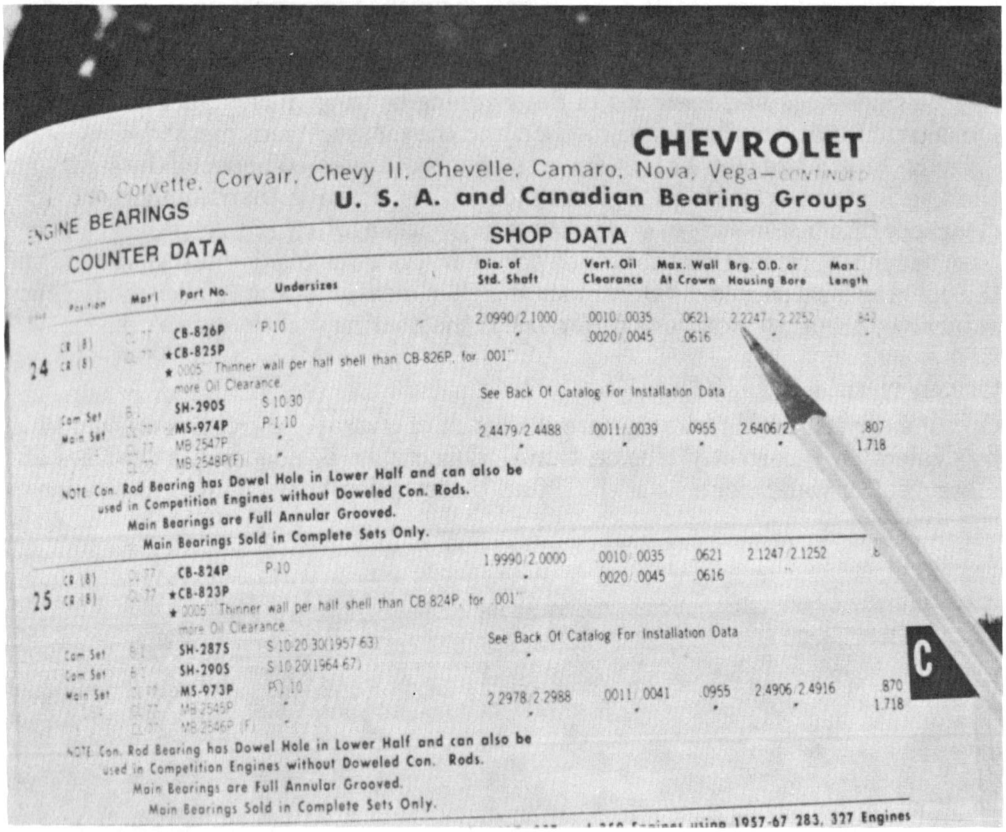

FIG 10-19 Bearing manual specifications.

CRANKSHAFT INSPECTION AND MEASUREMENT

Student Name ______________________ Date ______________

Make of Car ____________ Model ______________ Year ______________

Time Started ____________ Time Finished ____________ Total Time ____________

Flat Rate Time or Recommended Time ______________________

Specifications:

Refer to Job Report Sheet 3-A1.

Main Bearing Journal Diameter ______________________

Connecting Rod Journal Diameter ______________________

Checkup Questions:

1. Maximum main journal wear ______________________

   Maximum main journal taper ______________________

2. Maximum rod journal wear ______________________

   Maximum rod journal taper ______________________

3. Appearance of the front end vibration damper surface and bolt hole threads ____________

4. What type of crack detection was performed? ______________________

   ____

5. Were any other defects noted? ______________________

   ____

   ____

6. What service does the crankshaft need? ______________________

   ____

DATE COMPLETED ______________ INSTRUCTOR'S SIGNATURE ______________

OBJECTIVE 10-A2 (manipulative)

Working as a team with one other student and with the aid of Job Report Sheet 3-A1 and 10-A2, you will clean, inspect, and measure the crankshaft from an instructor-assigned vehicle engine.

Time Recommended:

Twice the flat rate time allowed in *Motors, Chilton,* or *National Service Data* manuals; approximately 40 minutes.

Preparation:

1. Review pages 299 through 312.
2. Study pages 315 through 320.
3. Study manufacturer's or general repair manuals for the engine being serviced.
4. Attend lectures and demonstrations on crankshaft and bearing service.

Tools and Materials Needed:

1. Micrometer
2. Slide hammer
3. Bearing puller

Procedure:

Note: Make a quick visual check of the journal surfaces. If the crank needs grinding, your instructor may want you to do the measurement work on the reground shaft. If the crank looks like it can be used without being reground, continue to step 1.

1. Remove pilot bearing or bushing if the cleaning process will cause them damage (Fig. 10-20).

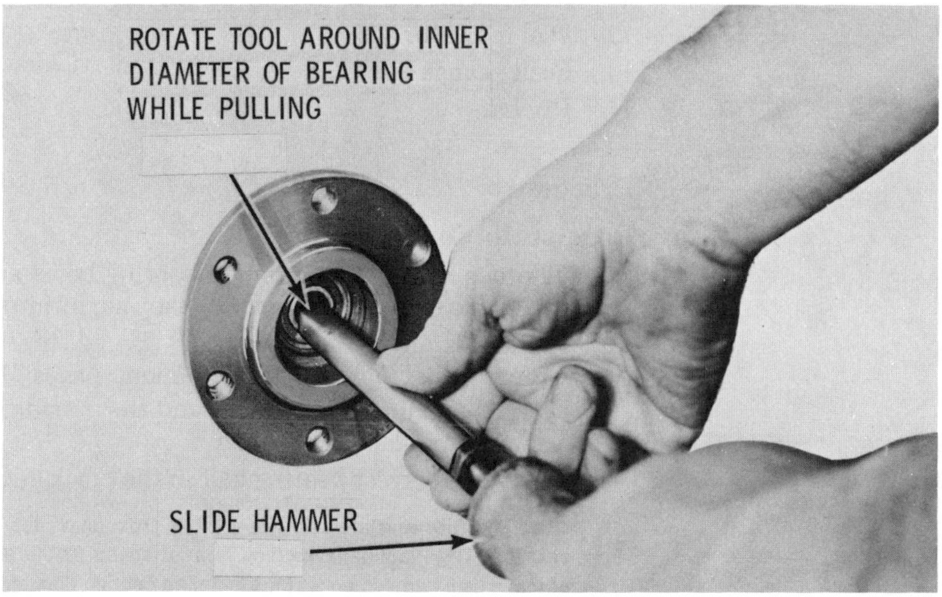

FIG 10-20 Slide hammer and bearing puller.
(Oldsmobile Division, GMC)

OBJECTIVE 10-A3 (manipulative)

Working as a team with one other student and with the aid of Job Report Sheet 10-A3, you will install main bearings and rear crankshaft seal, check bearing clearance and crankshaft end play on an instructor-assigned vehicle engine.

Time Recommended:

Twice the flat rate time allowed in *Motors, Chilton,* or *National Service Data* manuals; approximately 12 minutes per main bearing and seal.

Preparation:

1. Review pages 299 through 315.
2. Study pages 316 through 320.
3. Study manufacturer's or general repair manuals for the engine being serviced.
4. Attend lectures and demonstrations on crankshaft and bearing service.

Tools and Materials Needed:

1. Main bearing clearance specifications (Job Report Sheet 3-A1, Teardown and Assembly)
2. Crankshaft seal
3. File
4. RTV silicone sealer
5. Plastigage
6. Dial bore gauge or micrometer with a telescoping gauge
7. Torque wrench and appropriate sockets
8. Lacquer thinner
9. Grease
10. Soft-face hammer
11. Razor blade
12. Dial indicator
13. Feeler gauge
14. Pry bar

Procedure:

Preinstallation

1. Thoroughly inspect the main bearing bores and mating surfaces where the caps join the block. *Lightly* file any slight imperfections.
2. If there is any doubt about the size of the replacement bearings, refer to Crown Wall Thickness Measurement, pages 311–312.
3. Clean the main bearing bores and the outside surfaces of the bearing with a clean rag and lacquer thinner.
4. Install the correct bearing half in the block and in the caps (Fig. 10-21).

Note: The following are two acceptable procedures for checking the main bearing clearance. The Plastigage method is best suited to the auto mechanic. The second method uses measurements of the installed bearing diameter and crankshaft journal.

Plastigage Method—Clearance Check

1. With all the main bearings installed in the block, lay the *cleaned* crankshaft in position. There should be no oil on the crankshaft or bearings.
2. Cut a piece of Plastigage (green) to the length of each main bearing journal and place it on the journal slightly off center (Fig. 10-22).
3. Place the main bearing cap in position and torque to specified limit. *Do not* turn the crankshaft.
4. Remove the bolts and cap.
5. Compare the width of the flattened Plastigage with the color bands of the package (Fig. 10-23). The clearance should be approximately 0.0015 in. to 0.0025 in. (0.0381 to 0.0635 mm). Refer to Job Report Sheet 3-A1 for correct specifications.

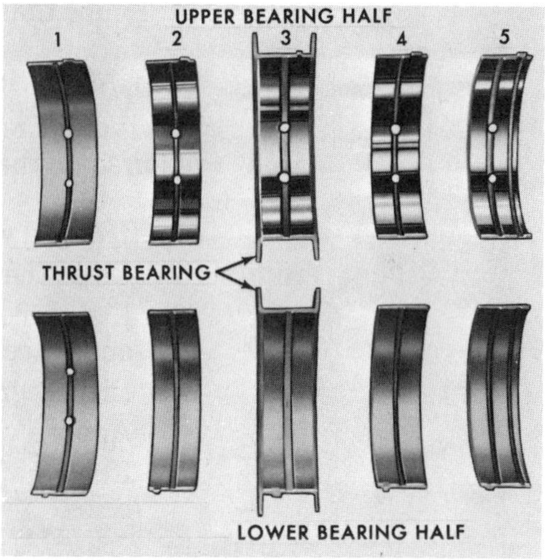

FIG 10-21 Main bearing installation locations.
(Cadillac Motor Car Division, GMC)

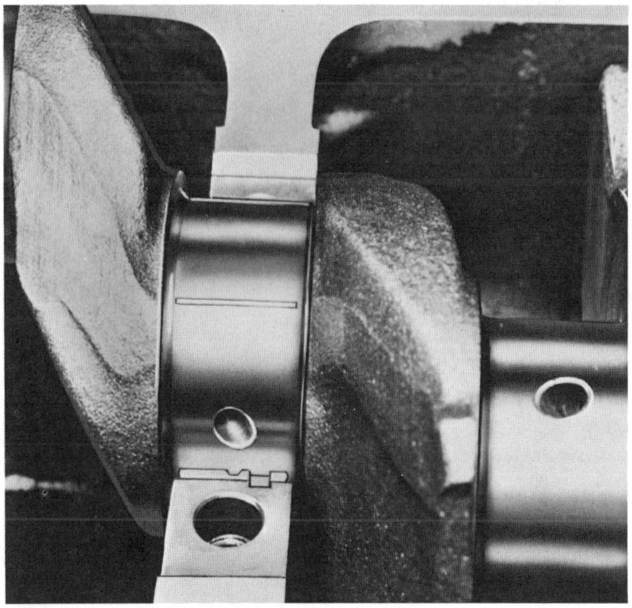

FIG 10-22 Lay Plastigage on the crankshaft journal.
(Oldsmobile Division, GMC)

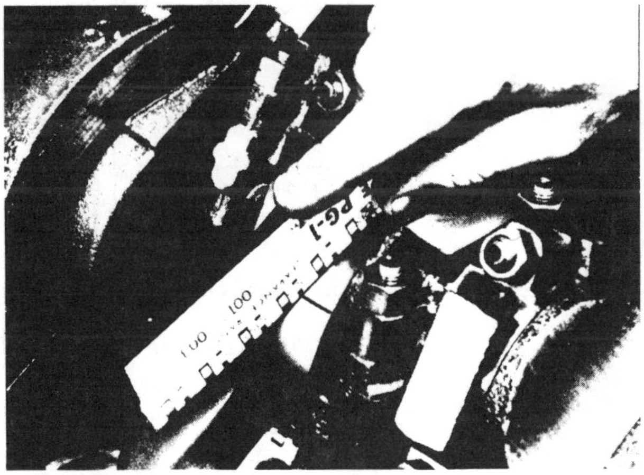

FIG 10-23 Compare the package to the flattened Plastigage.
(Oldsmobile Division, GMC)

6. Check all other main bearings in the same manner before lubing the bearings with grease and installing the rear crankshaft seal.
7. If the clearances are not proper, consult with your instructor.

Measurement Method—Clearance Check

1. Install all bearing caps with their correct bearings and torque to specifications.
2. With a dial bore gauge or a telescoping gauge and micrometer, measure and record the inside diameter of the main bearing bores.
3. Measure the main bearing journals and subtract those figures from the bore measurements to determine the clearance.
4. If the clearances are not proper, consult with your instructor. If they are correct, remove the caps and prepare to install the rear main seal.
5. Lube the main bearings with grease prior to installing the rear main seal.

Rear Main Seal Installation

Two types of rear main seals are commonly used. Follow the directions for the type required on the engine being repaired.

Neoprene Type

1. Clean the grooves in the block and cap (rear) that hold the seal in position.
2. Lightly coat the back (groove) side of the seal halves with RTV silicone sealer and place in the grooves with the lip facing forward (Fig. 10-24).
3. Lightly grease the seal surface touching the crankshaft.
4. Apply a very thin coat of RTV to the rear cap (Fig. 10-25).
5. Lube the journal and bearing with grease and install the cap before the RTV dries.

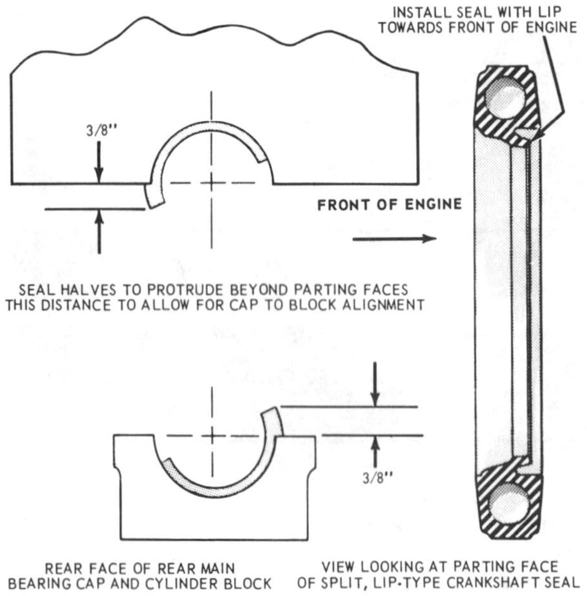

FIG 10-24 Installing rear main seal.
(Ford Motor Company)

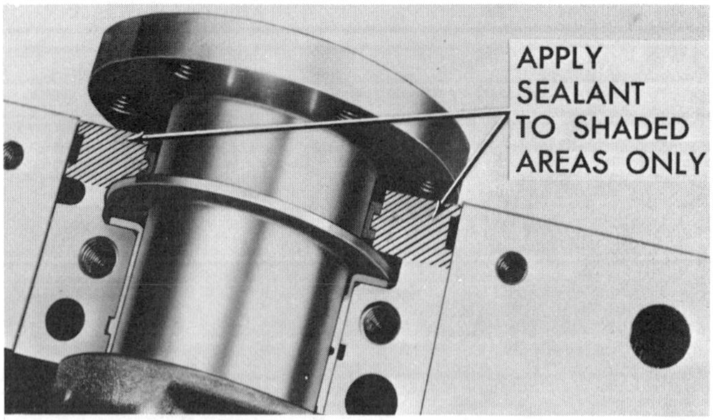

FIG 10-25 Apply a light coat of RTV to the parting surface as shown.
(Chevrolet Motor Division, GMC)

If the rear main bearing is also the thrust bearing, tighten the cap after the crankshaft has been pried forward to align the thrust surfaces and provide the correct thrust clearance. If the rear main is not the thrust bearing, merely tighten to specifications.

Rope Type

1. Remove the rear main bearing inserts from the cap and the block.
2. Clean the grooves in the block and cap where the seals are to be placed.
3. Lay the rope seals edge up in the grooves. Apply pressure by hand to form the seal to the circle of the groove.
4. With a large, round, smooth object (a 1½ in. socket with an extension works well), force the rope into the groove.
5. Place the crankshaft in the block and strike the rear flange with a soft-faced hammer to seat the seal.
6. Remove the crankshaft and cut the seal *flush* at the mating surface with a *sharp* razor blade.

Note: No part of the seal should be above or below the surface (Fig. 10-26).

7. Replace the rear main bearings in the cap and block and apply grease to it and the other bearings if this has not been done.
8. Apply grease to the seal area touching the crankshaft.
9. Replace the crankshaft.
10. Apply a thin layer of RTV silicone sealer to the rear main bearing parting surface of the cap or block and the ends of the rope seal.
11. If the rear main is also the thrust bearing, follow step 5 in the instructions under Neoprene seal.

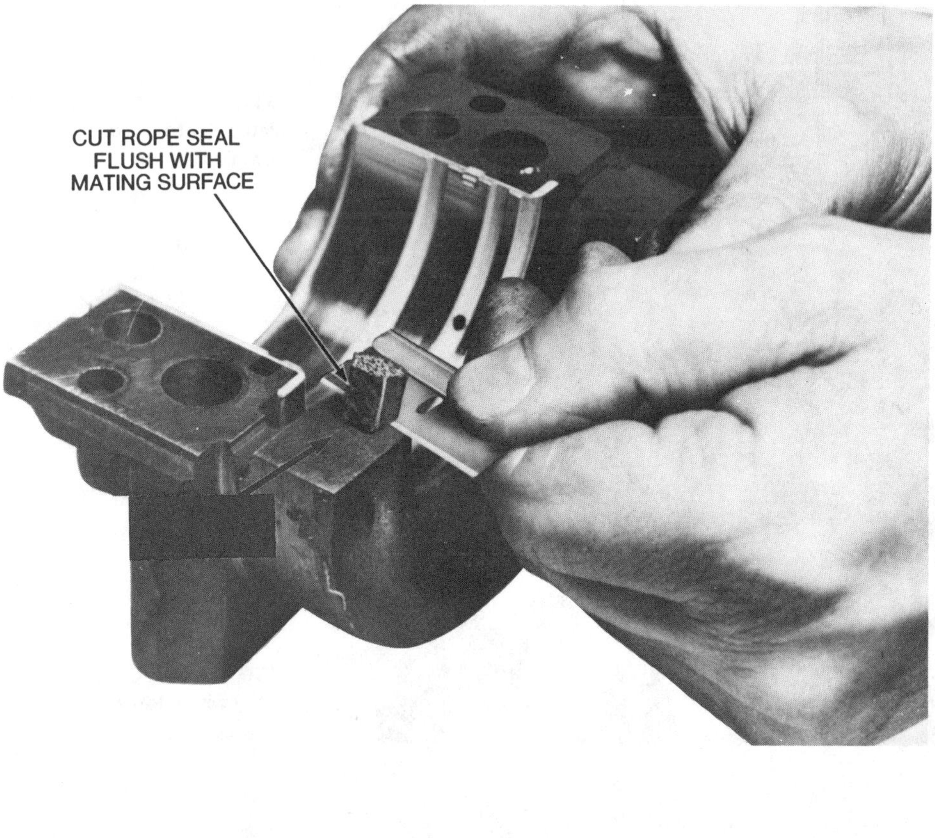

FIG 10-26 Cutting rope type seal flush with the surface. (Oldsmobile Division, GMC)

Crankshaft End Play—Clearance Check

Two methods are used to check crankshaft end play: a dial indicator to measure endwise movement of the crankshaft, and a feeler gauge to measure the gap between the crankshaft and its thrust bearing.

1. With the crankshaft in position and all the main bearing caps torqued, pry forward and backward on the crankshaft with a pry bar.
2. There should be roughly 0.004–0.008 in. (0.102–0.204 mm) movement.
3. Measure this movement with a dial indicator (Fig. 10-27) or a feeler gauge (Fig. 10-28).
4. If the movement is insufficient, loosen the thrust main bearing and pry forward to align it. Retighten and check again. If the clearance is still too small or too great, notify your instructor.
5. The engine should now be ready for the installation of the piston rod assemblies.

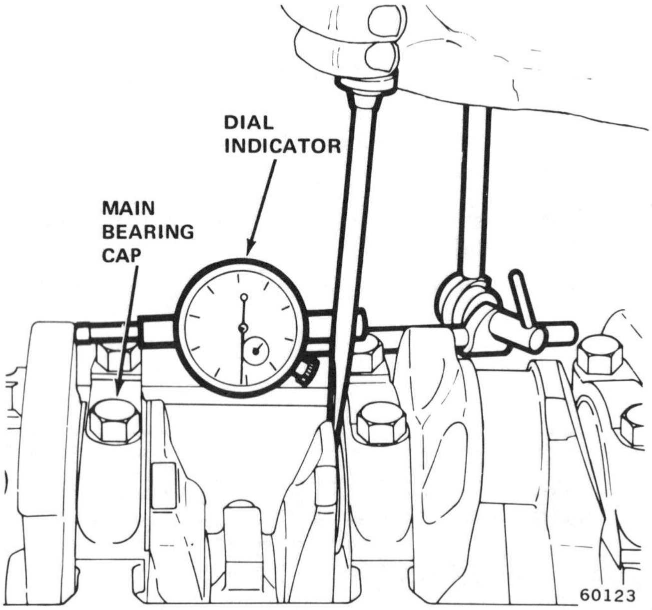

FIG 10-27 Using a dial indicator to check the crankshaft end play. (American Motors Corporation)

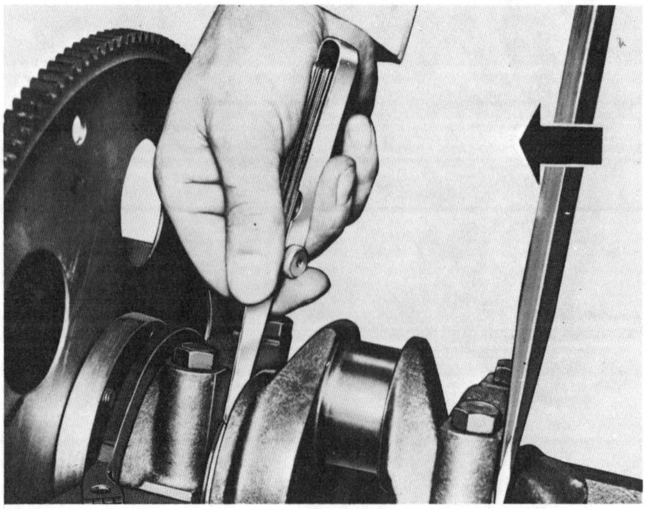

FIG 10-28 Checking the end play with a feeler gauge. (Oldsmobile Division, GMC)

MAIN BEARING AND SEAL INSTALLATION

Student Name ______________________ Date ______________

Make of Car ____________ Model ______________ Year ______________

Time Started ____________ Time Finished ____________ Total Time ____________

Flat Rate Time or Recommended Time ______________________

Specifications:

See Job Report Sheet 3-A1

1. Main bearing oil clearance

2. Crankshaft end play

Checkup Questions:

1. Why is it important to clean main bearing bores and outside surfaces of the bearings? ____________

2. Why do some main bearings have grooved upper halves? ____________

3. What color Plastigage is used to check bearings? ____________

4. Why should the rear main cap have RTV applied to the mating surfaces? ____________

5. Why is it necessary to lubricate the seal surfaces that contact the crankshaft? ____________

6. What is the most accurate method of measuring crankshaft end play? ____________

DATE COMPLETED ______________ INSTRUCTOR'S SIGNATURE ______________

FLYWHEELS

The type of flywheel found at the end of a crankshaft depends upon the type of transmission used. A flywheel from an engine with a standard transmission is a heavily weighted cast iron plate to which the clutch disc and pressure plate are attached. The clutch side of this part provides one-half of the friction surface which engages the clutch to make the vehicle move. The typical flywheel (sometimes called a drive plate) from an engine with an automatic transmission is a lightweight piece used to couple the torque converter to the engine.

The flywheel (and converter) serves to smooth out the power impulses of the crankshaft. Each flywheel is dynamically balanced, often with the crankshaft, to reduce unwanted vibration.

The ring gear common to most flywheels is located at the outer edge and is used with the starter motor to crank the engine. Service work on the flywheel consists of inspecting the ring gear for damage from the meshing of gear teeth of the engaging starter. It is possible to replace the ring gear alone on some flywheels. When an old one is cut off, the near gear is heated uniformly to 350°F, then slid into position on the cool flywheel.

The clutch type flywheel friction surface should be inspected and resurfaced if it is scored or grooved more than 0.005 (0.127 mm). Refinishing should also be done if there are excessive heat checks (cracks) or clutch surface runout of more than 0.005 in. (0.127 mm). Resurfacing is usually done by grinding.

VIBRATION DAMPERS

During engine operation, power impulses from the various power strokes in the cylinders apply a torsional twist to the crankshaft. As this happens, a harmonic frequency develops, causing harmful metal fatigue to the metal molecules. These molecules attempt to hold the crankshaft in one piece; however, if the twist and vibrations become severe and continue for a long period, the crankshaft can fail.

The amount of torsional twist and vibration a crankshaft experiences may be significantly reduced with a vibration damper. Since the flywheel acts, in part, to smooth out the crankshaft operation, the vibration damper is installed at the front of most engines. The damper is a small cast iron flywheel mounted in rubber. It is carefully balanced at the factory and should be treated with care, especially during its removal and replacement.

Name ______________________________ Date ______________

DIRECTIONS:

Circle the best answer to each question.

1. A modern engine relies upon lubrication
   I. supplied to its bearings under pressure.
   II. that is splashed by crankshaft agitation.
   A. I only B. II only C. Both I and II D. Neither I nor II
2. As bearing clearance increases
   I. the oil bypass in the pump picks up more oil.
   II. oil throwoff increases and some bearing may run dry.
   A. I only B. II only C. Both I and II D. Neither I nor II
3. Oil pumps
   I. are driven by the crankshaft.
   II. have some type of pressure relief valve.
   A. I only B. II only C. Both I and II D. Neither I nor II
4. Oil filter systems are classified in two groups
   I. the full flow and the bypass.
   II. the spin-on and the add-on.
   A. I only B. II only C. Both I and II D. Neither I nor II
5. The path of oil flow through a filter is
   I. top to bottom.
   II. outside to the center.
   A. I only B. II only C. Both I and II D. Neither I nor II
6. Oil pumps should be visually inspected and
   I. cleaned of any foreign matter.
   II. measured for wear.
   A. I only B. II only C. Both I and II D. Neither I nor II
7. Engine bearings are most often
   I. replaced rather than reconditioned.
   II. the precision insert type.
   A. I only B. II only C. Both I and II D. Neither I nor II
8. Engine bearings are designed
   I. with an extra distance across the parting faces called "crush."
   II. with a locking lip to allow additional oil flow.
   A. I only B. II only C. Both I and II D. Neither I nor II
9. A bearing with thrust surfaces would be a
   I. camshaft bearing.
   II. connecting rod bearing.
   A. I only B. II only C. Both I and II D. Neither I nor II

10. The common bearing backing metal is
    I. bronze.
    II. steel.
    A. I only   B. II only   C. Both I and II   D. Neither I nor II
11. Crankshafts have counterweights to
    I. provide bearing surfaces.
    II. offset the weights of the connecting rods and pistons during its rotation.
    A. I only   B. II only   C. Both I and II   D. Neither I nor II
12. Crankshaft service includes
    I. crack detection and grinding of the journal surfaces.
    II. heat treating.
    A. I only   B. II only   C. Both I and II   D. Neither I nor II
13. A crankshaft that is "10-20"
    I. has 0.010 in. oversize main bearing journals.
    II. has 0.020 in. oversize connecting rod journals.
    A. I only   B. II only   C. Both I and II   D. Neither I nor II
14. Plastigage is a soft wax string used to
    I. measure bearing spread.
    II. measure crankshaft end play or end clearance.
    A. I only   B. II only   C. Both I and II   D. Neither I nor II
15. The crown wall thickness of a modern bearing
    I. is ½ the difference in diameter of the housing bore and the shaft journal diameter plus ½ the oil clearance.
    II. is the thickness of a bearing measured halfway between the ends.
    A. I only   B. II only   C. Both I and II   D. Neither I nor II
16. When installing the Neoprene seal,
    I. coat the threads of the cap bolts with RTV to prevent loosening.
    II. lightly grease the seal where it makes contact with the crankshaft.
    A. I only   B. II only   C. Both I and II   D. Neither I nor II
17. Rope rear main seals
    I. must be cut flush with the mating surface of the block and cap.
    II. can be reused if they were carefully removed during engine disassembly.
    A. I only   B. II only   C. Both I and II   D. Neither I nor II
18. Which of the following tools can be used to measure crankshaft end play?
    I. Plastigage
    II. Dial bore gauge
    A. I only   B. II only   C. Both I and II   D. Neither I nor II
19. A flywheel
    I. for an automatic transmission car has the starter ring gear mounted on it.
    II. carries the clutch surface for the torque converter.
    A. I only   B. II only   C. Both I and II   D. Neither I nor II
20. A vibration damper
    I. absorbs torsional twist of the crankshaft.
    II. is mounted on the front of the crankshaft.
    A. I only   B. II only   C. Both I and II   D. Neither I nor II

DATE COMPLETED ____________ INSTRUCTOR'S SIGNATURE ____________________

unit 11

Piston and Connecting Rod Service

The piston and connecting rod assembly are presented in this unit. Operation fundamentals and complete service techniques are given in detailed objective procedures. The last portion of this unit includes the installation of the piston assembly in the cylinder block.

OBJECTIVE 11-A (written test)

After completing the preparation assignment, you will correctly answer a minimum of 14 out of 20 multiple choice questions in a written test, without reference material, on the following topics:

1. Pistons and their service
2. Piston pins and their service
3. Piston rings
4. Connecting rods and their service
5. Piston-connecting rod assembly installation

Preparation:

Study pages 327 through 365.

PISTONS

The piston is the engine part that is the heart of the power development. Along with the piston rings, it serves as the movable portion of the combustion chamber. This allows the expansion of the burning gases to push against the connecting rod and eventually to turn the crankshaft. To do its job properly, it must hold the piston rings against the cylinder wall while it moves smoothly back and forth.

PISTON NOMENCLATURE

The following terms are used to describe the design features of a common pair of pistons (Figs. 11-1 and 11-2). The majority of these terms have been adopted and approved by the Society of Automotive Engineers.

1. Land—that part of the piston above the top ring or between ring grooves. The lands confine and support the piston rings in their grooves.
2. Heat dam—a narrow groove cut in the top land of some pistons to reduce heat flow to the top ring groove. This groove fills with carbon during engine operation and reduces heat flow to the top ring.

3. Compression distance (or height)—the distance from the center of the pin hole to the top of the piston.
4. Ring belt—that area between the top of the piston and the pin hole where grooves are machined for the installation of piston rings.
5. Piston head—the top piston surface against which the combustion gases exert pressure. The piston head may be flat, concave, convex, or of irregular shape.
6. Piston pins (wrist pins)—connections between the upper end of the connecting rod and the piston. Pins may be held in one of three ways:
   a. Anchored in the piston with the bushing in the upper end of the connecting rod oscillating on the pin.
   b. Pressed (clamped) in the rod with the pin oscillating in the piston.
   c. Full floating in both connecting rod and piston with lock rings or other devices preventing the pin from contacting the cylinder wall.
7. Skirt—that part of the piston located between the first ring groove above the pin hole and the bottom (open end) of the piston. The skirt forms a bearing surface in contact with the cylinder wall.
8. Pin hole (bore)—an opening through the piston skirt to carry the piston pin.

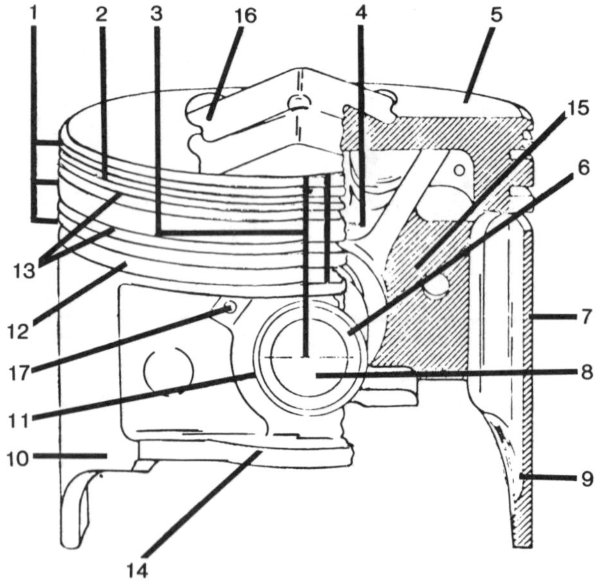

FIG 11-1 Slipper skirt piston.

(Dana Corporation)

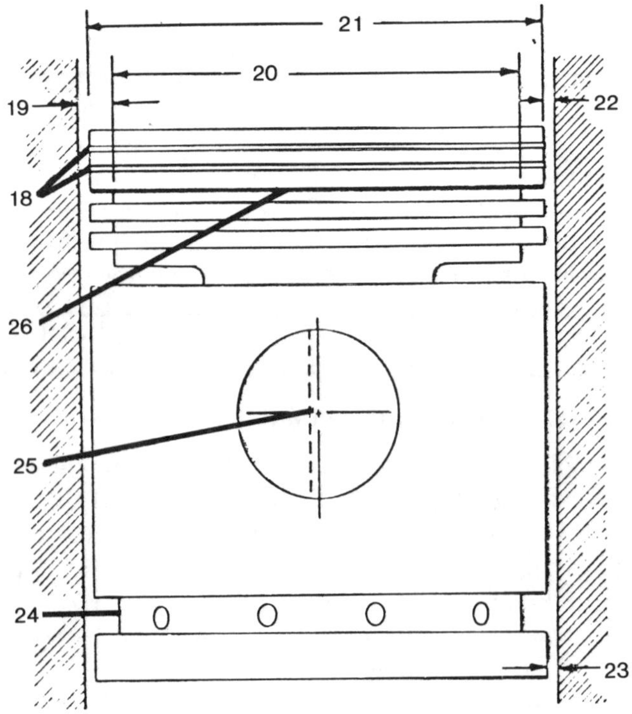

FIG 11-2 Full skirt piston.

(Dana Corporation)

9. Major thrust side—that portion of the piston skirt which carries the greatest thrust load. This is on the left side when viewing the engine from the front end with the crankshaft rotating clockwise.
10. Minor thrust side—that portion of the piston skirt which is opposite the major thrust face.
11. Piston pin bushing—a bushing fitted between piston pin and piston pin hole in some pistons to obtain a better bearing material. It is used rarely today, and only with iron pistons.
12. Oil ring groove—a groove cut into the piston around its circumference, at the bottom of the ring belt or at the lower end of the piston skirt. Oil ring grooves are usually wider than compression ring grooves and generally have holes or slots through the bottom of the groove for oil drainage to the interior of the piston.
13. Compression ring groove—a groove cut into the piston around its circumference, in the upper part of the ring belt. The depth of groove varies according to piston size and types of rings used.
14. Balance pads—extra aluminum generally cast on or below the wrist pin bosses. This material is machined off as necessary to "balance" the piston to a specified weight.
15. Struts—stamped, sheet steel reinforcements cast into the piston between the thrust faces on the inside of the wrist pin bosses. Their function is to control expansion and add strength.
16. Valve relief—a cast or machined depression in the head of the piston. This assures clearance between the piston and valve when the piston is at the top of its stroke and the valve is open.
17. Oil holes to pins—holes drilled from the outside or bottom of the wrist pin bosses to the inside of the pin hole which permit the entrance of oil for lubrication between the wrist pin and the wrist pin bosses.
18. Scuff band—a narrow, raised band or bands machined on the top land of some pistons to reduce the possibility of scuffing.
19. Groove depth—the distance from the cylinder wall to the bottom of the ring groove with the piston centered in the cylinder.
20. Groove root diameter—diameter of the piston measured at the bottom of the groove. The root diameter of each groove on any piston may differ, depending on the type of ring to be installed.
21. Land diameter—diameter of the land being measured. In some piston designs, all lands are of equal diameter; in others they increase from top to bottom.
22. Land clearance—the difference between the diameter of the land and cylinder diameter. (22 in Figure 11-2 shows ½ of total clearance)
23. Skirt clearance—the difference between piston skirt diameter (measured in a plane perpendicular to the piston pin) and cylinder diameter. (23 in Figure 11-2 shows ½ of total clearance)
24. Skirt groove—a groove cut into the piston around its circumference below the pin hole to carry an oil ring.
25. Offset pin hole—on most pistons, the pin hole is offset to the major thrust side of the piston centerline, approximately 0.060 in. (1.52 mm)
26. Top groove spacer—Badly worn top grooves must be remachined before installation of new rings. A steel spacer is installed above the ring in a reconditioned groove to reduce the side clearance to the recommended dimension.

27. Piston skirt taper—the difference between the diameter of the piston at the top of the skirt and at the bottom of the skirt with the diameters being measured in the thrust direction (not shown).
28. Piston cam—the circumferential shape in which a piston skirt is manufactured to provide proper cylinder contact and running clearance under all normal conditions of temperature and load (not shown).
29. Cast-in groove insert—a steel or cast iron insert, in an aluminum piston, either chemically or mechanically bonded during the manufacturing process, into which the top groove is machined to provide a longer wearing surface for the top ring (not shown).

Pistons in the modern engine are generally of the slipper design. There are other types used in engines designed for specific purposes. However, the slipper type has proven its superiority for passenger cars and light trucks through years of engine development. There are some variations in piston head design (Fig. 11-3).

Some pistons have steel alloy struts to improve their strength and control their expansion, but the majority of the material from which a piston is made is aluminum alloy. Aluminum alloy is used because it is a good conductor of heat and is lightweight. Combustion chamber temperatures can run as high as 4000°F. (2204°C.). The aluminum piston gives up this heat quite freely when it moves to the cooler part of the cylinder (BDC) and when the lubricating oil splashes against its interior side. Since it is lightweight, the piston can move back and forth more easily, allowing the engine to run at high speeds.

Pistons are further classified as to how they are manufactured. Cast pistons are made by pouring molten aluminum alloy into a mold and letting it solidify. The machining operations are then performed. Forged pistons are made by placing a slug (piece) of aluminum alloy between a male and a female set of dies. Extreme hydraulic pressure is used to forge the slug into the shape of a piston.

Most original equipment pistons are cast, but forged pistons are used in high performance engines. If the interior surface of a piston is very smooth and even, it is probably forged. An uneven surface with sharp, irregular edges is probably cast.

All automotive pistons are cam ground or cam turned. This is a machining operation which produces an oval piston skirt. This shape helps control the expansion of the aluminum alloy so that it closely fits the cylinder whether it is cold or warm. Before cam grinding was done, an aluminum piston had to be fitted rather loose in the cylinder, since the expansion rate of aluminum is roughly twice that of the cylinder. A piston that had the correct clearance at normal operating temperature had to be fitted so loose that it was very noisy during cold operating periods.

A further development of cam-ground pistons is the barrel skirt. This design has a cam-ground profile with its largest diameter at the bottom of the skirt. Because the lower skirt is not exposed to extremes in heat, it does not expand as much. For this reason, it can be fitted with as little as 0.0005 in. (0.0127 mm) clearance when cold.

PISTON REGROOVING

Several surfaces of every piston experience some wear. The ring grooves and the skirt are the parts of a piston that should be measured and serviced if necessary during every major reconditioning job.

As the engine operates, the top ring groove is subjected to high heat, shocking power impulses, and the expansion and contraction movements of the ring with little or no lubrication. All of these tend to wear the groove surfaces so they cannot hold the compression ring correctly against the cylinder wall. When this happens, there is compression loss and loss of control of the crankcase oil. The general trade-accepted wear limit for ring grooves is 0.006 in. (0.1524 mm) beyond the original groove width. The amount of wear should be checked at several places around the groove with a 0.006 in. (0.152 mm) feeler gauge and a twist drill shank of the correct size, or a ring groove wear gauge that is 0.006 in. larger than the stated width of the gauge (Fig. 11-4).

Assuming the piston has no irregularities that would prevent its reuse, the ring grooves with excess wear can be machined to an oversize width, usually 0.024 in. (0.610 mm) or 0.030 in. (0.762 mm). A steel spacer of that thickness is then installed *above* the new ring to fill the space removed.

If regrooving is to be completed, consult your instructor for specific operating instructions.

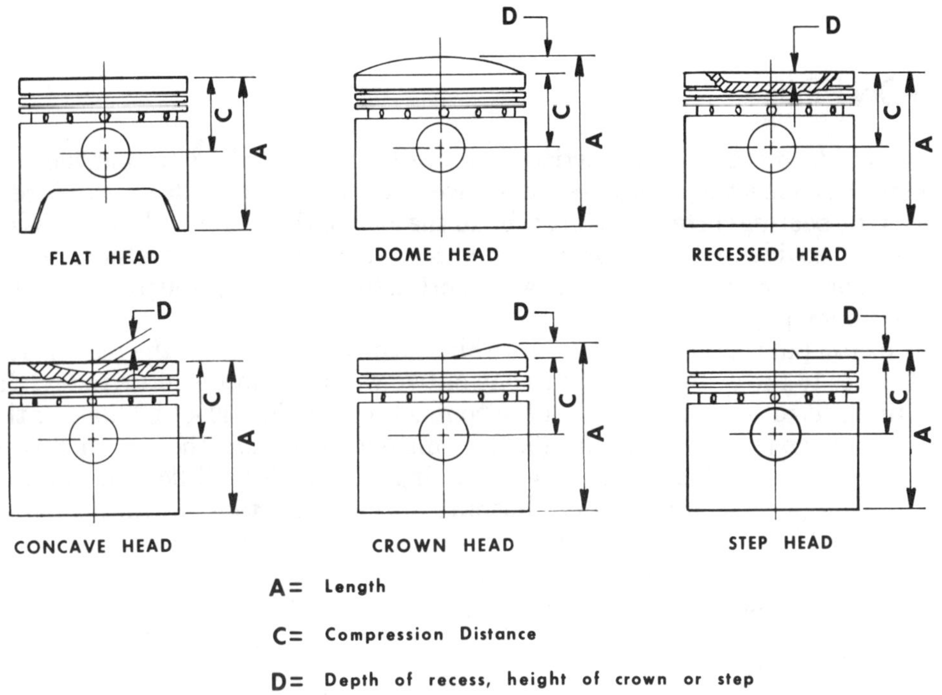

FIG 11-3 Common piston head designs.
(Dana Corporation)

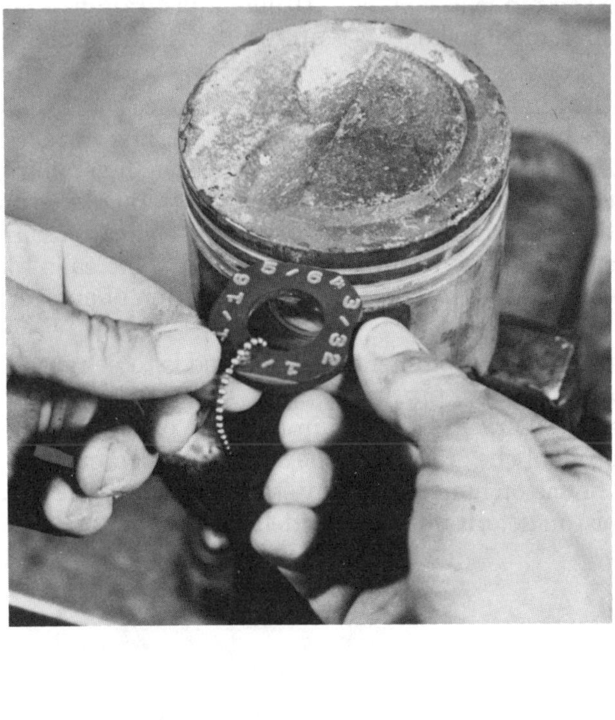

FIG 11-4 Piston ring groove wear gauge.

PISTON KNURLING

When piston skirts are worn less than 0.010 in. (0.254 mm) and the piston is to be reused, a machining operation known as knurling can be performed. Knurling is a means of reforming piston skirt material above the original surface (Fig. 11-5).

A ribbed wheel of hardened steel is pushed against either or both thrust surfaces of the piston. As the ribs of the wheel form grooves, they also push the surrounding metal upward. This raised metal increases the piston size to its original diameter or even larger. Excess knurling can be removed with a sanding block, using wet or dry abrasive paper.

Only general information on the operation of piston knurling equipment will be discussed in Objective 11-A1 because of the variety of machines in use today.

PISTON PINS

Piston pins (sometimes called wrist pins) are hard steel cylinders that join the piston to the connecting rod. These pins have a very smooth outside finish in order to act as the bearing surface that absorbs all the load of the power and compression strokes. To provide the exact size and smooth surface, each pin is carefully ground and then chromium plated. The pin is then fitted to the piston through a process of precision honing or reaming.

The piston pin is designed to move within the piston pin hole of the piston. The pin attaches to the connecting rod in one of two ways. In most engines today, the pin is press-fit into the small end of the connecting rod (Fig. 11-6) by making the rod hole approximately 0.001 in. (0.0254 mm) smaller than the pin diameter. Another means of joining the pin and connecting rod is with a full-floating pin (Fig. 11-7). This type of fit has a bronze bushing in the connecting rod that is honed 0.0003 in. to 0.0005 in. (0.0076 mm to 0.0127 mm) larger than the pin. This allows the pin to oscillate in the rod as well as the piston. Lock rings are used with the full-floating pin fit to keep the pin from rubbing the cylinder walls.

PISTON PIN SERVICE

"Fitting the pin" is providing the correct clearance (+ or −) in the piston and connecting rod openings where the pin is assembled. Although piston pins were fitted for many years with a reamer, modern engines require a piston and connecting rod with a precision fit.

During normal engine wear, the pin and its openings will experience some wear in the form of increased clearance between the piston pin and its bearing surfaces. Determining the amount of pin wear is almost impossible without disassembling the piston from the rod. Careful measurements must be taken at several places on the pin and in the piston pin bore. Compare these readings to specifications. In most cases of excessive pin wear, the piston bores are also worn. This calls for replacement of the piston, which comes with a new pin, or installation of an oversize pin.

If the press-fit pin shows wear beyond recommended limits and is to be reused, an oversize pin must be fitted to the rod and piston.

In the full-floating pin, excessive wear can be found in both the piston and the connecting rod. This wear will probably be greatest in the connecting rod bushing, which can be replaced and then precision honed to fit any standard or oversize pin. If there is excessive wear in the piston hole, it must be decided whether or not to use the piston.

An oversize pin is required if the old piston is to be reused. If a new piston is to be used, the pin should be the standard size, as originally equipped. Piston pins are available, separately, in standard, 0.003 in., 0.005 in., and 0.010 in. oversize.

FITTING PISTON PINS WITH A PRECISION HONE

Fitting piston pins means honing or reaming the inside surface of the piston bore and the small end of the connecting rod to the proper clearance specification. This process requires very careful measurements and exact use of equipment to obtain the correct fit. Improper fitting of a pin results, first, in a noisy engine and, later, in engine failure. The following are typical pin honing jobs performed on a precision hone:

a. Honing piston pin bores to oversize pins
b. Honing connecting rod bores (small end) to oversize pins (the rod may be a press-fit or full-floating type)
c. Honing connecting rod bores after the bronze bushing in a full-floating fit has been replaced (bushing replacement is described in the connecting rod service section of this unit)

FIG 11-5 Skirt pattern produced by the knurling process.
(Dana Corporation)

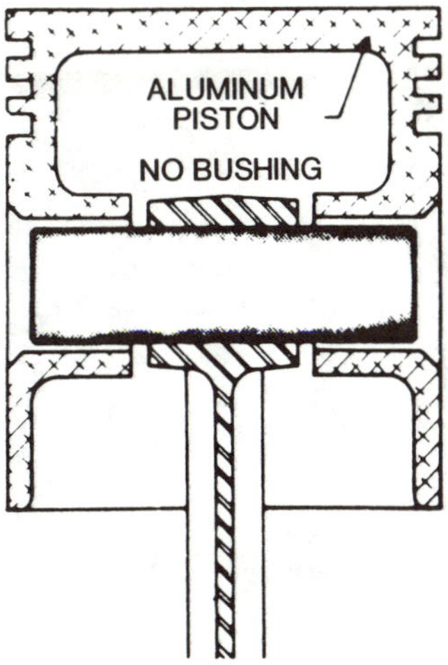

FIG 11-6 Press fit piston pin.
(Sunnen Products)

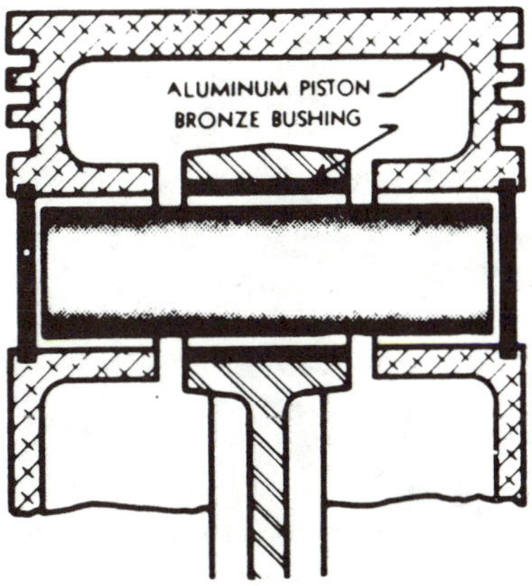

FIG 11-7 Full floating piston pin. (Sunnen Products)

PISTON-CONNECTING ROD DISASSEMBLY

Disassembly of the piston from the press-fit connecting rod normally requires a press and piston pin fixture (Fig. 11-8) to prevent damage to the piston or its pin. To determine the correct pin press tooling, locate the engine's size and year in the setup chart. Select the tooling and have your instructor assist you in removing and later replacing the pins. A press lube should always be used to prevent scoring or galling (severe scratching-gouging) of the pin bore or rod surfaces.

Full floating pins can be removed by removing both lock rings and pushing the pin out by hand. If a pin resists movement, the cause is usually a ring of oil sludge on the pin not covered by the rod or piston during normal operation. Remove this ring of sludge with a fast dissolving solvent such as lacquer thinner.

CONNECTING ROD BUSHINGS

In the full-floating pin fit, the small end of the connecting rod contains a bronze bushing. Bronze is an excellent bearing material for this location; however, it is somewhat soft and subject to wear and distortion over long periods of engine use. The following are service procedures required to renew the rod bushing:

a. Remove the old bushing (press-out).
b. Install the new bushing (press-in).
c. Expand the new bushing to the rod "eye."

The press work should be done with a bushing driver of the correct size. This will prevent damage to the rod opening when removing the old bushing and protect the new bushing from damage upon installation. Expanding the new bushing is necessary to prevent it from loosening and to assure good contact between the two metals. This can be done with a swaging tool or an expansion mandrel on the precision hone (Fig. 11-9).

FIG 11-8 Press fit piston pin removal. (Sunnen Products)

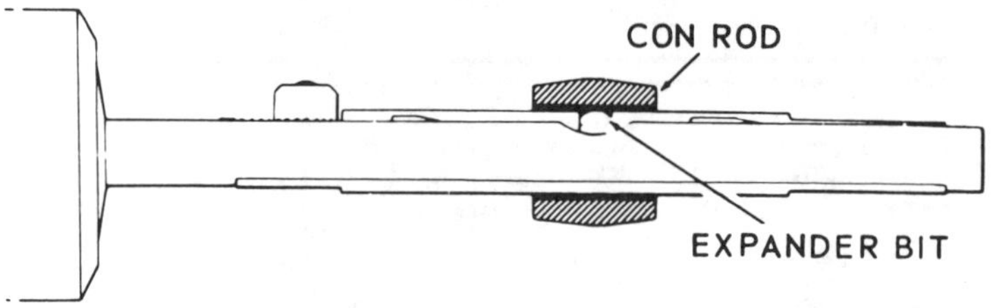

FIG 11-9 Expanding mandrel for seating pin bushing. (Sunnen Products)

FIG 11-10 Assorted piston pin press fixtures. (Sunnen Products)

OBJECTIVE 11-A1
(manipulative)
PISTON SERVICE

Working as a team with one other student and with the aid of Job Report Sheets 3-A1 and 11-A1, you will perform the following operations on 4 pistons from an instructor-assigned vehicle engine:

1. Clean and disassemble the piston and connecting rod.
2. Measure all dimensions.
3. Knurl and regroove if necessary.

Time Recommended:

Twice the flat rate time allowed in *Motors, Chilton,* or *National Service Data* manuals, or instructor's estimate.

Preparation:

1. Review pages 327 through 334.
2. Study pages 335 through 338.
3. Study manufacturer's or general repair manuals for the engine being serviced.
4. Attend lectures and demonstrations on piston and connecting rod service.

Tools and Materials Needed:

1. Cleaning materials or equipment, depending upon availability
2. Piston pin removal fixtures and/or press
3. Micrometer and telescoping gauge set
4. Precision gauge
5. Knurling equipment
6. Regrooving tools

Procedure:

1. Clean the piston surfaces with solvent to remove the oil.
2. Check for cracks or other irregularities on the pistons that would prevent their reuse.
3. Number the top of each piston according to its cylinder placement with a steel stamp or vibration marker.
4. Mark the beam of each rod by scratching the surface to indicate the front of the engine.
5. Remove the pistons from the connecting rods. For press-fit pins, a press with fixtures for that style of piston or a special pin puller is required (Figs. 11-10 and 11-11). Seek your instructor's help in completing this operation. For full-floating pins, remove both lock rings and push the pin out by hand. If the pin resists, apply a small amount of lacquer thinner to the ring of oil sludge on the pin.
6. Each piston and its pin should now be measured in the following places and compared to specifications:
   a. Maximum diameter (Fig. 11-12). Use an outside micrometer at 90° from the piston ring on the lower skirt to find the maximum diameter.
   b. Ring grooves (usually top groove only). Measure grooves with a feeler gauge and drill shank or a groove gauge.

c. Piston pin bore in the piston and connecting rod. Measure both sides of each piston at several positions. A precision gauge is the most accurate device; however, if careful measurements are taken, a micrometer and telescoping gauge will work (Figs. 11-13 and 11-14).
d. Piston pin. Measure with a micrometer with a ten-thousandths vernier scale. Check areas of the pin where the piston would make contact.
e. Measure each connecting rod (large end) with no bearing inserts and the cap torqued to specifications (Fig. 11-15). Measure each rod for out-of-roundness in two places as shown in Figures 11-16 and 11-17. Record measurements on Job Report Sheet 3-A1.
The measurement tools for this two-step operation should be one of those listed below:
1. Micrometer and telescoping gauge
2. Dial bore gauge (Fig. 11-18)
3. Precision gauge

Record your measurements on Job Report Sheet 3-A1. Consult with your instructor about the service work to be completed on the pistons and connecting rods or about whether a new piston set would be most appropriate.

Note: The following are common automotive machine shop service operations completed on pistons. The description of procedure steps is general because of the variety of equipment in use. All surfaces of the piston should be cleaned before further service work begins.

7. Piston knurling
   a. With a power knurler, clamp the piston in the chuck, holding it at the ring land area (ring belt), not at the skirt.
   b. Set the rotational stroke if so equipped.
   c. Bring the knurling wheel(s) to the skirt area just below the oil control ring groove.
   d. Adjust the knurling tool to apply pressure to the skirt for the appropriate amount of expansion.
   e. The knurl pattern should cover one or both of the thrust surfaces without going over an edge of the skirt.
8. Piston regrooving
   a. Purchase the spacers before cutting the grooves to be certain the cutter will remove the correct amount from the piston.
   b. The ring grooves should be thoroughly cleaned. (Fig. 11-19).
   c. The regroover, whether hand- or power-operated, will require careful positioning of the cutter (Fig. 11-20).
   d. Using motor oil, lubricate the guides that maintain the cutter's position (usually located in the second ring groove).
   e. The cutter should remove a slight (0.002 in. or 0.0508 mm) amount from the lower part of the groove and cut the major amount from the upper (most worn) side of the groove (Fig. 11-21).

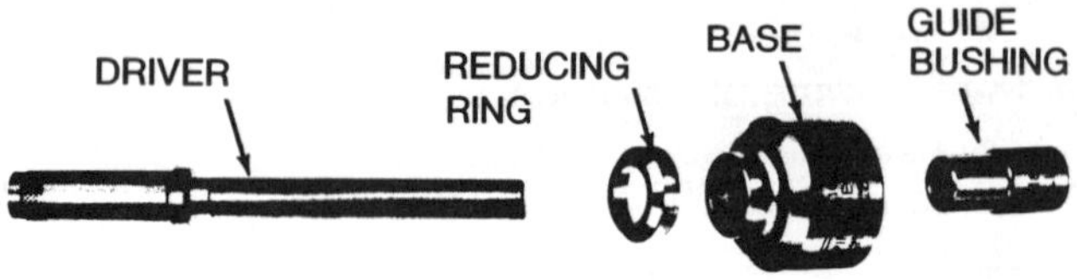

FIG 11-11 Pin press fixtures assembled. (Sunnen Products)

FIG 11-12 Measuring maximum piston diameter.

FIG 11-13 Measuring pin end with a precision gauge.
(Sunnen Products)

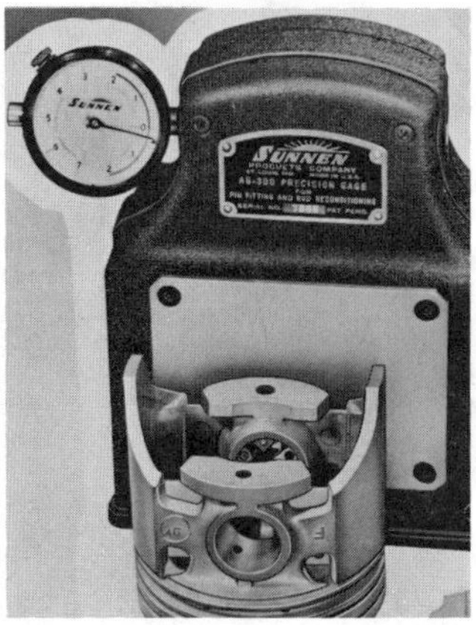

FIG 11-14 Measuring the piston pin bore with a precision gauge.
(Sunnen Products)

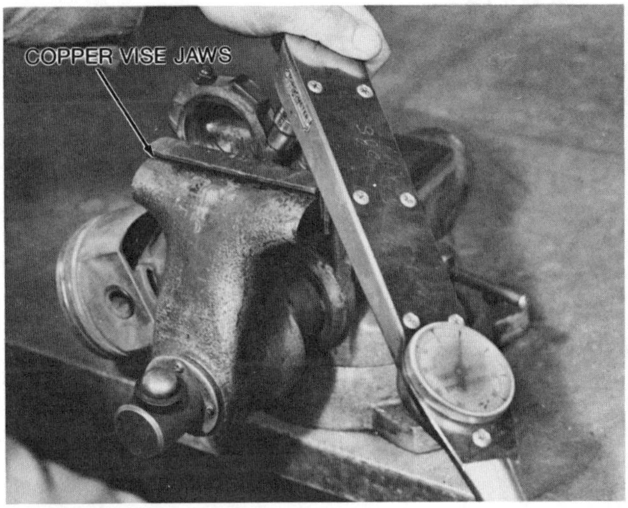

FIG 11-15 Torquing rod cap to specifications.

FIG 11-16 Measuring the vertical distance of the rod bore.

FIG 11-17 Measuring the horizontal distance of the rod bore.

FIG 11-18 Using a dial bore gauge to measure the rod bore.

FIG 11-19 Ring groove cleaner.

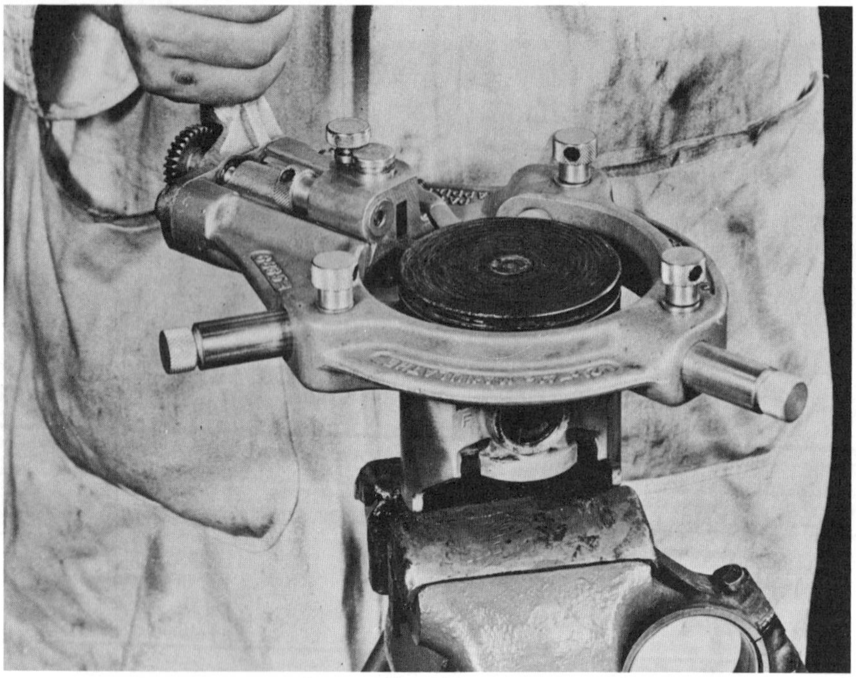

FIG 11-20 Hand operated piston regroover.
(Dana Corporation)

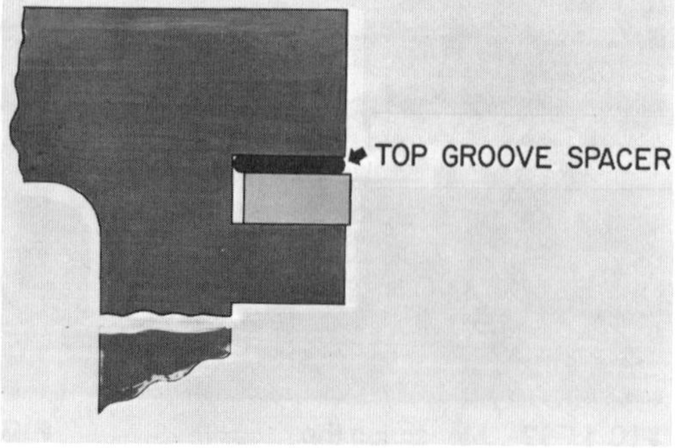

FIG 11-21 Top groove spacer installed above the ring.
(Dana Corporation)

JOB REPORT SHEET 11-A1

PISTON SERVICE

Student Name ______________________ Date ______________

Make of Car ______________ Model ______________ Year ______________

Time Started ______________ Time Finished ______________ Total Time ______________

Flat Rate Time or Recommended Time ______________________

Specifications:

1. Piston diameter (standard or oversize) ______________________
2. Piston pin clearance ______________________
3. Original piston pin diameter ______________________
4. Piston ring groove width: Top ______________ 2nd ______________ Oil ______________

Checkup Questions:

1. What part of the skirt showed the greatest diameter? ______________________

   Why? ______________________
2. What type of pin fit was used? ______________________
3. What area of the piston pin showed the greatest wear? ______________________
4. How many connecting rods showed excessive bushing wear? ______________________

   How many were out-of-round (large end)? ______________________

DATE COMPLETED ______________ INSTRUCTOR'S SIGNATURE ______________

Working as a team with one other student and with the aid of Job Report Sheet 3-A1 and 11-A2, you will fit the pins of 4 piston-connecting rod assemblies from an instructor-assigned vehicle engine, using a precision hone.

OBJECTIVE 11-A2 (manipulative)

Time Recommended:

Twice the flat rate time allowed in *Motors, Chilton,* or *National Service Data* manuals; approximately 20 minutes per piston.

Preparation:

1. Review pages 327 through 338.
2. Study pages 341 through 342.
3. Study manufacturer's or general repair manuals for the engine being serviced.
4. Attend lectures and demonstrations on piston and connecting rod service.

Tools and Materials Needed:

1. Piston-connecting rod assemblies
2. New piston pins (if necessary)
3. Precision hone
4. Precision gauge

Procedure:

The following steps are general because of differences in models of machines and accepted shop procedures. The typical precision gauge and hone described here is shown in Figure 11-22. Your instructor is responsible for the preliminary setup

FIG 11-22 Precision honing machine. (Sunnen Products)

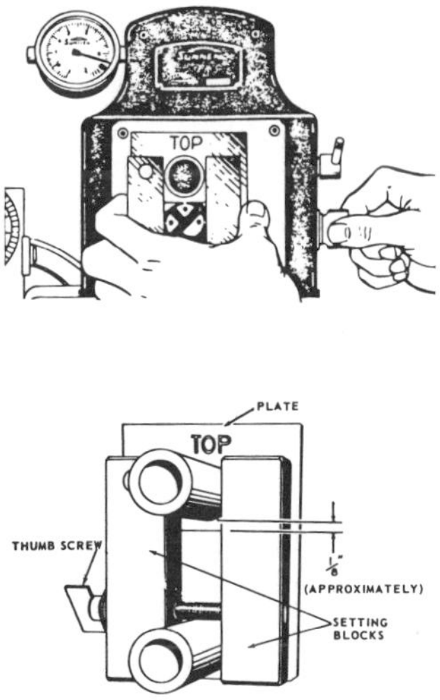

FIG 11-23 Setting fixture.
(Sunnen Products)

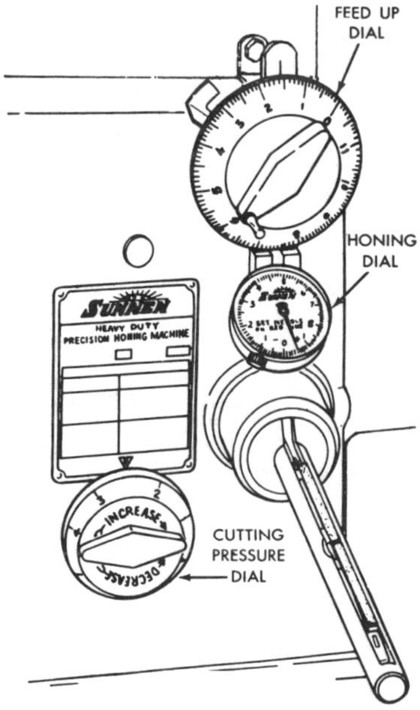

FIG 11-24 Precision hone controls. (Sunnen Products)

of the gauge and honing machine and will assist you in completing the operations.

1. Whenever piston pins (new or used) are to be fitted in the pistons, they should be on hand before the operation begins.
2. Place 2 of the pins in the gauge setting fixture and tighten the clamp (Fig. 11-23).
3. Place the setting fixture on the precision gauge and adjust the gauge to read within the range of the new pin bore size. Rotate the dial level to zero and add the specified pin clearance for piston and full-floating rods. Subtract the interference for press-fit rods.
4. Place the piston pin or connecting rod bore over the gauge and determine the amount of material to be removed by the hone.
5. With the proper size mandrel (honing unit) installed in the machine, slip the piston or rod over the mandrel. Adjust the support(s) for connecting rods.
6. Turn the micrometer feedup dial until the honing dial begins to move. This dial shows both the stock removal and the stone wear (Fig. 11-24).
7. Set the dial on zero and turn on the machine's motor.
8. With a good flow of coolant directed at the stones, begin honing by depressing the floor pedal all the way.
9. Advance the micrometer feed knob until the honing dial advances several divisions, but no more than the amount to be removed.
10. Hone the pin bore by stroking back and forth the full length of the stone (Fig. 11-25). Reverse the work frequently and continue honing until the honing dial reads zero. If there appears to be too much or too little honing, have your instructor change the cutting pressure.
11. Release the foot pedal and remove the piston or rod. Gauge the bore. It will be undersize by the amount of stone wear.
12. Continue steps 10 and 11 until the bore reaches the size desired (Fig. 11-26).
13. All pistons or connecting rods are to be honed in the same manner.

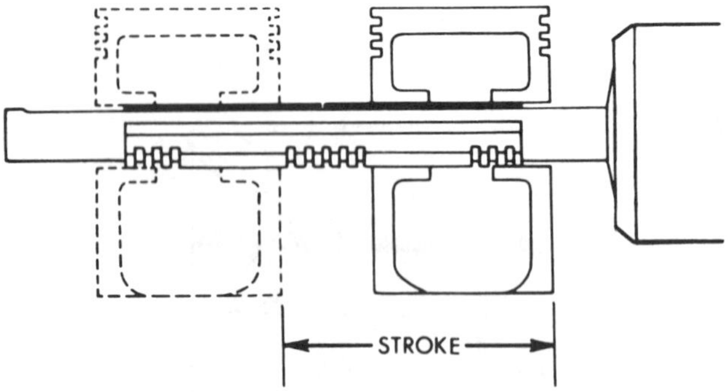

FIG 11-25 Proper stroking motion while honing.
(Sunnen Products)

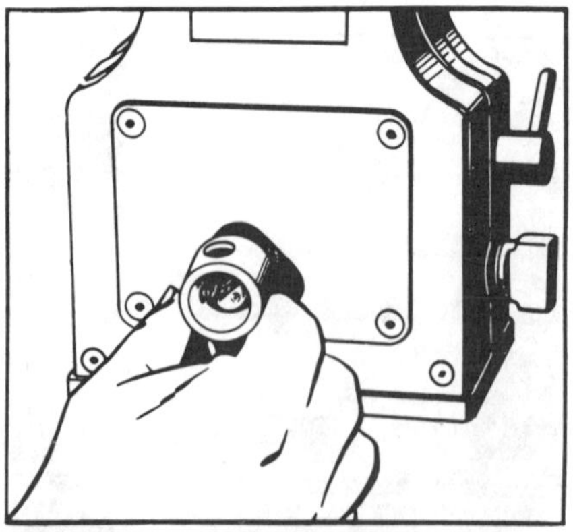

FIG 11-26 Measuring the piston pin bore of a connecting rod with a precision gauge.
(Sunnen Products)

JOB REPORT SHEET 11-A2

PISTON PIN FITTING

Student Name ______________________ Date ______________

Make of Car ____________ Model ______________ Year ______________

Time Started ____________ Time Finished ____________ Total Time ____________

Flat Rate Time or Recommended Time ______________________

Specifications:

Piston pin diameter (new or used) ______________________

Specified clearance:

Piston ______________________

Connecting Rod ______________________

Checkup Questions:

1. The honing dial shows the amount removed and ______________________
2. What is the function of the foot pedal? ______________________
3. Why is it important to stroke the part the full length of the stone? ______________________
4. Why is it important to reverse the work frequently? ______________________

DATE COMPLETED ______________ INSTRUCTOR'S SIGNATURE ______________

CONNECTING RODS

The connecting rod serves as the link between the piston and the crankshaft. Its major purpose is to convert the reciprocating (up-and-down) motion of the piston into the rotary motion of the crankshaft.

In vee engines, the connecting rods have a slight offset designed into the large end. This is done because each connecting rod journal holds two rods and the cylinder banks have a slight stagger from one another. If the rods were not offset, there would have to be more stagger between cylinder banks to compensate.

The connecting rod is manufactured of high quality steel shaped by drop forging. This method hammers the steel into the rough shape of a connecting rod while compressing it to form strong lines of metal.

A connecting rod typically has an I-beam, cross-sectional shape through the mid-area to give it high strength and light weight. The rod and cap are forged separately and assembled after the bolt holes and mating surfaces are machined.

Most connecting rods today go through a process of peening (usually with steel shot) to relieve the stresses built up during the forging operation. A rod that has been shot-peened has a fine "cratered" surface finish except in the bore openings. Careful finish machining is then done to give it circular bores at both ends that are the correct distance between one another.

CONNECTING ROD WEAR

A connecting rod in the average engine, with press-fit piston pins, will experience very little wear. Full-floating pin connecting rods have bronze bushings which usually require service during each engine reconditioning. This is done by pressing out the old bushing and pressing in a new one with the help of a bushing driver.

If serious bearing failure occurs, the large end surface of the rod may be affected. The bearing insert begins to turn within the bore under marginal conditions of lubrication. As the bearing failure accelerates, the bearing insert backing (steel) and the connecting rod bore begin scoring. An engine that has had this problem is often referred to as having "spun a bearing." If neglect continues, the rod is "thrown"—the rod leaves the crankshaft journal. In most cases, a thrown rod is not serviceable and must be replaced. Very often extensive engine damage occurs as a result of a thrown rod.

CONNECTING ROD RECONDITIONING

Reconditioning of a connecting rod is done to restore accuracy of the large-end bore. A precision hone can be used to recondition a rod to meet O.E.M. (Original Equipment Manufacturer) specifications for roundness, bore size, and surface finish (Fig. 11-27). Assuring the straightness or alignment of the two bore openings of a rod is a part of rod reconditioning, but is usually a physical check made after the honing process.

The large end of the connecting rod becomes out-of-round, or stretched, as a result of the high pressure of combustion and the inertia forces (moving a weighted material back and forth) during periods of both hot and cold operation. The stretch is usually in a vertical direction. However, because of the rod's position during the maximum pressure of the power stroke and its position during the beginning of the compression and exhaust strokes, the longest measurement can sometimes be found up to 35° from the vertical (Fig. 11-28).

Reconditioning the rod involves two major steps (Fig. 11-29):

a. Reduce the rod bore slightly below its standard size through precision grinding of the cap and rod (0.002 in. or 0.0508 mm each).

b. Remove just enough material from the rod bore, by precision honing, to return it to original size.

These two steps will correct out-of-roundness and size and will also renew the surface finish so the bearing insert will be held correctly and be able to give up its excess frictional heat to the connecting rod.

The straightness or alignment of any connecting rod should be checked during an engine overhaul, whether or not the large end has been honed. A device called a rod aligner is used to determine the straightness of a connecting rod with or without the piston attached (Fig. 11-30). The rod (preferably with its piston attached) is placed with its bore contacting a surface that is exactly 90° from the piston vertical surface.

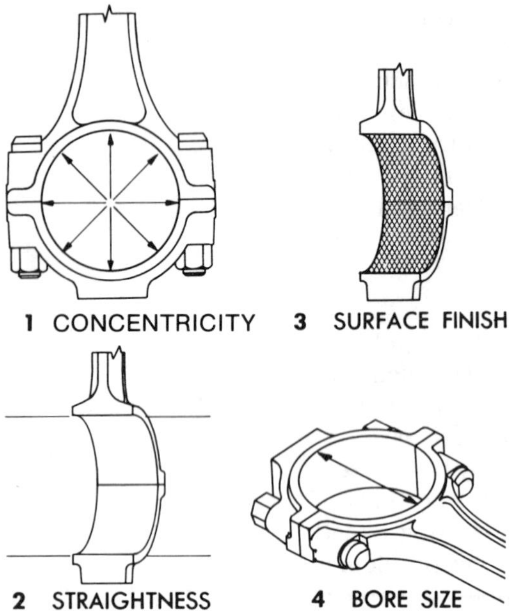

FIG 11-27 Requirements of a reconditioned rod. (Sunnen Products)

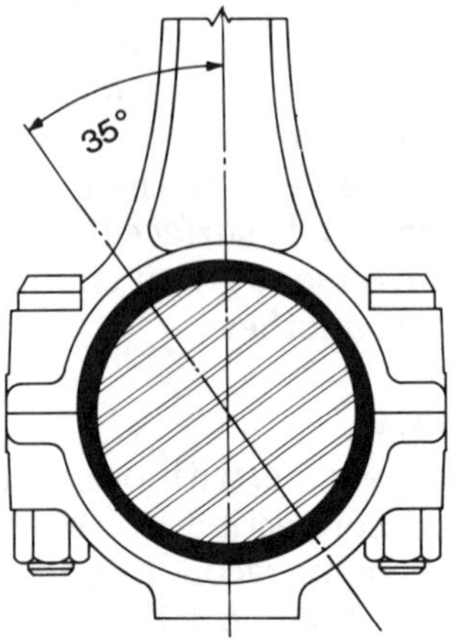

FIG 11-28 Greatest stretch in a rod bore may be up to 35° from vertical. (Sunnen Products)

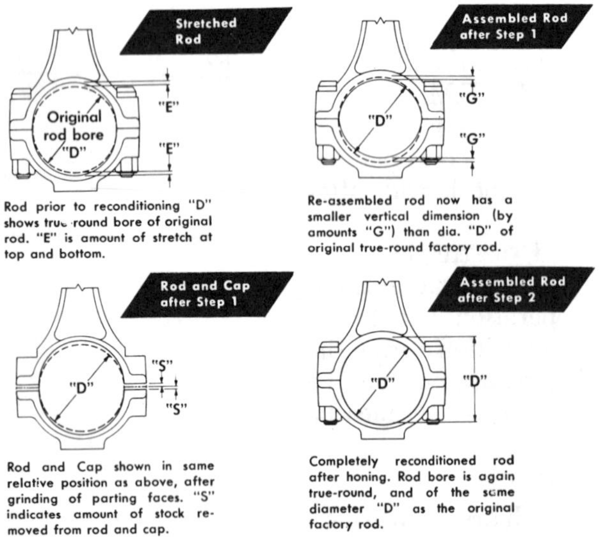

FIG 11-29 Steps in reconditioning a rod. (Sunnen Products)

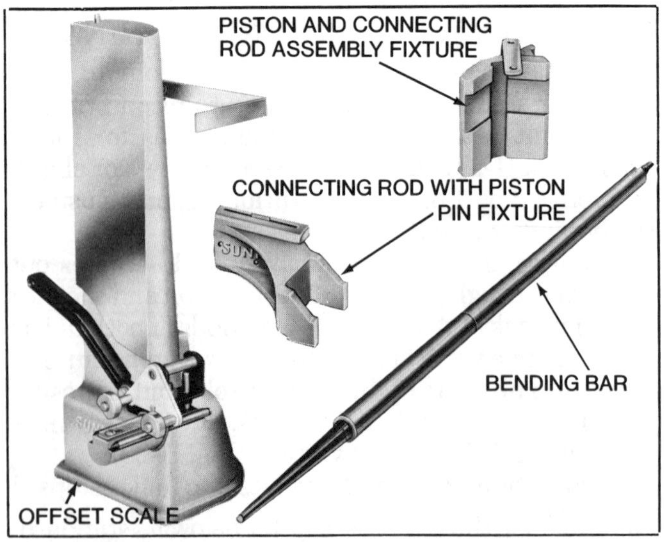

FIG 11-30 Typical rod alignment checking tool. (Sunnen Products)

Connecting rods lose their straightness in one or more of the following ways:

a. Bend
b. Twist
c. Offset

Bend and twist can be corrected easily through the use of a bending bar while the rod is clamped in a vise with protected jaws. Brass or copper plates are often used between the jaws and the rod. Offset is a problem best rectified by a replacement rod.

OBJECTIVE 11-A3 (manipulative)

Working as a team with one other student and with the aid of Job Report Sheet 11-A3, you will recondition the large (crankshaft) end and reassemble 4 pistons and connecting rods from an instructor-assigned vehicle engine.

Time Recommended:

Twice the flat rate time allowed in *Motors, Chilton,* or *National Service Data* manuals.

Preparation:

1. Review pages 327 through 347.
2. Study pages 347 through 349.
3. Study manufacturer's or general repair manuals for the engine being serviced.
4. Attend lectures and demonstrations on connecting rod reconditioning.

Tools and Materials Needed:

1. Pistons and connecting rods
2. Cap and rod grinder
3. Precision gauge
4. Precision hone

Procedure:

Connecting rod reconditioning involves these major steps:

a. Measurement of the bore (Job Sheets 3-A1 and 11-A1)
b. Disassembly and grinding
c. Honing and assembly

The further breakdown which follows remains general because of differences in models of machines and accepted shop procedures.

Fill in all specifications on Job Report Sheet 11-A3.

1. When it has been determined that the connecting rod bore is out-of-round, misaligned, or distorted, the unit should be disassembled.
2. Drive the rod bolts out of their holes.
3. Grind the rod and cap on a precision grinder (Fig. 11-31). Remove approximately 0.002 in. (0.0508 mm) from each side of each half of the rod and cap. This operation reduces the rod bore approximately 0.004 in. (0.102 mm).

More material would have to be removed if the connecting rod was badly damaged by a turning bearing insert.

4. Reassemble the bolts into the rod and replace the cap, torquing to specifications.
5. With the proper honing unit installed on the machine, place 1 or 2 connecting rods in position (Fig. 11-32). V-8 and other narrow bearing rods should be honed 2 at one time to maintain a true honing stone.
6. Have your instructor set the cutting pressure and adjust the honing unit stones.
7. Back off the feed dial until the honing dial reads zero with the foot pedal fully depressed.
8. Release the pedal and check the amount of stock to be removed.
9. Turn ON the machine and begin stroking as the pedal is being depressed. Push slowly until it is all the way down.
10. Direct the oil from the honing nozzles directly at the rods.
11. While stroking the rods back and forth over the stone (without going off either end), advance the feed dial slightly to keep the stone in contact.
12. Check the bore size frequently until it reaches the specified diameter as set on the precision gauge.
13. Have your instructor assist you in the use of the honing dial to bring the remaining rod bores to the correct size.
14. Thoroughly clean all rods, pistons, and their pins. All the rod caps should be disassembled and any burrs removed from the parting faces.
15. For press-fit pins, locate and assemble the correct piston pin press fixtures or puller.
16. Lubricate the pin and pin bores with press lube (Sunnen B-200) or substitute SAE 90 gear oil.
17. Press or pull the pins into position while noting that the major thrust side is always to the left when viewed from the front (Fig. 11-33). The pin should be equidistant from both sides of the pin bore when the rod is centered between the interior pin bore ends.
18. For full-floating pins, lubricate and assemble the pins in the piston and rod. Carefully set *both* pin lock rings in position with the solid part up on each piston.
19. Assemble remaining pistons and rods.

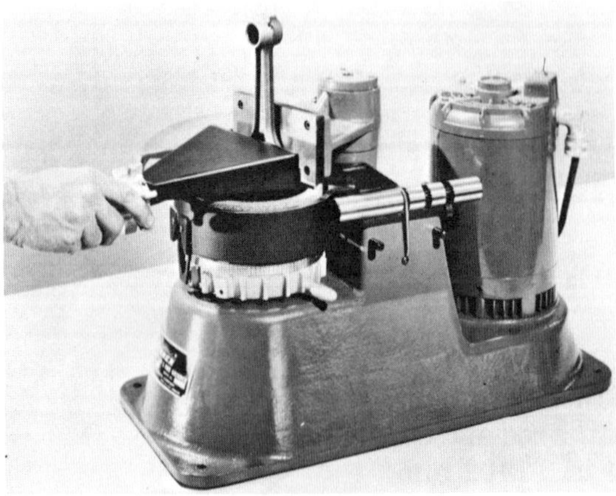

FIG 11-31 Precision grinder.
(Sunnen Products)

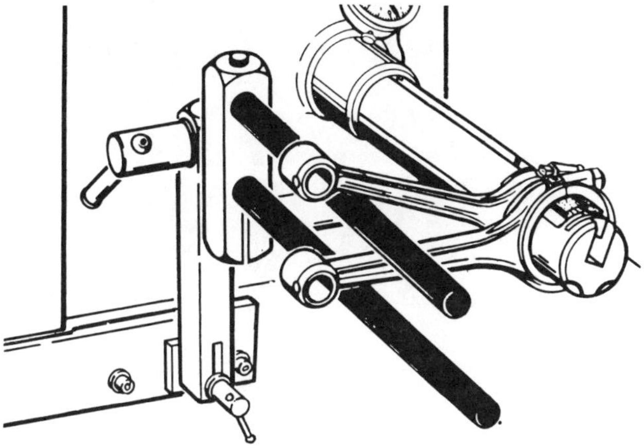

FIG 11-32 Small width rods being honed two at one time.
(Sunnen Products)

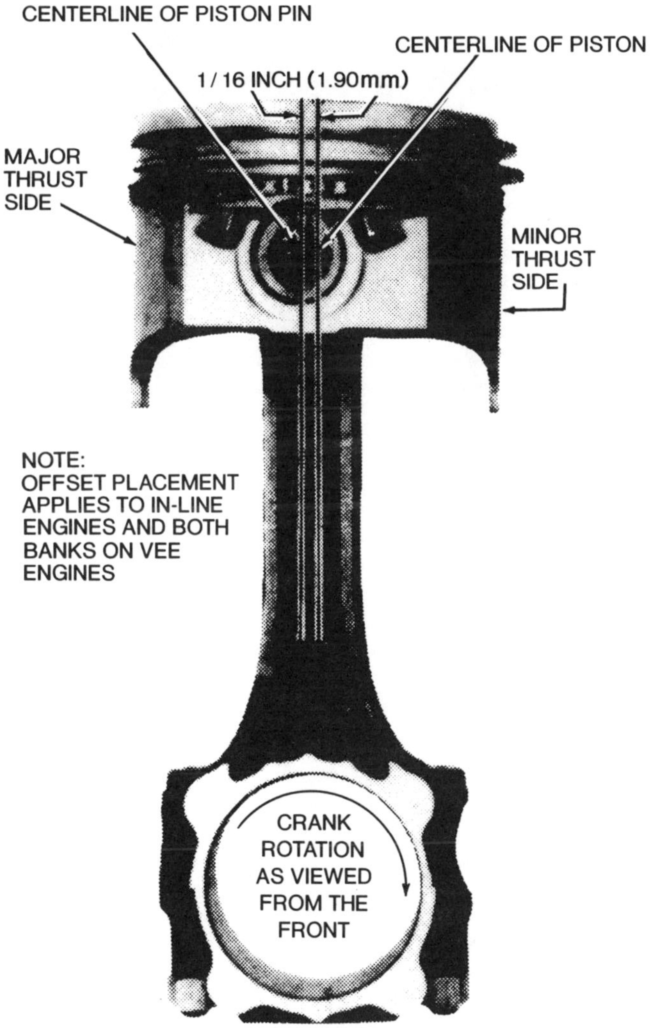

FIG 11-33 Assemble rods and pistons according to notch on piston or offset to the left as viewed from the front.
(Buick Motor Division, GMC)

JOB REPORT SHEET 11-A3

CONNECTING ROD RECONDITIONING

Student Name ______________________ Date ____________

Make of Car ____________ Model ____________ Year ____________

Time Started ____________ Time Finished ____________ Total Time ____________

Flat Rate Time or Recommended Time ______________________

Specifications:

1. Connecting rod bore diameter ______________________
2. Connecting rod cap torque ______________________

Checkup Questions:

1. What is the recommended amount of material to be removed from

   each cap? ____________ each rod? ____________
2. Why is it important to hone two narrow connecting rods at one time? ____________

   ____
3. Why is it important not to allow the rod to move off stones while stroking? ____________

   ____
4. What type or brand of press lube was used during the assembly operation? ____________

DATE COMPLETED ____________ INSTRUCTOR'S SIGNATURE ____________

PISTON RINGS

The piston ring's primary purpose is to form a seal between the piston and the cylinder wall. This seal helps prevent the escape of expanding combustion gases (blow-by) during the power stroke and also prevents lubricating oil from entering the combustion chamber from the crankcase. To do these two jobs, the piston ring must conform to the shape of the cylinder wall during its rapid movements up and down.

TYPES OF PISTON RINGS

Piston rings can be classified in two general types: compression rings and oil control or oil rings (Fig. 11-34). The top two are considered compression rings and the lower one the oil ring. In actual operation, the compression rings also serve as the final oil control rings. Likewise, the oil control ring serves as the last compression ring. The middle ring, usually classed as a compression ring, serves as a combination of the other two. The function of each ring depends, in part, on which of the four strokes the piston is moving in at the time.

COMPRESSION RINGS

Compression rings are classified by their metal composition, cross-sectional shapes, and type of applied coating. The majority of passenger car engines use compression rings made from highly refined cast iron. Some require rings made from ductile iron, which are less apt to break during abnormal combustion periods.

The cross-sectional shapes of common compression rings are shown in Figure 11-35. Each shape has features which help it seal the combustion gases and control the lubricating oil. Today, the most popular shape is the torsional twist design. During the power stroke, it twists slightly in the groove to apply high unit pressure

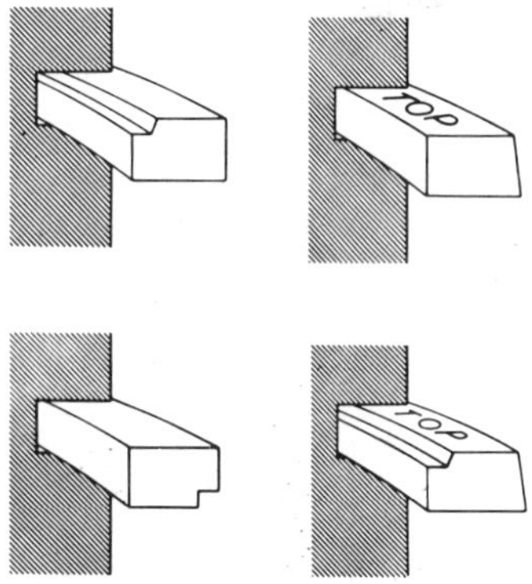

FIG 11-35 Typical compression ring shapes.
(Dana Corporation)

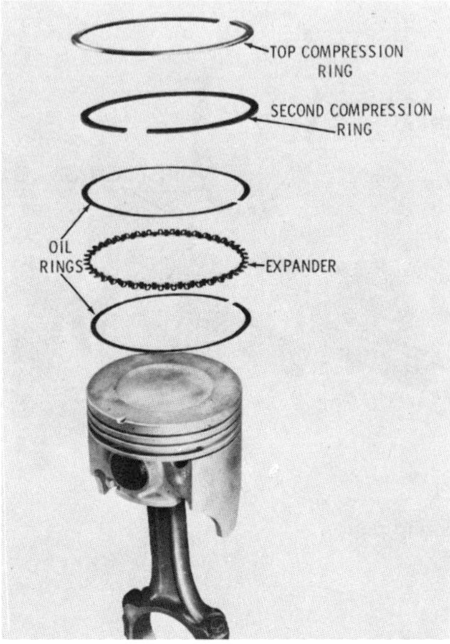

FIG 11-34 Placement of piston rings.
(Oldsmobile Division, GMC)

to the cylinder wall and piston groove (Fig. 11-36). This assists in preventing excessive oil movement into the combustion chamber. The twist action also provides a very tight seal to reduce loss of combustion gases. To provide this twist action, the top ring must be chamfered or counter-bored on the inside of the upper edge. The second compression ring may also have a counter-bore; however, its face is often stepcut to further assist in oil control by functioning as a scraper.

PISTON RING SERVICE

Two measurement operations are normally required when new piston rings are installed in an engine: the ring end gap (clearance), and the ring side clearance measurements.

Piston ring end gap measurements are made to be certain there is a minimum gap. With too little gap, the combustion heat during operation would cause the ends to butt, and scoring and seizure would occur in the cylinder. The measurements are taken with each ring in its cylinder bore. The exact position within the bore is determined by the cylinder's reconditioning. Measuring the gap within a rebored cylinder can be accomplished at any position. If a cylinder is *not* rebored, the gap must be checked at the bottom of that ring's normal travel.

Each ring must be checked in the cylinder in which it will be installed. This will take care of any differences in cylinder diameter that could cause problems during engine operation.

An inverted piston is the most convenient device for placing the ring perpendicular in the cylinder at the proper location. Look ahead to Figure 11-42. A feeler gauge is then used to measure the gap (Fig. 11-43). The amount of gap is considered a minimum measurement. This means the ends should have *no less than* the specified clearance. However, no ring should have more than 0.040 in. (1.016 mm) end gap. If it does, consult with your instructor for advice. When the end gap clearance is not specified, the following general trade-accepted specification may be used:

Top Compression Ring	0.003–0.004 in. (0.076–0.102 mm) per inch of cylinder bore
Second Ring	0.0015 in. (0.0381 mm) per inch of cylinder bore
Oil Control Ring	0.001 in. (0.0254 mm) per inch of cylinder bore

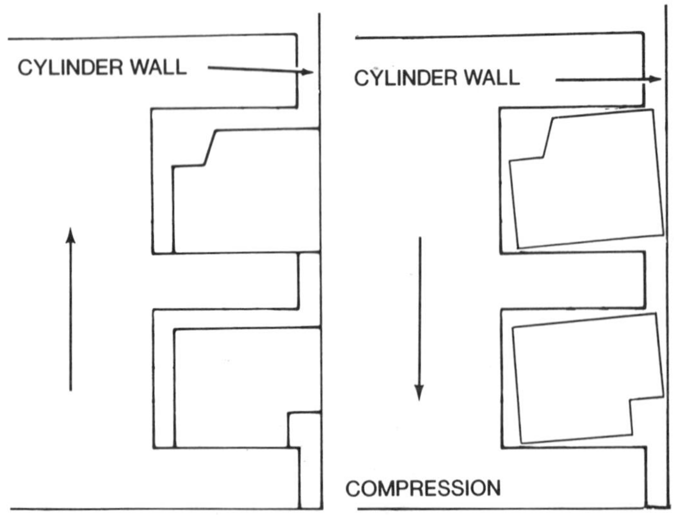

FIG 11-36 Twisting action of a piston ring.

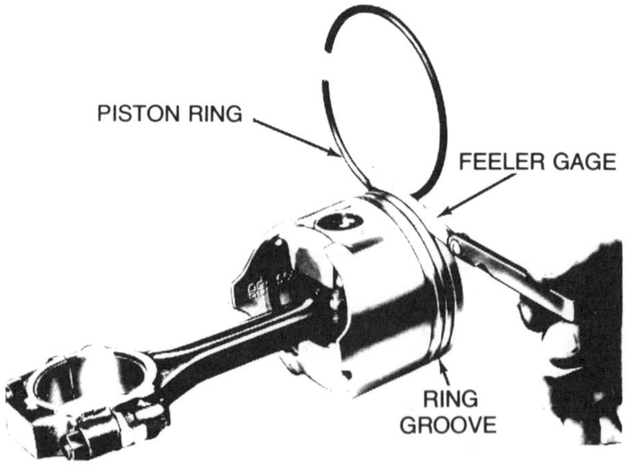

FIG 11-37 Measuring piston ring side clearance.
(Chrysler Corporation)

Piston ring side clearance is also a minimum measurement. The ring must have enough space to expand its width as it is heated by combustion gases. This clearance is normally measured with a feeler gauge and a new ring placed in its groove (Fig. 11-37). Follow specifications listed in a shop manual or the ring package instructions. If no side clearance specifications are available, use the following:

Top Compression Ring	0.002 in. (0.0508 mm)
Other Rings	0.0015 in. (0.0381 mm)

OBJECTIVE 11-A4 (manipulative)

Working as a team with one other student and with the aid of Job Report Sheets 3-A1 and 11-A4, you will align the piston and connecting rod, measure the ring end gap (clearance), side clearance, and install the rings on 4 pistons from an instructor-assigned vehicle engine.

Time Recommended:

Twice the flat rate time allowed in *Motors, Chilton,* or *National Service Data* manuals.

Preparation:

1. Review pages 327 through 355.
2. Study pages 355 through 357.
3. Study manufacturer's or general repair manuals for the engine being serviced.
4. Attend lectures and demonstrations on piston and connecting rod service.

Tools and Materials Needed:

1. Connecting rod aligner and bending bar
2. Piston and rod assemblies
3. Piston rings and their installation instructions
4. Torque wrench

Procedure:

1. Check the alignment of each rod and piston by torquing the rod cap to specified tightness and placing it on the aligner with the major thrust side toward the operator (the major thrust side is always on the left when viewing from the front of the engine).
2. With the V-block (piston fixture) held firmly in the right hand (Fig. 11-38), check for gaps between the fixture and the vertical checking surface. A gap at the top or bottom when the piston is held vertically indicates the rod is bent. If this is the case, complete step 2A to straighten it.

2a. Remove the rod from the aligner and place the connecting rod thrust side up in a vise. Straighten the rod by bending it toward the gap seen in the previous step (Fig. 11-39).

2b. Return the rod to the aligner and check your work. There should be no gaps between the V-block and the vertical surface.

3. Check for rod twist by moving the piston on its pin. Look for gaps at the top or bottom (Fig. 11-40).
3a. If the gap is at the top, correct the twist as shown in Figure 11-41A. To correct the opposite twist, bend as shown in Figure 11-41B. The piston in the vise is shown with its major thrust surface up.
4. Locate the piston rings and the cylinder block.
5. Read *all* the directions included in the ring package. Handle all the rings with care, especially cast iron rings, which are subject to breakage.
6. Place the top ring in the cylinder bore (any convenient place for rebored cylinders and at the bottom of the ring travel for cylinders not rebored).
7. Use a piston to push the ring into place. Then measure the clearance between the two ends with a feeler gauge (Fig. 11-42).
8. Compare the measured gap with clearance specifications in the ring installation instructions, shop manual, or in this text (Fig. 11-43).
9. If the gap is too small, carefully file the ends with a device designed for that purpose. Be certain the ends remain square and that the moly or chrome is

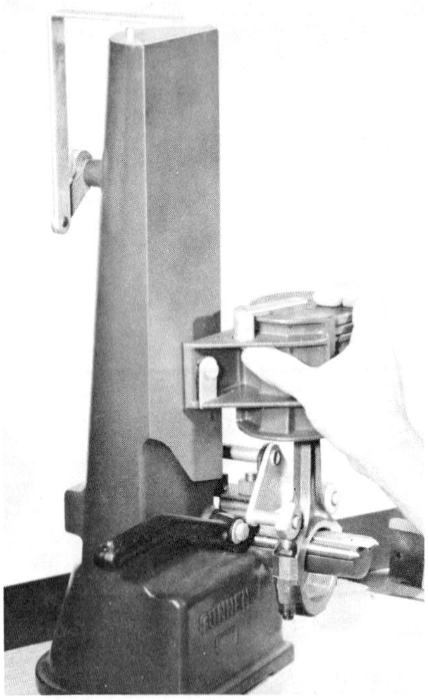

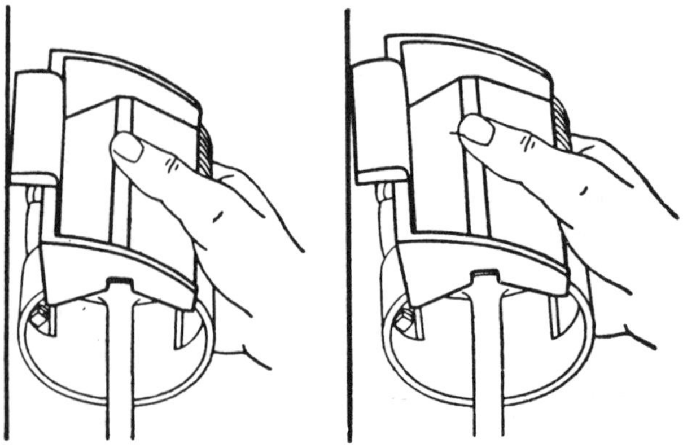

FIG 11-40 Indications of misalignment of a connecting rod assembly.
(Sunnen Products)

FIG 11-38 Checking connecting rod alignment.
(Sunnen Products)

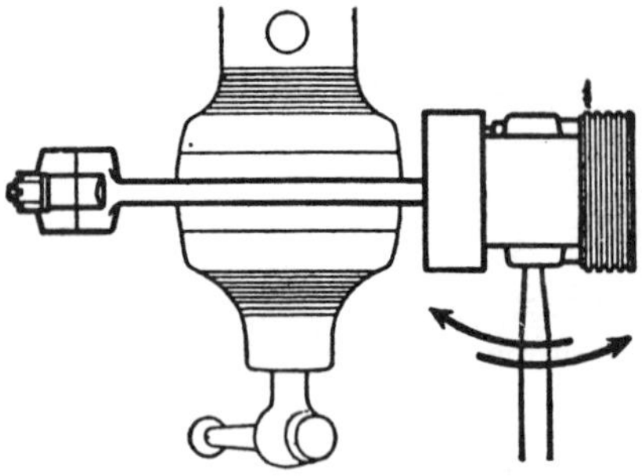

FIG 11-39 Bending a rod to correct misalignment. (Sunnen Products)

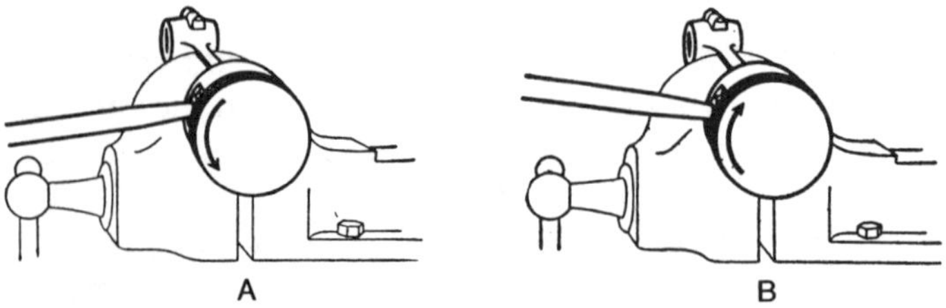

FIG 11-41 Correcting for rod twist. Bend according to gap found while rocking the piston on aligner. (Sunnen Products)

not chipped away. This can be avoided by operating the file to move from the outside toward the inside of the ring.

10. Make all end gap measurements and adjustments.
11. Measure the side clearance with a feeler gauge and a new ring. Use specifications in the shop manual, ring package instructions, or this text.
12. Install the oil control ring expander (spacer), placing its ends as directed.
13. Next, install the *top* rail by inserting one end to right of the expander joint *first,* then setting in the rest of the rail (Fig. 11-44).
14. Install the lower rail by inserting one end to the left of the expander joint. Set in the rest of the rail, placing its gap according to directions.

Note: Double check to see that expander ends are butted and *not* overlapped.

15. Using a piston ring expander tool, install the second groove ring, noting any TOP, dot, or other directions (Fig. 11-45).
16. Install the top ring according to directions with a ring expander tool (Fig. 11-46).

FIG 11-42 Pushing the piston ring into position for end gap checking.

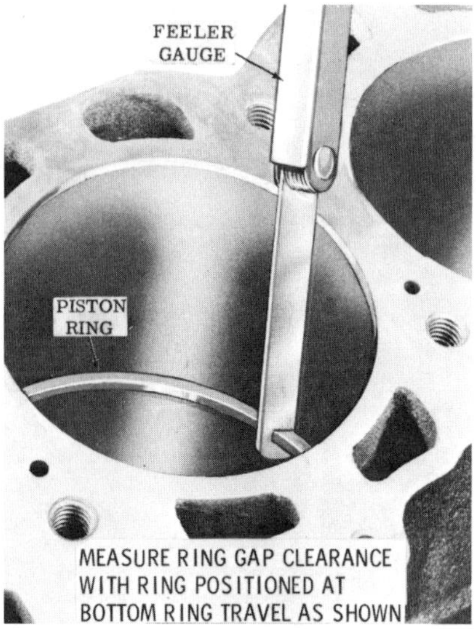

FIG 11-43 Measuring ring end gap (clearance) with a feeler gauge. (Oldsmobile Division, GMC)

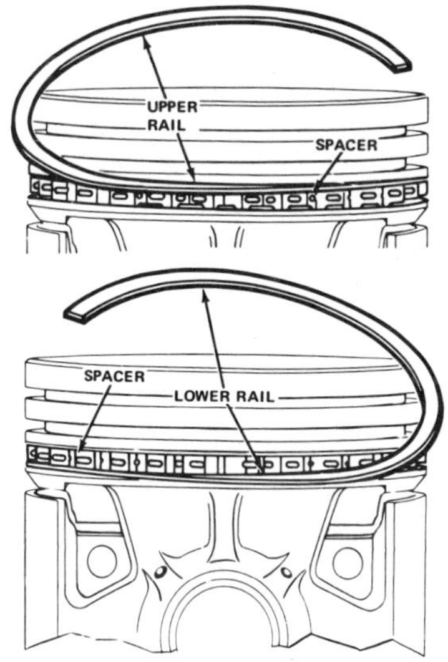

FIG 11-44 Oil control ring rail installation. (American Motors Corporation)

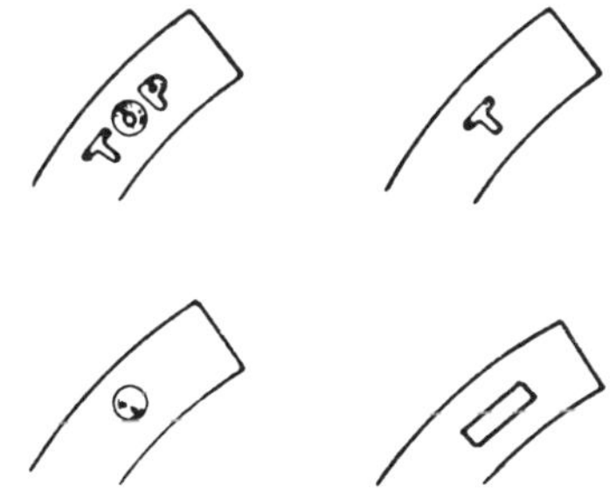

FIG 11-45 Typical compression ring indicator marks.

(American Motors Corporation)

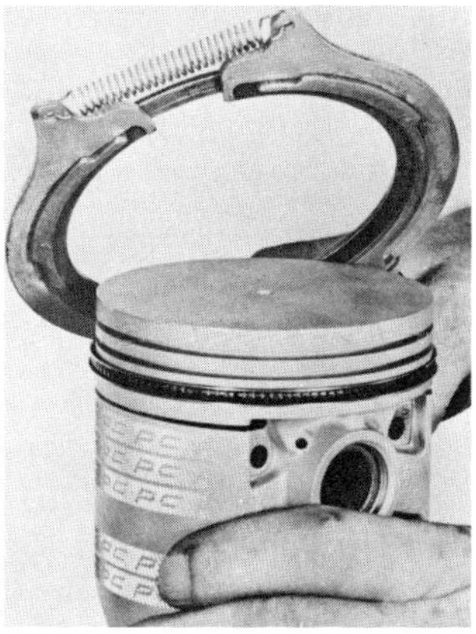

FIG 11-46 Recommended type of ring installation tool.

(Dana Corporation)

CONNECTING ROD ALIGNMENT AND PISTON RING INSTALLATION

Student Name ______________________ Date ______________

Make of Car ____________ Model ______________ Year ______________

Time Started ____________ Time Finished ____________ Total Time ____________

Flat Rate Time or Recommended Time ______________________

Specifications:

	TOP	SECOND	OIL
Ring end gap specifications			
Ring side clearance			

Checkup Questions:

1. What effect would a bent connecting rod have on a piston and cylinder bore? ______________

   on the connecting rod bearing? ______________________

   ____

2. What effect would piston rings with insufficient end gap have on the cylinder walls? ______________

   ____

   ____

3. Why should the oil ring top rail be installed before the bottom rail? ______________

   ____

4. Did the top ring have a chamfer or bevel? ______________________

   If so, which way was it facing when installed? ______________________

DATE COMPLETED ______________ INSTRUCTOR'S SIGNATURE ______________

Working as a team with one other student and with the aid of Job Report Sheets 3-A1 and 11-A5, you will place the ring end gaps, install in the block and check the connecting rod bearing oil clearance and side clearance of 4 piston and connecting rod assemblies from an instructor-assigned vehicle engine.

OBJECTIVE 11-A5 (manipulative)

Time Recommended:

Twice the flat rate time allowed in *Motors, Chilton,* or *National Service Data* manuals; approximately 20 minutes per assembly.

Preparation:

1. Review pages 327 through 357.
2. Study pages 361 through 363.
3. Study manufacturer's or general repair manuals for the engine being serviced.
4. Attend lectures and demonstrations on piston and connecting rod service.

Tools and Materials Needed:

1. Piston rod assemblies
2. Connecting rod bearings
3. Piston ring compressor
4. Plastigage
5. Torque wrench
6. Assorted socket wrenches and handles
7. Grease

Procedure:

1. Blow off any foreign matter from the piston.
2. Place the ring gaps according to the shop manual specification or as shown in Figure 11-47. Be certain the oil control ring expander is not overlapped.
3. Remove the rod cap and clean the connecting rod bore and the backside of the bearing with lacquer thinner (Fig. 11-48).

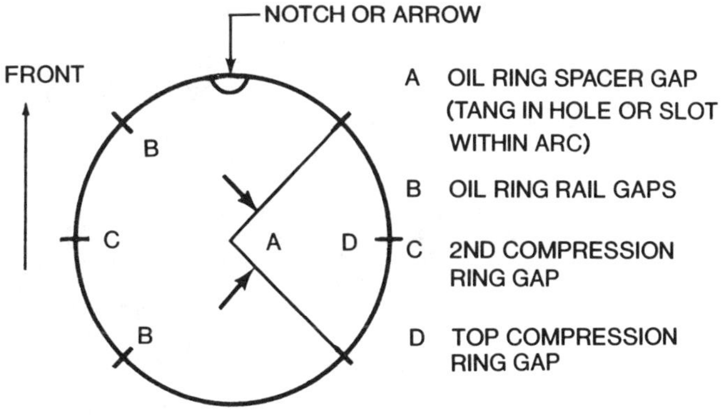

FIG 11-47 Ring end gap placement locations prior to piston installation. (Oldsmobile Division, GMC)

FIG 11-48 Cleaning rod bore with lacquer thinner prior to bearing installation.

4. Snap the rod bearing in place. Double check to see if there are differences between upper and lower bearing halves (Fig. 11-49).
5. Lubricate the ring area of the piston with motor oil and the bearing insert with grease.
6. Note the up and down side of the ring compressor band and tighten the band over the rings. No end gaps should be exposed.
7. Attach rubber hose or spark plug boots over the rod bolt ends (Fig. 11-50).
8. With the connecting rod journal of the crankshaft in the BDC position, place the piston in the cleaned cylinder bore with the notch, hole, or arrow pointing toward the front.
9. Tap the piston into position with the butt end of a soft-handled hammer (Fig. 11-51).
10. Clean the rod cap bore and bearing insert backside with lacquer thinner before snapping bearing into place.
11. Cut a strip of green Plastigage the width of the bearing, and lay it across the journal.
12. Install the rod cap, noting location of the stamped numbers.
13. Torque the rod cap to specifications; then remove it. Compare Plastigage package stripes with the compressed Plastigage (Fig. 11-52).
14. If the clearance is within limits, apply a light coat of grease to the journal and retighten the cap to proper torque. Incorrect clearance can have many causes:
    a. Wrong bearings
    b. Dirt between bearings and rod bore or cap
    c. Improper rod bore size, wrong diameter, or out-of-round. If clearance cannot be brought within limits, consult with your instructor.
15. Install the remaining piston-rod assemblies by following steps 1–13. As each piston is installed, turn the crankshaft slightly. An increasing resistance should be felt as each new piston is installed.
16. Measure the side clearance (gap between the connecting rod and side surface of bearing journal or a companion connecting rod in the case of vee engines where two rods share the same journal) (Fig. 11-53).
17. Turn the crankshaft with a pry bar against bolts inserted in the flywheel flange. Some resistance to movement should be felt, but it should not be extreme.

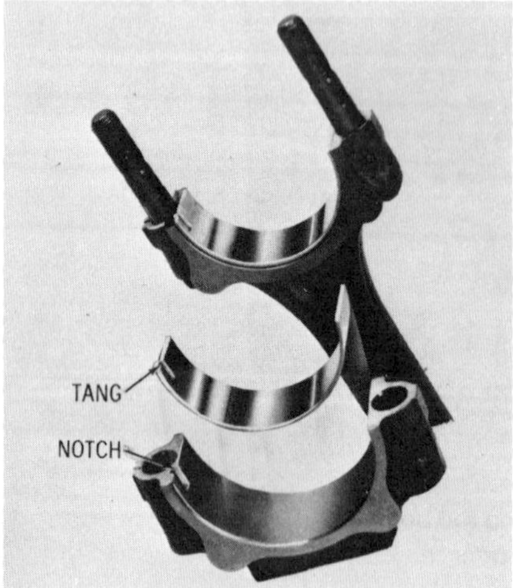

FIG 11-49 Bearing tang (lip) should fit into the rod groove.
(Oldsmobile Division, GMC)

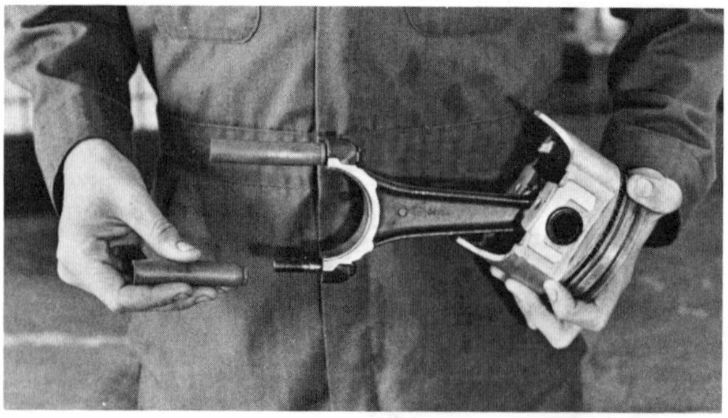

FIG 11-50 Spark plug boots placed over rod bolts to prevent damage to the crankshaft journals.

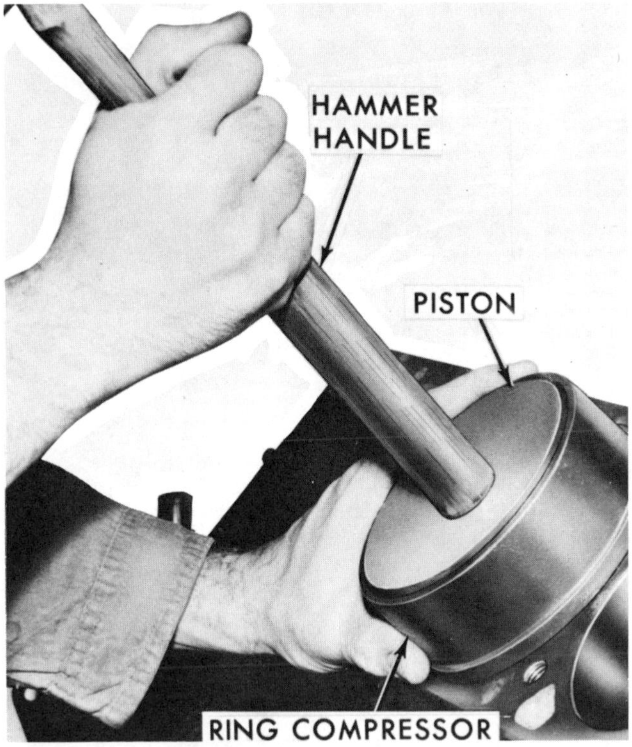

FIG 11-51 Installing a piston in the cylinder block.
(Cadillac Motor Car Division, GMC)

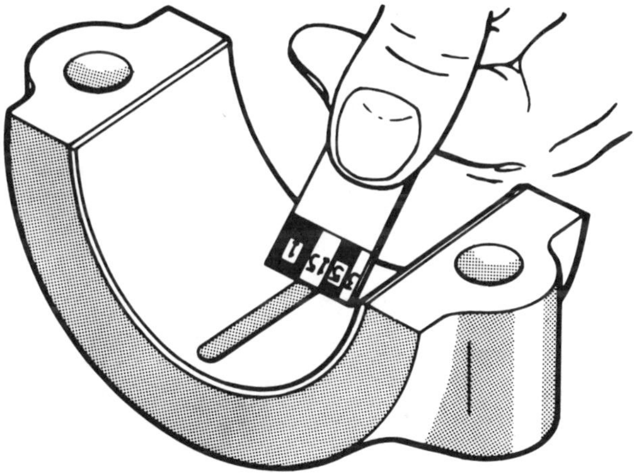

FIG 11-52 Comparing compressed Plastigage with package stripes.
(Chrysler Corporation)

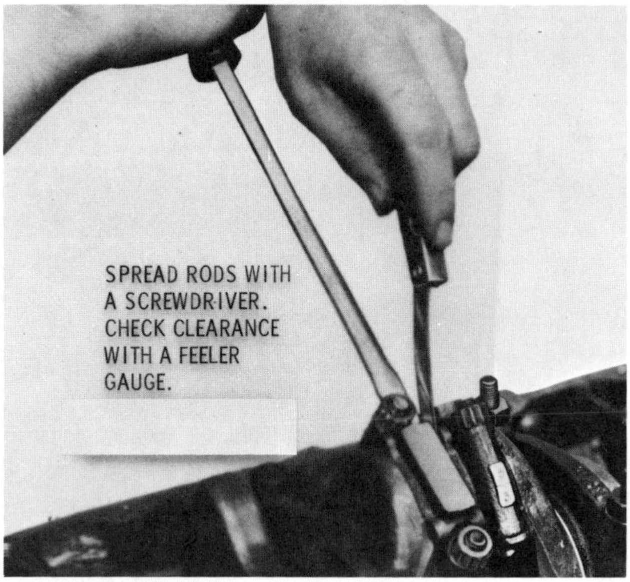

FIG 11-53 Checking connecting rod side clearance.
(Oldsmobile Division, GMC)

JOB REPORT SHEET 11-A5

PISTON-CONNECTING ROD ASSEMBLY INSTALLATION

Student Name ______________________________ Date ________________

Make of Car ____________ Model ________________ Year ________________

Time Started ____________ Time Finished ____________ Total Time ____________

Flat Rate Time or Recommended Time ______________________________

Specifications:

1. Ring gap placement diagram
2. Connecting rod torque
3. Connecting rod oil clearance
4. Connecting rod side clearance

Checkup Questions:

1. Why should the rod bore and backside of each bearing be thoroughly cleaned? ____________

   ____

2. What function do the hose pieces or spark plug boots on the rod bolts serve? ____________

   ____

3. Were the pistons clearly marked as to their forward direction?

   ____

4. Why use a soft-handled hammer rather than a wooden one? ____________

   ____

5. Were any bearing clearances found to be outside recommended limits? ____________

   Explain: ______________________________

   ____

   ____

DATE COMPLETED ________________ INSTRUCTOR'S SIGNATURE ________________

Name ______________________________ Date ______________

DIRECTIONS:

Circle the best answer to each question.

1. Cam-ground pistons
   I. have major and minor thrust sides.
   II. are used exclusively in automobile engines today.
   A. I only B. II only C. Both I and II D. Neither I nor II
2. The shape of a piston head has an effect on
   I. compression ratios
   II. compression pressures
   A. I only B. II only C. Both I and II D. Neither I nor II
3. When the top ring groove of a piston wears excessively
   I. the piston must be replaced.
   II. the groove can be enlarged and a steel spacer is placed below the new ring.
   A. I only B. II only C. Both I and II D. Neither I nor II
4. Knurling of a piston skirt
   I. is done to expand the overall diameter.
   II. should not be done to a used piston.
   A. I only B. II only C. Both I and II D. Neither I nor II
5. Press-fit design piston pins
   I. require lock rings to keep the pin from slipping.
   II. do not require honing.
   A. I only B. II only C. Both I and II D. Neither I nor II
6. Disassembly of a press-fit piston and connecting rod assembly
   I. requires a driver tool and hammer
   II. requires a press and special pin fixtures.
   A. I only B. II only C. Both I and II D. Neither I nor II
7. Full-floating piston pins
   I. require a press and special pin fixtures to remove the pin.
   II. have a bronze bushing surface within the connecting rod.
   A. I only B. II only C. Both I and II D. Neither I nor II
8. A connecting rod
   I. should be measured for out-of-roundness at the large end.
   II. with a full floating pin should be measured for lateral runout.
   A. I only B. II only C. Both I and II D. Neither I nor II
9. A ring groove cleaner
   I. should remove only the carbon buildup within the top groove.
   II. can be used to regroove the top ring groove.
   A. I only B. II only C. Both I and II D. Neither I nor II

10. Fitting piston pins with a precision hone
    I. requires two pins for the gauge setting fixture.
    II. requires stroking the piston or rod back and forth over the turning mandrel.
        A. I only B. II only C. Both I and II D. Neither I nor II

11. Connecting rods
    I. are manufactured by the drop forging process.
    II. require reconditioning if their beam areas become rusty.
        A. I only B. II only C. Both I and II D. Neither I nor II

12. Connecting rod reconditioning requires
    I. heat treating.
    II. enlarging the bore size to the next closest bearing diameter.
        A. I only B. II only C. Both I and II D. Neither I nor II

13. When assembling a press-fit piston and rod
    I. lubricate all surfaces making contact with the pin.
    II. the pin should be centered within the piston and rod.
        A. I only B. II only C. Both I and II D. Neither I nor II

14. When assembling a full-floating piston pin and rod
    I. use a press and special tool.
    II. place the pin lock rings with the solid part facing upward.
        A. I only B. II only C. Both I and II D. Neither I nor II

15. The major thrust side of a piston
    I. should be on the left when viewing the engine from the front.
    II. is on the right for engines that have full-floating pins.
        A. I only B. II only C. Both I and II D. Neither I nor II

16. Piston rings
    I. are larger than the cylinder before they are installed.
    II. are coated and plated to prevent rusting before they are installed.
        A. I only B. II only C. Both I and II D. Neither I nor II

17. The markings on compression rings
    I. are trade marks.
    II. indicate the UP side.
        A. I only B. II only C. Both I and II D. Neither I nor II

18. Connecting rod alignment must be done
    I. before the rings are installed on the piston.
    II. during each engine overhaul.
        A. I only B. II only C. Both I and II D. Neither I nor II

19. Piston ring end gap
    I. is the minimum clearance necessary to prevent the loss of compression at wide open throttle.
    II. should be measured while the piston is in the cylinder.
        A. I only B. II only C. Both I and II D. Neither I nor II

20. After the piston and connecting rod are installed in the cylinder
    I. the ring gaps should be aligned with one another.
    II. the bearing and rod side clearance should be measured on all rods.
        A. I only B. II only C. Both I and II D. Neither I nor II

DATE COMPLETED ______________ INSTRUCTOR'S SIGNATURE: ______________________________

unit 12

Gaskets, Sealants, and Engine Assembly

Gaskets, seals, adhesives, and sealants are often overlooked in providing engine repair information. These topics, along with engine assembly procedures, are discussed in detail in this unit.

OBJECTIVE 12-A (written test)

After completing the preparation assignment, you will correctly answer a minimum of 7 out of 10 multiple choice questions in a written test, without reference material, on the following topics:

1. Gaskets and their service
2. Sealants and adhesives and their service
3. Engine assembly procedures

Preparation:

Study pages 369 through 379.

GASKETS AND SEALS

Within the engine are fluids (liquids and gases) which at times have high pressure and high temperature. Since the engine is assembled from many parts, a method of sealing in these fluids is required. There are also locations that require sealing to prevent foreign matter from entering the engine.

Gaskets are materials used to seal two or more nonmoving surfaces. A seal is typically used to contain liquids between a moving and a nonmoving surface (Fig. 12-1).

The materials from which gaskets are made are many and varied. The material selected for a particular gasket depends on the type of fluids which it is to seal or control.

Asbestos, a natural mineral fiber, is used where there will be extremes in heat. Most often, it is used in combination with one of the metals, especially steel, to assist it in holding its shape. Asbestos gaskets are used in the cylinder head, intake and exhaust manifold, and carburetor mounting flange (Fig. 12-2).

Cork, which comes from the bark of the cork oak tree, is used most often to seal surfaces containing oil. Cork gasket material is made by grinding the cork into small bits and using an adhesive to bind it together. It is then rolled into sheets of various thicknesses and cut to the desired shape. It has the advantage of being highly compressible, which can help to seal relatively uneven surfaces, such as an oil pan or valve cover (Fig. 12-3).

Since cork is brittle, it must be handled with care. It is subject to shrinkage during storage, since it is cut to shape while moist. To expand a shrunken cork gasket, soak it in water for three to five minutes. Most often it will return to its cut size and fit the surface for which it was intended.

Caution: While soaking the cork, check its fit each minute or two so it does not overexpand. When cork is coated or mixed with rubber, it is less apt to shrink or break as easily.

Synthetic (man-made) rubber is a highly resilient (able to return to its original shape after being distorted) and compressible material. Neoprene is the most common rubber compound used in making automotive gaskets and seals. This material is not affected by oils, greases, coolants, or water and does not shrink or expand during storage. Rubber is never used where there are high temperatures such as combustion or exhaust gases.

Rubber is frequently used in combination with cork, paper, or asbestos to bind these materials together. It is also the most common material used as the sealing element of a lip seal (Fig. 12-4), in which rubber is molded to a steel casing and a spring (not always used) applies pressure to the turning shaft. This type of seal is typically found on timing covers. Note how the lip faces the liquid (oil).

Rubber is used by itself in making ring seals (Fig. 12-5). The cross-sectional shape of a ring seal often determines its name.

Paper is the name used to describe a number of paper-like gaskets which are made from vegetable fibers. One dark gray or black "paper" gasket is actually rubber-impregnated (soaked) asbestos fiber. The vegetable fiber is usually treated

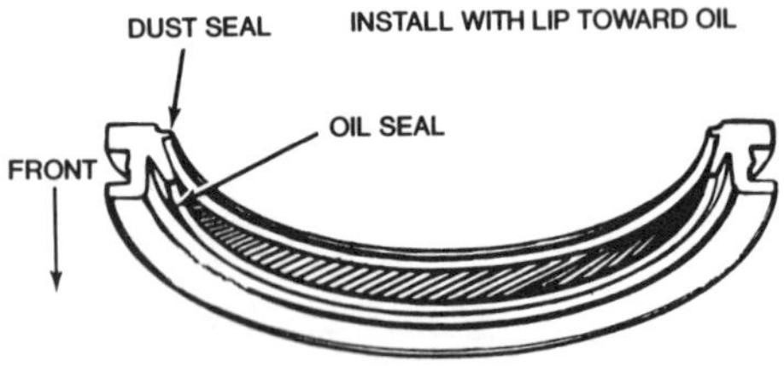

FIG 12-1 Rear main bearing oil seal. (Chevrolet Motor Division, GMC)

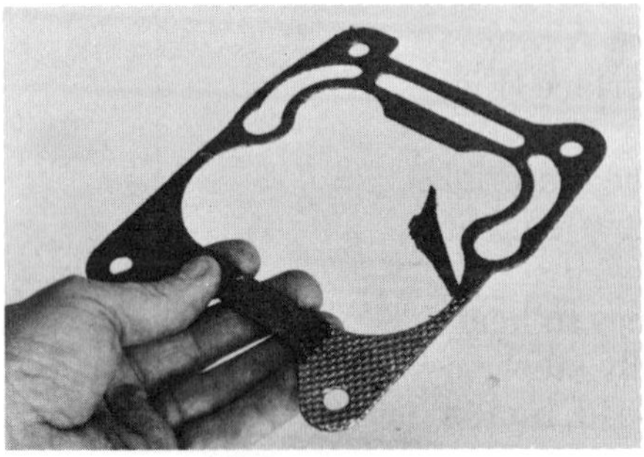

FIG 12-2 Asbestos and steel gasket.

FIG 12-3 Cork (side gaskets) and rubber (front and rear seal) gasket materials. (Chevrolet Motor Division, GMC)

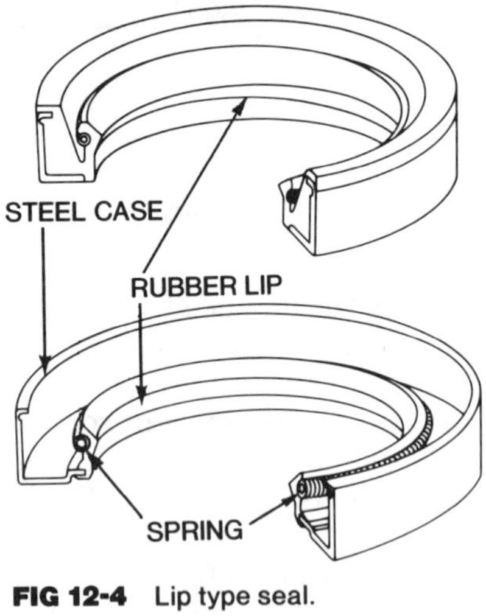

FIG 12-4 Lip type seal.

with resin to keep it from breaking apart after it comes into contact with engine liquids. This type of gasket is most often used in sealing coolant at places like the water pump and thermostat housing.

Steel (iron plus a small amount of carbon) is used in many gasket and seal locations on the engine. Cylinder head gaskets are most often a combination of steel and asbestos or plain steel with embossed sealing surfaces. As the head is torqued in place, the embossed ridges crush down to make a tight seal (Fig. 12-6).

The steel and asbestos head gasket is the most popular for engine repair application, since it is able to withstand more surface and torquing irregularities. Also, if a combination gasket is used to replace the embossed type after the head has been resurfaced, the chances of bolt hole alignment with the intake manifold on a vee engine are much greater.

Aluminum is used in combination with asbestos to seal areas of high heat or high pressure.

Copper is used alone or in combination with asbestos in locations with a high unit pressure. This pressure can be from fluids to be sealed or the required tightness of the fastener.

Gaskets for engine service are sold in varying numbers depending on the extent of the work being done.

Note: The major gasket manufacturers list the gaskets enclosed in each package. Read the label before leaving the parts counter to be certain all needed gaskets are included (Fig. 12-7A).

1. Full set (complete set)—all gaskets needed to completely recondition an engine. This set is recommended for any engine removed from the vehicle for service (Fig. 12-7B).
2. Head set (valve grind set)—all gaskets needed to grind the valves.
3. Oil pan set—gaskets needed to replace oil pan. This set, along with a head set, could be used for an in-car ring and valve job.

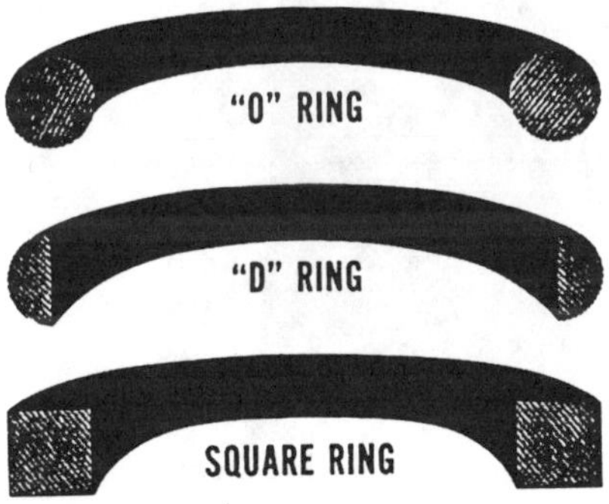

FIG 12-5 Cross-sectional shapes of ring seals. (Chrysler Corporation)

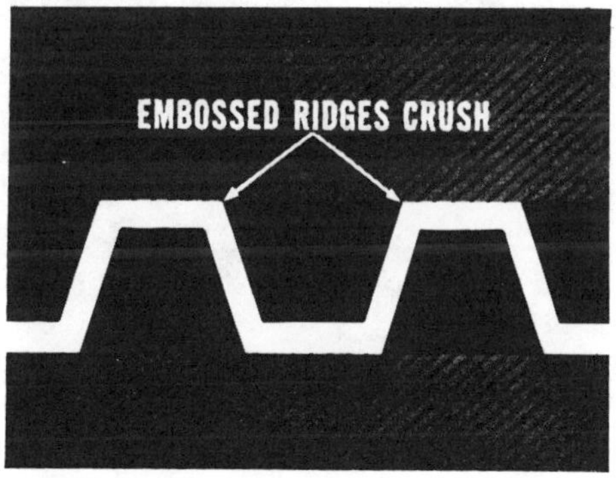

FIG 12-6 Gasket crush. (Chrysler Corporation)

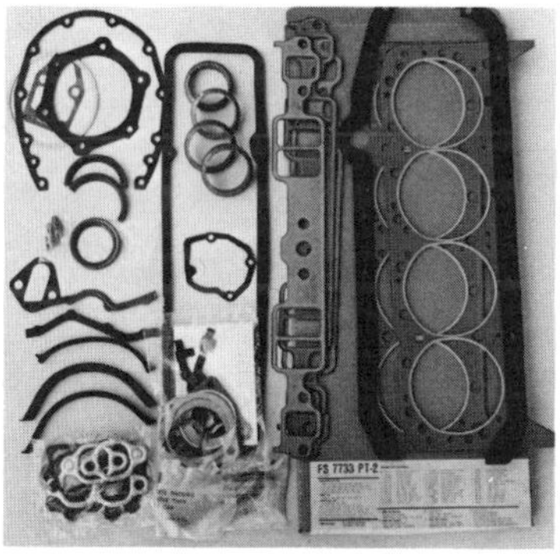

FIG 12-7(B) Gasket set.

FS 7733 PT-2

Full Set

CHEVROLET V8
Pass. & Truck, Engine 265, 283 (Exc. FI), 302, 307, 327 (Exc. Sp. H/Perf.), 350 (Exc. Sp. H/Perf.): 1957-76; Truck 1½ Ton, 2 Ton Engine 265 after 9-1-55 1955-56
GMC V8
Engine 283, 307, 327, 350, 1967-76
PONTIAC V8
Engine 307, 1971-72
Victor FS 1178 VF, FS 1178 VH, FS 1178 VE
McCord OH 6631 M-5, OH 6631 M, OH 6631 M-6
Fitzgerald RS 0710 L-1, RS 0723 L-2, RS 0861 L
Application coverage may be greater than shown on package. Consult Fel-Pro catalogs for latest data.

Gasket Set Includes:

16- R 5112	Valve Stem Seals	1- K 5198	Air Cleaner to Carb.	1- 13032	Water Pmp. Body
1- T 5124	Timing Cover	2- 7733 PT-2	Head Gasket	1- 13835	Oil Pmp. Cover
1- K 5142	Water Outlet	2- LA 8194	Exh. Pipe Packing	1- 14842	Timing Cover Seal
1- D 5146	Oil Pan Right	2- LA 8592	Exh. Pipe Packing	1- 30328	Oil Pan Front
1- R 5147	Oil Pan Rear	2- LB 9274	Exh. Mfld. Center	1- 60083	Exh. Mfld. Vlv. Heat Riser
1- R 5148	Oil Pan Front	4- LB 9275	Exh. Mfld. Ends	1- 60099	Carb. Mounting
1- R 5151	Int. Mfld. Sl. Rear	1- 9288	Carb. Mounting	1- 70194	Distributor
1- D 5152	Wtr. Pmp. to Blk.	1- K 11528	Air Cleaner to Carb.	1- 90102	Mfld. Seal Front
2- A 5155	Rear Main Brg. Sl.	2- R 11829	Rear Main Brg. Sl.	1- 90133	Mfld. Seal Front
1- P 5181	Fuel Pmp. Mtg. 2-hole	1- 70822	Oil Pan Drain Plug	1- 90134	Mfld. Seal Rear
1- P 5182	Fuel Pmp. Mtg. Plate	1- K 12569	Water Outlet	1- 90238	Int. Mfld. Side
1- D 5197	Oil Pan Left	2- R 12869	Valve Cover	2- 90314	Int. Mfld. Side

1. **Does Not Contain:** Valve Stem Seals 1959 Pass. Eng. 283 - SS 12434
2. **Does Not Contain**
2. Continued Exh. Pipe Gasket 1955-56 Truck - LB 9272
3. **Does Not Contain**
3. Continued 60002,60021,60043, 60110,60111,60153 For Specific Applications See

FIG 12-7(A) List.

SEALANTS AND ADHESIVES

Nearly all engine mating components require a gasket or seal between their surfaces. If the surfaces are correctly prepared and the proper gasket is used, no sealant or adhesive should be necessary. When two surfaces are joined without a gasket and there is to be a liquid- or gas-tight seal, a sealant must be used. Most engine technicians prefer to use small amounts of these chemicals during assembly to ensure against leakage or slippage of any sealing device. Many times adhesives are used merely to hold the gasket in place while the parts are assembled. In places where there is excessive vibration, an adhesive may be used to prevent loosening of a nut or bolt.

Note: Many head and other gaskets are coated with materials to improve their sealing ability. These applied chemicals should *not* be covered with any additional sealant or adhesives.

RTV

The foremost sealant and adhesive combination is a multipurpose silicone rubber substance usually referred to as RTV (Fig. 12-8). These initials stand for room temperature vulcanization, which means the material will cure (harden) when exposed to average room temperatures (70° F., 21°C.). This material is the most forgiving of all the sealants because it does not totally harden for 24 hours unless it is heated—running the engine, for instance. This means that if it is necessary to disassemble a recently installed part, chances are the gasket will not be damaged. Even when a disassembly is completed many miles later, this sealer is relatively easy to remove. The temperature range of effective sealing is −80° to 450° F. (−62° to 232° C.).

FIG 12-8 RTV (Room Temperature Vulcanizing) sealant.

FIG 12-9 Popular gasket sealants and adhesives.

CEMENTS

Gasket cements have been used in the automotive trade for many years. (Fig. 12-9). Most of these sealers-adhesives have a resin base with varying amounts of filler material, oils, and solvents. These can usually be grouped according to their setting nature: hard, flexible, and soft. The name Permatex, a brand name of the Permatex company, is most often used for this group of sealers. Select the sealant most suitable to the location on the engine.

A third group of chemicals in popular use today is the thread sealants, most often referred to as Loctite, a brand name of the Loctite Corporation (Fig. 12-10). This material comes in three strengths: Blue is for sealing threads that require disassembly. Red is for high strength applications that would not require later disassembly. Green is used for permanent placement of press-fit parts.

GASKET AND SEAL INSTALLATION

Check all gaskets by placing them in position before attempting to install the part. Look for directions like FRONT, THIS SIDE UP (Fig. 12-11). *When in doubt, compare new gaskets with old ones* or consult your instructor.

Make certain all surfaces are properly cleaned and all thread holes have been chased and chamfered, if necessary. All bolts and nuts should be cleaned. Assemble close at hand all parts to be installed and check all bolts and nuts for placement.

Use a torque wrench and the correct tightening sequence where required.

Use no more sealant or adhesive than necessary. When in doubt, ask your instructor.

Aluminum paint may be used on head gaskets that are *not* coated by the manufacturer (Fig. 12-12). *Use spray application only.*

Use a seal driver or suitable flat surface to replace lip seals. Do not strike the seal directly with a hammer.

FIG 12-10 Locking chemicals.

FIG 12-11 Installation direction label on a gasket

Note: The lips of all seals should face the liquid. Lightly coat the outer edge of a steel-cased lip seal with RTV or No. 2 Permatex before driving into position. Lubricate all rubber or fabric seals to prevent burnouts when the engine is started.

OBJECTIVE 12-A1 (manipulative)

Working as a team with one other student and with the aid of Job Report Sheets 3-A1 and 12-A1, you will assemble the remaining components of an instructor-assigned vehicle engine prior to installation.

Time Recommended:

Twice the flat rate time allowed in *Motors, Chilton,* or *National Service Data* manuals.

Preparation:

1. Review pages 369 through 373.
2. Study pages 374 through 378.
3. Study manufacturer's or general repair manuals for the engine being serviced.
4. Attend lectures and demonstrations on engine assembly.

Tools and Materials Needed:

1. Engine parts and gaskets
2. Engine oil and filter
3. RTV sealant
4. Moly grease or molykote
5. Aluminum paint (spray can only)
6. Feeler gauge set
7. Drill motor (reversible)
8. "Dummy" distributor
9. Torque wrench
10. Assorted sockets and handles
11. 2 hammers (1 ball peen, 1 soft-faced)

Procedure:

Note: Just prior to final assembly, gather all parts, fasteners, and gaskets to review their placement. Also, be certain all machining and cleaning work has been completed. Flatten the gasket surface of all sheet metal parts with two hammers as shown in Figure 12-13.

1. Place the oil pan gasket on the block. Set the end pieces (usually rubber) on after the edge gaskets (cork or composition). This is necessary because most pan gasket pieces are designed with interlocking joints to prevent slippage.
2. Seal all joints with RTV sealant just prior to bolting on the pan. First tighten the bolts that are closest to the joints, then those remaining, to the specified torque. Be sure pan plug is tight.
3. Revolve the engine to bring it right side up and pour 2 to 3 quarts of oil into the crankcase.
4. Install the oil filter after filling it with oil if possible.

5. Install the lifters after coating the faces with grease or molykote.

Note: Some engines with nonadjustable and/or semi-permanent rocker arms require installation of the pushrods at this time.

6. Place all cleaned cylinder head bolts in their openings in the head. Check for correct lengths of bolts.
7. Install the cylinder heads after placing the gaskets in position. Aluminum paint may be sprayed on both sides of the gaskets before installation if they are *not* seal-coated by the manufacturer.
8. Apply a light coat of RTV sealant to the thread area and the underside of each cylinder head bolt (Fig. 12-14).
9. Torque the head bolts to the specified tightness following the order recommended (Fig. 12-15).

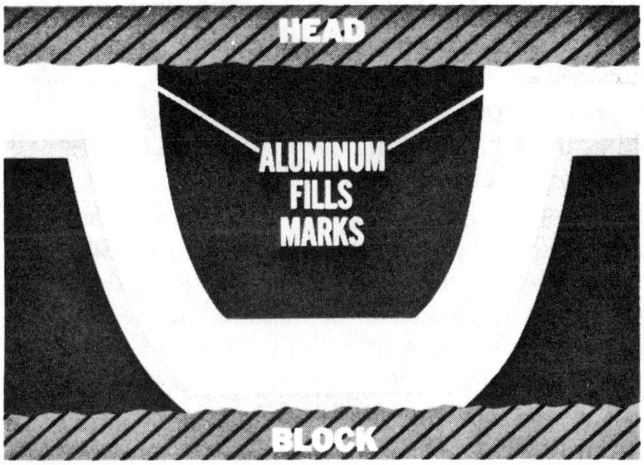

FIG 12-12 Aluminum coating fills in irregular surfaces.
(Chrysler Corporation)

FIG 12-14 Sealant applied to a head bolt to prevent coolant leakage.
(Buick Motor Division, GMC)

FIG 12-13 Straightening sheet metal distortion around the bolt holes.

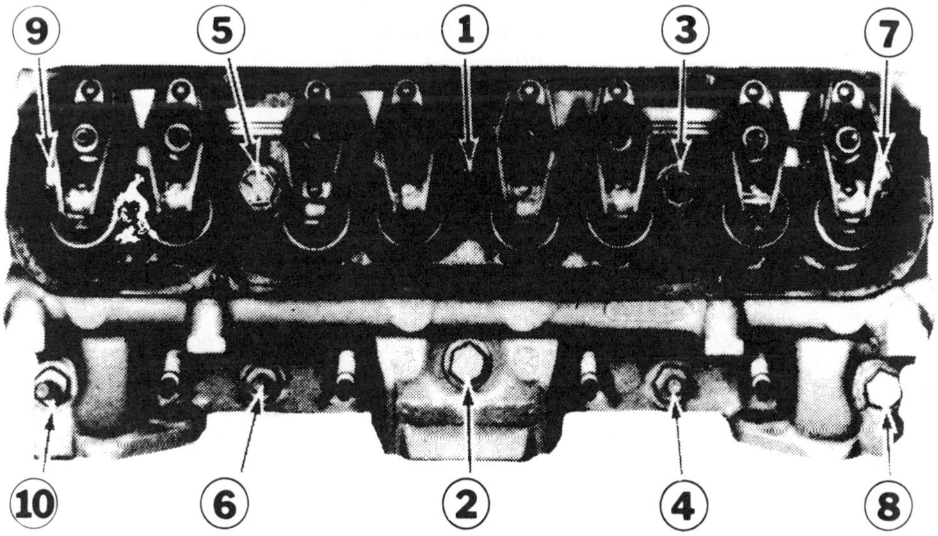

FIG 12-15 Cylinder head bolt tightening sequence.
(Buick Motor Division, GMC)

10. Install the pushrods and tighten the rocker arms just enough to take up most of the gap between the rocker arm and the valve stem.
11. Pre-adjust the valve clearance for both types of lifters (hydraulic or mechanical) in the following manner:
    a. Rotate the crankshaft to bring the piston to TDC compression stroke in #1 cylinder. Place 2 or 3 flywheel bolts in the crankshaft flange and use a pry bar to turn the crank.
    b. For hydraulic lifters, tighten both intake and exhaust rocker arms of #1 until the plunger part of the lifter just begins to move. Further tighten these two down until the specified number of turns has been reached. For mechanical lifters, tighten the two rocker arms of #1 to the specified lash (clearance) with a feeler gauge of the specified thickness. If "hot" clearance is specified, increase feeler thickness by 0.003 in. (0.076 mm).
    c. Rotate the crankshaft clockwise (viewed from the front) until the next cylinder is in the TDC position of its compression stroke. Adjust these two rocker arms.
    d. Continue step c until all rocker arms have been adjusted.
12. Install the pushrod cover if so equipped.
13. Install the intake manifold gaskets interlocking each corner. Apply a small amount of RTV to each joint and to the underside of each long piece. RTV will help hold the gasket's position (Fig. 12-16).
14. Prerun the oil pump, using a "dummy" distributor and a reversible drill motor (Fig. 12-17). Make the pump turn in the same direction as the distributor normally rotates. A pressure gauge can be installed at the oil pressure sending unit opening to check the pressure relief valve of the oil pump. Run the pump long enough to bring oil to each rocker arm and lifter.
15. Install the oil pressure and temperature indicator sending units. Seal only the upper portion of the threads with RTV.
16. Install the valve cover(s).
17. Install the water pump and thermostat. Be certain the thermostat is placed correctly. Look for arrows indicating the direction of flow (Fig. 12-18).
18. Cover all engine openings with tape or paper; then spray exterior surface with factory color paint.
19. Install exhaust manifolds, watching for differences in right and left side designs.
20. Install the fuel pump.
21. Attach motor mounts, noting right and left side differences.
22. Install the flywheel and bellhousing (clutch housing) (Fig. 12-19). Torque the bolts or nuts to specifications. If it is necessary to remove the engine from the assembly stand, set it on the floor supporting the sides of the pan-block surface with wood blocks.
23. Install the pilot bearing or bushing as required (Fig. 12-20).
24. Install the clutch disc and pressure plate with a torque wrench. Be sure clutch disc is facing the proper direction and is centered. Use a clutch aligning tool or an input shaft from a similar transmission.
25. Install the starter motor if it will not interfere with the engine installation.
26. Install any additional parts which are easier to attach to the engine prior to its installation in the vehicle.
27. Cover the engine with a plastic bag until just before installation (Fig. 12-21).

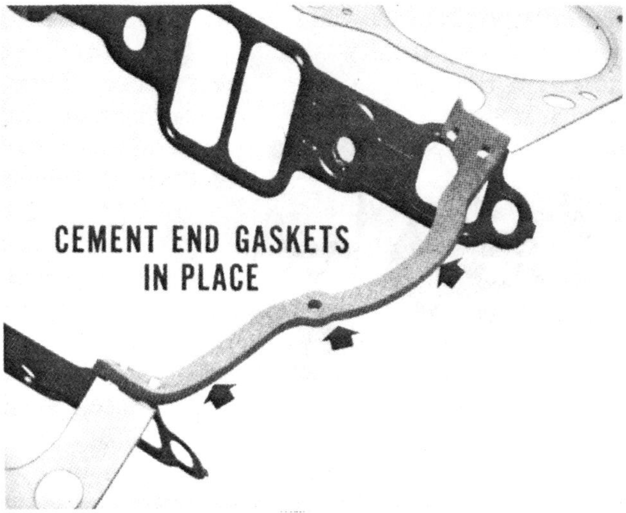

FIG 12-16 Use a sealant on gasket ends.
(Chrysler Corporation)

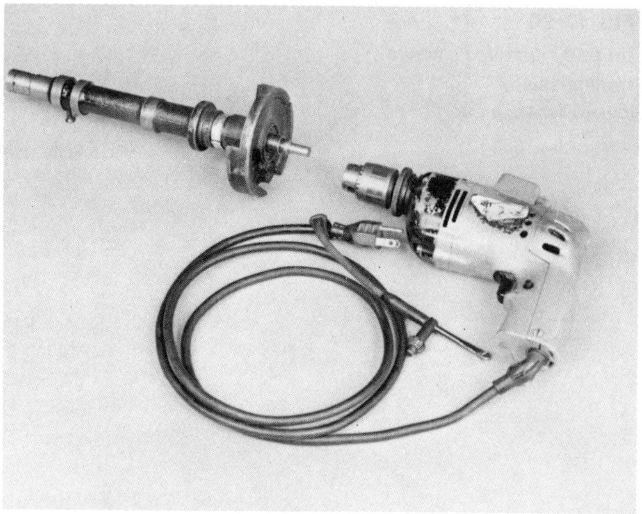

FIG 12-17 "Dummy distribution and drill motor used to drive the oil pump.

FIG 12-18 Place the thermostat according to directions on its housing.
(Oldsmobile Division, GMC)

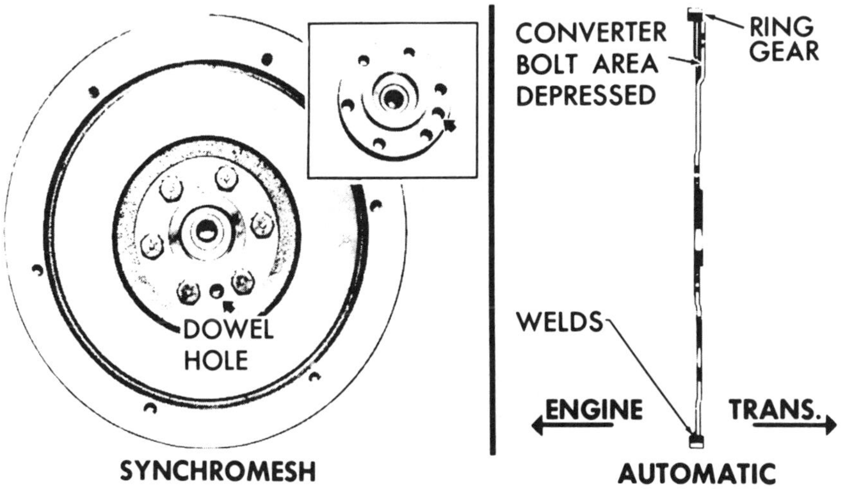

FIG 12-19 Flywheel types and their permanent location.
(Buick Motor Division, GMC)

FIG 12-20 Pilot bearing properly installed (manual transmission).
(Oldsmobile Division, GMC)

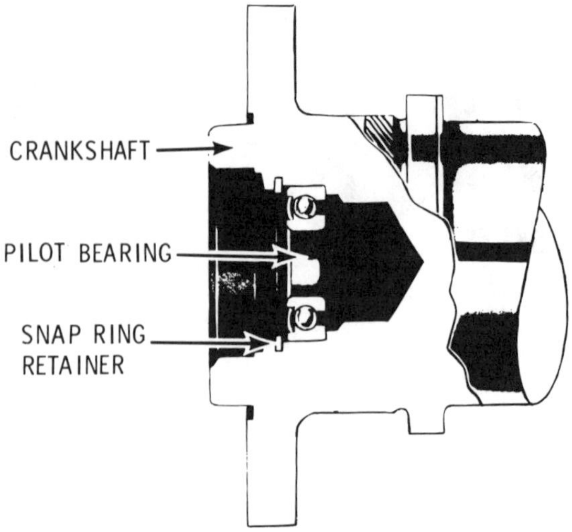

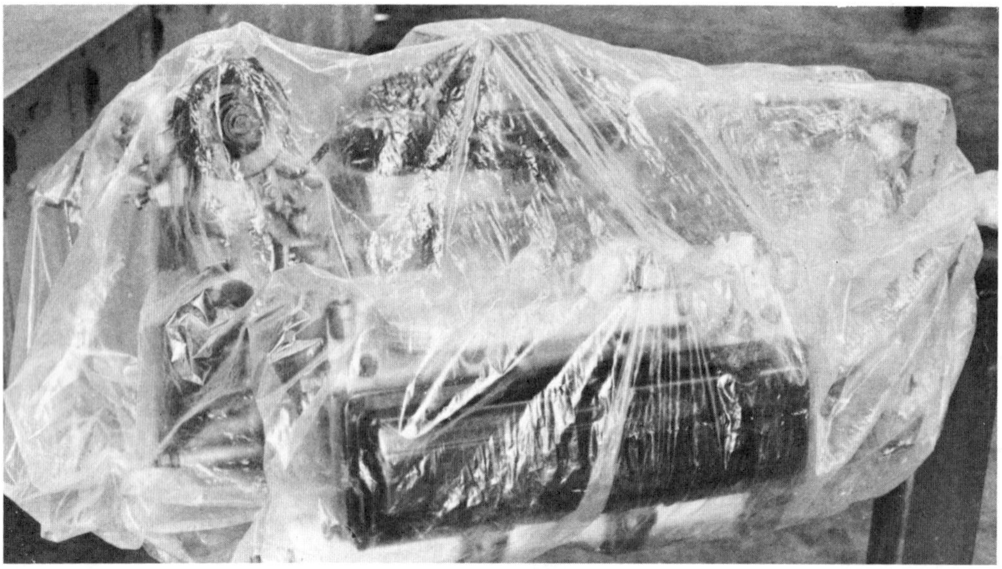

FIG 12-21 Cover the engine with a plastic bag during storage prior to installation.

ENGINE ASSEMBLY

Student Name ______________________ Date ______________

Make of Car ______________ Model ______________ Year ______________

Time Started ______________ Time Finished ______________ Total Time ______________

Flat Rate Time or Recommended Time ______________________

Specifications:

Engine tightening specifications (Job Report Sheet 3-A1)

Checkup Questions:

1. Explain how the timing marks should be aligned on this engine.

   ____

   ____

2. How many different lengths of cylinder head bolt did the engine have? ______________

3. Why is it necessary to follow a particular order when tightening the cylinder head bolts? __________

   ____

4. Why is RTV recommended on the threads and underside of the bolt head? ______________

   ____

5. How many turns are required past zero lash, *or* what was the specified valve clearance? __________

6. What problems may result if no oil reaches the rocker arms after 5 minutes of oil pump operation?

   ____

   ____

DATE COMPLETED ______________ INSTRUCTOR'S SIGNATURE ______________

Name ______________________________ Date ______________

DIRECTIONS:

Select the best answer to each question and circle the correct letter at the left.

1. Asbestos materials are used for gaskets
   I. to control oil flow and hot coolant.
   II. to seal fuel lines.
   A. I only
   B. II only
   C. Both I and II
   D. Neither I nor II
2. Lip seals
   I. are made from synthetic rubber and leather.
   II. should be installed so the lip points toward the liquid to be sealed.
   A. I only
   B. II only
   C. Both I and II
   D. Neither I nor II
3. Gasket sealants
   I. of RTV will cure to a flexible, rubber-like material.
   II. are necessary on asbestos gaskets to keep them from decomposing.
   A. I only
   B. II only
   C. Both I and II
   D. Neither I nor II
4. Cork gaskets
   I. can be expanded by placing in warm water.
   II. are used where exhaust gases are to be sealed.
   A. I only
   B. II only
   C. Both I and II
   D. Neither I nor II
5. Sealing the corners where two gaskets are joined
   I. should be done after the gasket is compressed.
   II. is good insurance against leakage.
   A. I only
   B. II only
   C. Both I and II
   D. Neither I nor II
6. Head gaskets without factory applied sealer
   I. should be coated with aluminum paint on both sides.
   II. should be preheated before installing.

A. I only
B. II only
C. Both I and II
D. Neither I nor II

7. Sealant should be applied
   I. to head gaskets.
   II. to all asbestos gaskets.
   A. I only
   B. II only
   C. Both I and II
   D. Neither I nor II

8. Hydraulic lifters
   I. can be adjusted before the engine is run.
   II. should be filled with oil after the prelube operation.
   A. I only
   B. II only
   C. Both I and II
   D. Neither I nor II

9. The clutch disc
   I. must be installed with the correct side toward the transmission.
   II. requires alignment with the pilot bearing.
   A. I only
   B. II only
   C. Both I and II
   D. Neither I nor II

10. Which of the following should be used on an engine which will be stored for any length of time?
    I. Tape over any openings
    II. Plastic bag over the entire engine
    A. I only
    B. II only
    C. Both I and II
    D. Neither I nor II

DATE COMPLETED ______________ INSTRUCTOR'S SIGNATURE ______________

unit 13

Engine Installation and Operation

Before installing the engine, the vehicle must be properly prepared. Preparation, engine installation, and break-in procedures are covered in this unit.

After completing the preparation assignment, you will correctly answer a minimum of 7 out of 10 multiple choice questions in a written test, without reference material, on the following topics:

1. Vehicle preparation
2. Break-in procedures
3. Engine installation

OBJECTIVE 13-A (written test)

Preparation:

Study pages 383 through 390.

VEHICLE PREPARATION

When the engine is removed, there is an excellent opportunity to service such parts as the exhaust system, the front seal of an automatic transmission, and the radiator and/or transmission cooler.

The engine compartment and front suspension areas should be thoroughly cleaned before the engine is reinstalled. Be careful not to destroy the labels attached to the wires, cables, and hoses during this cleaning process.

BREAK-IN PROCEDURES

Break-in is the term used to describe the operating conditions an engine should experience during its initial running following major or minor reconditioning. Since many of the parts and their surfaces are new, they will frequently have more than the normal amount of friction to overcome. If the engine is properly operated, the parts should wear with one another until this high initial friction is gone. Piston rings offer the most friction, so the break-in process is designed primarily for them.

The slight irregularities of the ring face of new rings and the reconditioned surface of the cylinder wall must be worn away before break-in is considered complete. The amount of break-in running time and the speed of operation is the subject of much discussion. A prominent piston manufacturer states it may take as many as 2,500 miles (4 022 km) for piston rings to become fully seated.

The actual engine operation should involve high idle speed (2500 rpm) and periods of acceleration and deceleration. The high idle is necessary to keep the crankcase oil agitated so it amply covers the cam lobes, lifters, rocker arms, valve stems, and other areas that rely on splash lubrication. Acceleration and deceleration periods of operation are needed soon after the engine first runs so that combustion gases will force the rings to the cylinder wall. This speeds up the ring seating process. Deceleration pulls oil into the upper ring area of the piston and past the valve guides to assist in their initial break-in.

The vehicle should not be driven at sustained high speed for the first 100 miles (161 km).

OBJECTIVE 13-A1 (manipulative)

Working as a team with one other student and with the aid of Job Report Sheets 3-A1 and 13-A1, you will prepare the vehicle and install the engine in an instructor-assigned vehicle.

Time Recommended:

Twice the flat rate time allowed in *Motors, Chilton,* or *National Service Data* manuals.

Preparation:

1. Review pages 383 through 384.
2. Study pages 384 through 387.
3. Study manufacturer's or general repair manuals for the vehicle being serviced.
4. Attend lectures and demonstrations on engine installation.

Tools And Materials Needed:

1. All remaining engine and vehicle parts and gaskets
2. Steam cleaner
3. Battery charger
4. Fender covers
5. Engine hoist and sling
6. Vehicle hoist
7. Floor jack
8. Guide bolts
9. Assorted sockets and handles
10. Assorted wrenches

Procedure:

1. Steam clean the engine compartment area of the vehicle. Paint the surfaces if desired.
2. Charge the battery at a slow rate (5–10 amps) for at least 12 hours (Fig. 13-1).

3. Bring the vehicle to the engine installation area, preferably over a vehicle lift. Tape fender covers to the body areas that can be scratched during engine installation procedures.
4. Remove any transmission support members installed after the engine was removed.
5. Place a floor jack under the transmission and jack it up to a position slightly above normal.
6. Check the position of the torque converter to see that it is fully engaged in the front pump. Ask for help from your instructor if you are uncertain (Fig. 13-2).
7. Check the exhaust pipes and replace the gaskets.
8. Attach the lifting sling to the engine as suggested in Unit 4. Use high-strength bolts (class 5) and washers or spacers.
9. If not previously done, lift the engine off the stand and lower it to the floor to remove the engine stand holding fixture.
10. Attach the transmission with its converter or clutch assembly if it was removed from the vehicle, or install guide bolts to assist in lining up the transmission during engine replacement (Fig. 13-3).

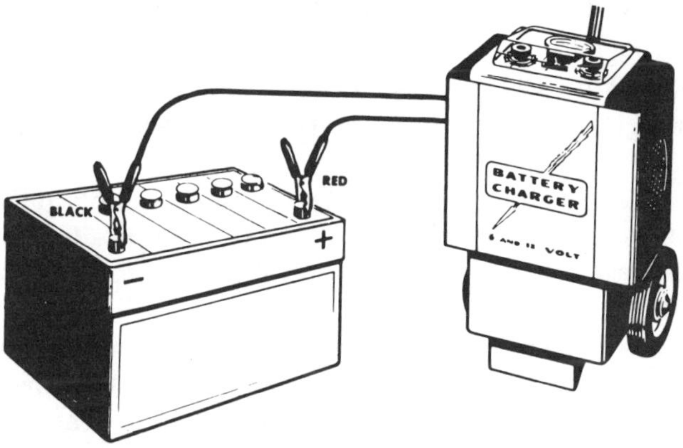

FIG 13-1 Charge the vehicle's battery to the full charge level. (Chrysler Corporation)

FIG 13-2 The torque converter must be fully engaged in the front oil pump of the transmission. (Chrysler Corporation)

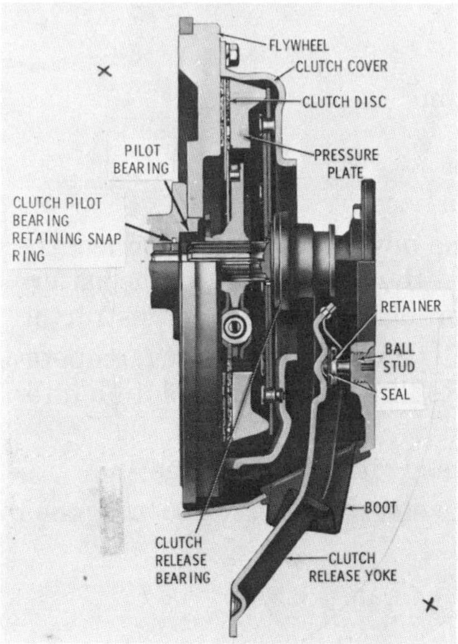

FIG 13-3 Clutch assembly. Install the cover and disc with an alignment tool. (Oldsmobile Division, GMC)

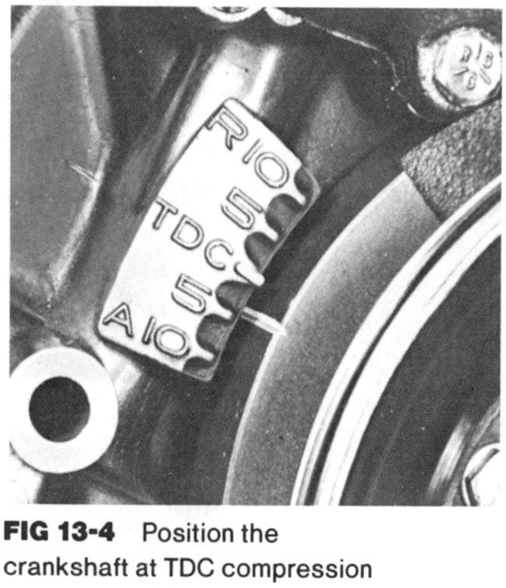

FIG 13-4 Position the crankshaft at TDC compression stroke in #1 cylinder. (American Motors Corporation)

11. Raise the engine in a slightly "front up" position and lower it slowly into the engine compartment. Watch the position of the exhaust pipe and other pieces. The mating surfaces of engine and transmission should be brought to a parallel position.
12. Lower the engine so that the guide bolts at the rear may slide into the bolt holes in the transmission. When the two surfaces have come together properly, insert two or more bolts to firmly attach the transmission to the engine.

Note: There should now be some distance (approximately ¼ in. or 6.35 mm) between the flywheel and the torque converter. If not, the converter is not fully engaging the front pump of the transmission, which will cause serious damage to the pump. Remove the engine and reposition the torque converter to engage the front pump.

13. Lower the engine all the way after installing the front motor mount bolts. Remove the jack from beneath the transmission.
14. Remove the engine sling and place the engine lifting device out of the way.
15. Raise the vehicle to remove the engine guide bolts and replace them with the bolts for those openings.
16. Install the remaining transmission bellhousing bolts and attach the torque converter and dust (flywheel) cover.
17. Install the exhaust manifold(s) if not done earlier. Attach the exhaust pipes.
18. Install any remaining parts, such as transmission linkage or lines, speedometer cables, starter or its wires, while the vehicle is raised.
19. Lower the vehicle and install the ignition system parts. Replace any part in questionable condition.
20. The internal engine parts should still be in position from the assembly process, so #1 piston is in TDC of the compression stroke. If it is not, position it now. Turn the crankshaft until the timing marks are aligned according to each engine's specifications (Fig. 13-4).
21. Install the distributor with the rotor pointing toward #1 spark plug wire. The ignition points should be close to breaking (separating), or the armature, reluctor, light gap, etc., should be aligned if it is the breakerless ignition system (Fig. 13-5). The final steps in static ignition timing will be discussed in steps 27 through 29.
22. Install the following components in the listed order:
    a. Air conditioning compressor
    b. Power steering pump
    c. Carburetor, linkage, and all fuel lines
    d. Fan and all drive belts—check belt tension
    e. Radiator, shroud, and oil cooler lines
    f. Hoses—radiator and heater
    g. All previously labeled wiring and vacuum hoses
23. Install the charged battery after cleaning the posts and cables. Connect the positive side and lightly touch the cable end of the negative side to the battery post. If it sparks, there may be an electrical problem. Consult your instructor.
24. Fill the radiator with plain water unless it will be subject to freezing temperatures immediately. Then check for leaks. Refill with coolant solution after road testing.
25. Secure a charged fire extinguisher. Have it ready in case of emergency.
26. Add oil to the engine, transmission, and power steering reservoir to bring their levels to the full mark.

27. To static time the ignition system, the engine must be in TDC compression stroke in #1 cylinder, and the distributor rotor should be pointing toward #1 spark plug wire. If necessary, check the rotor position by removing the cap.
28. Turn the distributor *with* rotation (retard), moving the cap so the rotor is halfway to the last firing plug wire. Now, remove the #1 spark plug wire from the plug and hold it near a ground.
29. Turn the distributor *against* rotation (advance) until the coil discharges. Tighten the distributor clamp down at the position where the spark occurred while advancing (turning distributor *opposite* normal rotor rotation). Replace #1 spark plug wire. Turn the key switch OFF.
30. Prime the carburetor by filling the bowl with gasoline through the internal vent opening (Fig. 13-6).
31. The engine should now start. Crank it over and run it at a fast idle, 2000–2500 rpm. Check the oil pressure gauge or light immediately. Watch for any liquid leakage or unusual noises. Stop the engine if there is anything unusual; otherwise run it until it reaches operating temperature (3–5 minutes).
32. Recheck the radiator level, adjust the carburetor and ignition timing, and install the hood.
33. Road test the vehicle by driving it normally at less than 55 miles per hour (88 kph). Accelerate and decelerate at least 10 times between 30 and 55 mph (48–88 kph). Return to the shop and check for liquid leakage or other irregularities.
34. Your instructor will want to give the vehicle a final road test before its release to the customer.
35. Inform the customer of his/her responsibility in maintaining the engine.

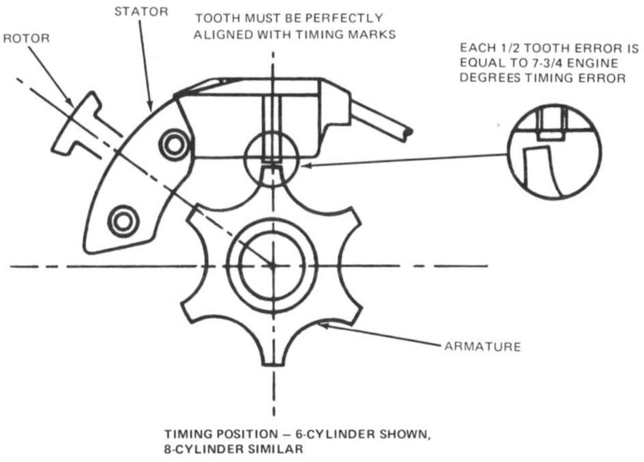

FIG 13-5 Static timing procedures for magnetic breakerless distributors.
(Ford Motor Company)

FIG 13-6 Prime the carburetor bowl by filling with gasoline through the internal vent opening.

JOB REPORT SHEET 13-A1

ENGINE INSTALLATION

Student Name ______________________ Date ______________

Make of Car ____________ Model ______________ Year ______________

Time Started ____________ Time Finished ____________ Total Time ____________

Flat Rate Time or Recommended Time ______________________________

Specifications:

Belt tension ______________ Alternator

______________ Power Steering

______________ Air Pump

______________ Air Conditioner Compressor

Ignition Timing ______________ Degrees BTDC-ATDC

Idle Speed ______________ RPM in N or Drive

Checkup Questions:

1. A/T Was the torque converter double checked for engagement with the front pump? ______________

   M/T Was the clutch disc properly aligned for installation of the engine? ______________

2. Were guide bolts used during the installation? ______________________________

3. Why is plain water used for the cooling system during the initial operating period? ______________

   ____

   ____

4. What was used to support the transmission during engine installation?

   ____

5. Why is it necessary to be cautious about connecting the negative battery cable (step 23)? ______

______

______

6. Why is it important to run the freshly started engine at 2000–5000 rpm? ______

______

DATE COMPLETED ______ INSTRUCTOR'S SIGNATURE ______

Name ______________________ Date ______________

DIRECTIONS:

Select the best answer to each question and circle the letter at the left.

1. Before the engine is reinstalled,
   I. check the torque converter position to make certain it is fully engaged with the front pump.
   II. steam clean the underhood areas.
   A. I only
   B. II only
   C. Both I and II
   D. Neither I nor II

2. Break-in is a term used to describe
   I. ring seating.
   II. engine operation following a major overhaul.
   A. I only
   B. II only
   C. Both I and II
   D. Neither I nor II

3. A reconditioned engine
   I. should not be run above 2500 rpm for the first 100 miles (161 km).
   II. must have a fully charged battery for starting.
   A. I only
   B. II only
   C. Both I and II
   D. Neither I nor II

4. Guide bolts are used
   I. to line up the front engine mounts.
   II. to hold the engine sling.
   A. I only
   B. II only
   C. Both I and II
   D. Neither I nor II

5. The clutch release bearing
   I. is attached to the clutch disc.
   II. applies pressure to the pressure plate.
   A. I only
   B. II only
   C. Both I and II
   D. Neither I nor II

6. The distributor is installed
   I. with #1 cylinder in the TDC compression stroke position.
   II. with its rotor pointing to #1 spark plug wire in the cap.
      A. I only
      B. II only
      C. Both I and II
      D. Neither I nor II
7. Before starting the engine for the first time,
   I. fill the radiator with coolant.
   II. static time the distributor.
      A. I only
      B. II only
      C. Both I and II
      D. Neither I nor II
8. Static timing involves
   I. rotating the distributor (advance direction with the ignition switch on until a spark snaps at #1 spark plug).
   II. placing the piston of #1 cylinder in the TDC position
      A. I only
      B. II only
      C. Both I and II
      D. Neither I nor II
9. When the engine first starts,
   I. it should be run at 2000–2500 rpm.
   II. the oil pressure light should be on until it is road tested.
      A. I only
      B. II only
      C. Both I and II
      D. Neither I nor II
10. All reconditioned engines
   I. should be road tested.
   II. should be watched carefully by the owner.
      A. I only
      B. II only
      C. Both I and II
      D. Neither I nor II

DATE COMPLETED ______________ INSTRUCTOR'S SIGNATURE ______________________

unit 14

In-Car Engine Repair And Maintenance

Many engine repair and maintenance operations are completed without removing the engine. These jobs are covered in this unit with special emphasis on cooling and lubricating system service objectives.

OBJECTIVE 14-A (written test)

After completing the preparation assignment, you will correctly answer a minimum of 14 out of 20 multiple choice questions in a written test, without reference material, on the following topics:

1. General repairs
2. Minor overhaul
3. Cooling system service
4. Lubricating system service

Preparation:

Study pages 393 through 419.

GENERAL REPAIRS

In-car service refers to repair or maintenance operations that are performed while the engine remains in the vehicle. Sometimes this work is called minor repair, as opposed to major repair, which involves removal of the engine from the chassis. Engine work that falls within the category of in-car service can be as simple as replacing a gasket or as complex as a ring and valve job.

On many occasions, an engine is partially disassembled in the vehicle for minor repairs only to reveal major engine damage. The engine must then be removed to complete the needed repairs. Through careful diagnosis as explained in Unit 3, many such situations can be avoided.

Minor repairs should be done in an area that has plenty of light available and is free from blowing dust. For repairs in which the valve covers, oil pan, or other protective parts are removed, a dirt free area is absolutely essential to prevent engine contamination. The most convenient place to work on repairs beneath the vehicle is the vehicle hoist.

The following text material will outline common service work performed on an engine while it remains in the chassis.

VALVE LIFT–CAM LOBE MEASUREMENT

When an engine fails to perform as it should although it has good compression and no electrical or fuel system problems, the problem may be worn cam lobes. This can be checked in the following manner:

1. Remove the valve cover(s).
2. Locate the cam lobe *or* valve lift specifications.
3. Depending upon the specifications, attach a dial indicator to a valve that will measure the valve lift or cam lobe lift. (Fig. 14-1).
4. Remove the spark plugs.
5. Crank the engine while watching the dial indicator.
6. Measure the lift of all valves and record the readings.

If any cam lobe lift readings are less than 0.006 in. (0.154 mm) or any valve lift readings are less than 0.009 in. (0.229 mm) of the specifications, the camshaft should be replaced.

CAMSHAFT REPLACEMENT

Camshaft replacement in an OHV engine requires the following steps:

1. Disconnect the battery.
2. Drain the coolant and remove the radiator and oil cooler lines if the vehicle is equipped with an automatic transmission.
3. Remove the alternator, power steering, and air conditioning drive belts.
4. Remove the water pump (most engines).
5. Remove vibration damper and timing cover. The fuel pump may have to be removed.
6. Remove the intake manifold with all its linkage, wiring, and tubing and distributor.

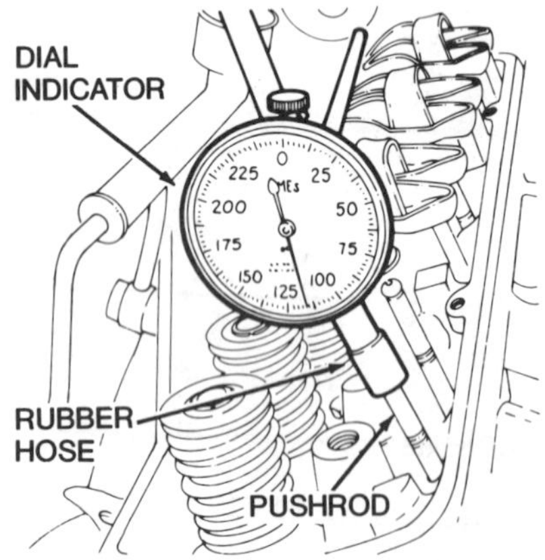

FIG 14-1 Dial indicator attached to the cylinder head and pushrod to measure the cam lobe lift.

(American Motors Corporation)

FIG 14-2 Raising the valve lifters.

(Cadillac Motor Car Division, GMC)

7. Remove the valve covers and loosen all rocker arms.
8. Remove the pushrods, keeping them in order to ensure their replacement exactly as removed.
9. Raise or remove the lifters, keeping them in order (Fig. 14-2).
10. Remove the camshaft sprocket and chain or gear.
11. Remove thrust plate bolts or screw if so equipped.
12. Install at least one long bolt in the camshaft end to assist in its removal (Fig. 14-3).
13. Remove the camshaft by pulling gently forward. Avoid gouging the cam bearings.
14. Install the new camshaft after coating the cam lobes and journal with moly grease.
15. Align the timing marks as indicated in a shop manual (Fig. 14-4).
16. Complete this job by following the reverse order of disassembly. Be certain gasket surfaces, bolts, and bolt holes are cleaned. Seal gasket joints with RTV sealant. Follow recommended torque patterns and specifications.

VALVE SEAL-SPRING REPLACEMENT

If diagnosis reveals excessive valve guide oil consumption, there may be defective valve stem oil seals or a broken spring. The procedures following will assist in their replacement.

1. Remove the valve cover(s).
2. Remove the rocker arms. On engines with hydraulic lifters and pivot rocker arms, count the number of turns to remove each nut and record this number. On some engines, rocker arm removal is not possible; refer to a shop manual for further directions.

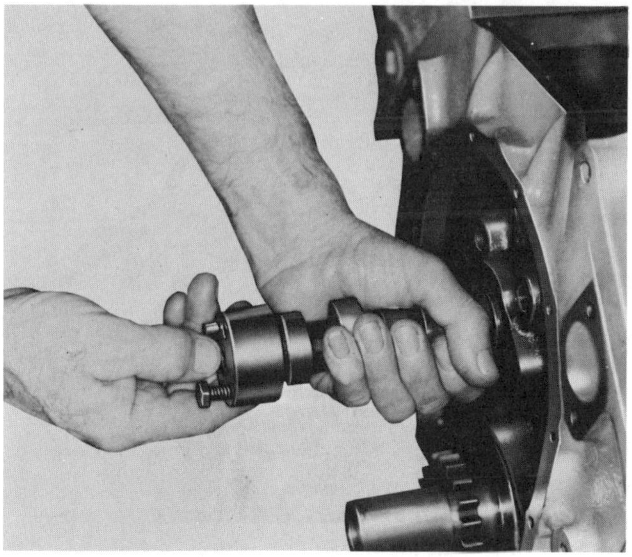

FIG 14-3 Removing the camshaft.
(Chevrolet Motor Division, GMC)

FIG 14-4 Typical valve timing marks. (Chevrolet Motor Division, GMC)

3. Remove the spark plug and install a shop air adapter.
4. Pressurize the cylinder with shop air (Fig. 14-5).
5. Compress the valve spring, remove the locks (keepers), and carefully release the spring.
6. Replace the seal or spring.
7. Reassemble in reverse order.
8. Adjust valve clearance (Fig. 14-6). Hydraulic lifter pivot rocker arms are adjusted by turning the nut down the number of turns past zero lash required in the specifications or the same number required for removal. Mechanical lifter pivot rocker arms for each cylinder should be adjusted to specifications with the piston at TDC compression stroke.

   Shaft rocker arm assemblies can be replaced as a unit. Tighten each stand bolt only slightly until all stands are seated on the head.

OIL PAN REMOVAL AND REPLACEMENT

Oil pan removal is usually required for bearing replacement. Some vehicles are designed so the pan cannot be removed without engine removal. Before attempting pan removal, check a repair manual to be certain this work can be completed as desired. The following are general procedures for oil pan removal and replacement:

1. Disconnect the battery.
2. Remove the fan shroud holding screws on the fan blade.
3. Raise the vehicle on a hoist or support it with jack (safety) stands.
4. Drain the oil.
5. Loosen or remove clutch and transmission linkage.
6. Remove the crossover exhaust pipe.
7. Remove the front motor mount bolts.

FIG 14-5 Compressing the valve spring with the cylinder pressurized with shop air.
(Chevrolet Motor Car Division, GMC)

FIG 14-6 Adjusting the valve lifter clearance.
(Chevrolet Motor Car Division, GMC)

8. Raise the engine by jacking up on the vibration damper or oil pan. Place a wood block between the jack and the damper or pan to prevent damage. Set wood blocks between the engine and its mounting points to hold it in the raised position (Fig. 14-7).
9. Remove the clutch or torque converter dust cover.
10. Remove the oil pan bolts and loosen the pan. If it will not loosen, check for remaining bolts and strike the side with a rubber mallet. Occasionally, the crank may have to be turned to allow for the oil pan's removal.
11. Reassemble in reverse order. Be certain all gasket and seal surfaces are clean and the gaskets are in their proper position. Apply RTV to gasket joints and the block surface to assist in holding the gasket placement.

OIL SEAL REPLACEMENT

Most engines have oil seals at the front and the rear of the block to prevent the oil from passing by the crankshaft as it turns. These seals should not be confused with the front and rear pieces of the oil pan gasket.

The front seal (timing cover seal) is usually a lip seal made of rubber (neoprene). To replace this seal, the timing cover may have to be removed (Fig. 14-8).

Timing Cover Removed:

1. Review the steps listed for Camshaft Replacement (pages 394–395) to remove the cover.
2. With the timing cover removed, drive the old seal out of the opening in the cover. Using a punch or old screwdriver set just behind the lip from the backside, strike with a hammer.
3. Apply a light coat of RTV on the outside seal surface and the inside of the seal opening.

FIG 14-7 Wooden blocks set between the motor mounts and the cross member to keep the engine raised.

(Chevrolet Motor Car Division, GMC)

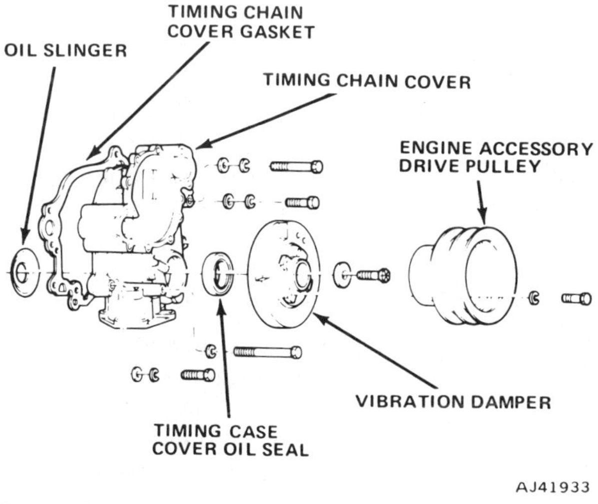

FIG 14-8 Disassembled view of a timing cover and related parts.

(American Motors Corporation)

4. Drive or press the seal into position while supporting the seal opening area.

Note: The lip always faces toward the oil.

Timing Cover in Place:

1. If the seal can be removed without removing the front cover, the drive belts and vibration damper must first be removed.
2. Using a slide hammer puller or any other durable tool with a hook jaw, slip the hook just behind the lip and pull out the seal.
3. Clean the seal opening area and apply a light coat of RTV to the seal case and the opening in the cover.
4. Drive the seal in place with a piece of pipe that will contact the steel casing of the seal and will slip over the crankshaft.

The rear seal is either rope (fabric) or rubber (neoprene). When the engine is in the vehicle, it is nearly impossible to successfully install a new rope seal in the block. A new seal can be installed in the rear main bearing cap; however, since this is only one-half of the seal, it may not cure an oil leak. Several gasket manufacturers have neoprene replacement seals designed to fit some engines originally equipped with the rope seal. If one of these is used, the upper half of the rope seal is removed and discarded along with the lower half. Follow the manufacturer's directions for installation.

MINOR OVERHAUL

Minor overhaul—not necessarily minor in effort and time needed to complete the job—involves completing any one or combination of the following operations while the engine remains in the vehicle:

1. Valve Job
2. Ring or re-ring job
3. Bearing replacement (main or connecting rod bearings or both)

VALVE JOB

Of the three repair operations listed, the valve job is most often performed in all segments of the auto repair industry.

RING AND VALVE JOB

A ring and valve job requires removal of the cylinder head(s) and the oil pan, valve service as outlined in Unit 7, and preparation of piston and cylinder walls as discussed in Units 9 and 11. It should be understood that an in-car ring and valve job is not the best method of engine reconditioning. It should be performed only when the engine and/or vehicle condition does not justify the cost of a major overhaul (engine removed).

The ring job and bearing replacement work is still being done today; however, it is becoming increasingly difficult to perform these operations while the engine is in the vehicle, for more vehicles each year are designed so the oil pan cannot be removed without engine removal. Improved piston rings and bearing materials, along with significant developments in lubricating oils, have reduced but not eliminated the need for ring and bearing service. Extra engine heat resulting from emission control systems that have lean air-fuel mixtures, retarded timing, high coolant temperatures, and the use of unleaded fuel are all factors which cancel out these improvements to some degree.

BEARING REPLACEMENT

While the engine is in its chassis, the bearing (connecting rod and main) can be replaced. With the oil pan removed, the caps are loosened and removed. The connecting rods would be removed during a ring job so bearing replacement could take place at that time. Connecting rods that remain in the block must be pushed upward far enough to allow access to the upper bearing half. Clean the rod bore surfaces, and replace the bearing. The oil clearance should be checked; then the cap should be assembled by torquing to specifications.

Main bearings are removed by loosening all the main bearing caps and bolts ½–1 turn. Remove one cap all the way. Using a roll-out tool placed in the oil hole, roll-out the upper main bearing by slowly turning the crankshaft so the locking lip moves out of its groove in the block (Fig. 14-9). This will push the upper half out so it can be removed.

To install, select the upper bearing for that opening and start it into position. Insert the roll-out tool in the crankshaft and oil hole. Slowly turn the crankshaft until the locking lip is seated in its groove in the block. Wipe all the surfaces clean and check the oil clearance. Lubricate the bearing surface. Install the cap by torquing to correct specifications. Replace all upper main bearings in the same manner.

Note: When torquing the thrust main bearing, be certain the crankshaft end play is within specifications.

COOLING SYSTEMS AND THEIR SERVICE

The engine's cooling system is designed to *control* the temperatures of all components. Removing excess heat and providing fast warmups are the two major jobs of the cooling system. When temperatures are not controlled, problems such as the following can occur:

1. Abnormal combustion (detonation-preignition)
2. Scuffled or scored pistons and rings
3. Warped or burned valves
4. General lubrication failure

OPERATION

The cooling system functions like an endless conveyor belt. Heat that develops from the combustion process is transferred to the coolant in the water jackets. The pump forces the coolant out of the engine past the thermostat to the top side of the radiator. The radiator "takes away" the heat in the air passing over its tubes and fins (Fig. 14-10). This is an application of one of the basic physical principles of heat: "Heat travels from a warmer to a cooler substance." In this situation, the passing air reduces the heat level (temperature) of the metal parts of the radiator below the temperature of the coolant. The coolant continues its path of travel through the radiator and lower hose until it reaches the pump. The term *coolant* rather than *water* has been used in this text because modern cooling systems require a mixture of water and other fluids to provide efficient heat exchange and cold protection.

The internal parts of the engine are designed with extreme care so the coolant will flow to the areas of greatest heat. Head gaskets are often designed to allow the correct coolant flow to the critical heat areas around the combustion chambers. A continuous flow of coolant is maintained within the block and head(s) by the pump. It does this whether the thermostat is open or closed, using a bypass recirculation passage between the pump and the block.

Typically, the coolant temperature causes changes in the opening of the thermostat valve which controls the coolant flow to maintain efficient operating temperatures. Depending on the type of driving and ambient (outside) temperature, the thermostat may be in any position from totally closed (light driving or cool weather) to wide open (heavy driving or hot weather) (Fig. 14-11).

COOLING SYSTEM PROBLEMS

Cooling system troubles result from both overheating and underheating. Overheating is easier to detect through dash warning lights and gauges, but is often more difficult to correct than underheating. Overheating can cause a vehicle to stop altogether as a result of extensive engine damage; underheating can take place for weeks or months with no engine stoppage. Underheating is normally caused by a thermostat stuck in the open position or not having a thermostat installed. This causes a slow warmup, which results in poor combustion, excessive fuel consumption and exhaust emissions and rapid wear of internal engine parts. For each gallon of gasoline burned in the engine, more than a gallon of water is formed. This water is normally removed from the engine as a vapor with the exhaust gases. In cold operating engines, a portion of the vapor and exhaust gas "blows by" the piston rings into the oil pan area (Fig. 14-12) where it condenses (turns into a liquid) and forms corrosive acids that damage engine parts, especially bearing materials. The water vapor in the blow-by gases is whipped into the lubricating oil to form what is called sludge. This is the black, mayonnaise-like material that builds up in all areas where the lube oils flow.

Since underheating is almost always the fault of the thermostat, replace it if there is any doubt about its operation. To detect an underheating engine, operate the passenger compartment heater. If the heat output is insufficient, replace the thermostat. A rough test of the thermostat can be made by removing the radiator

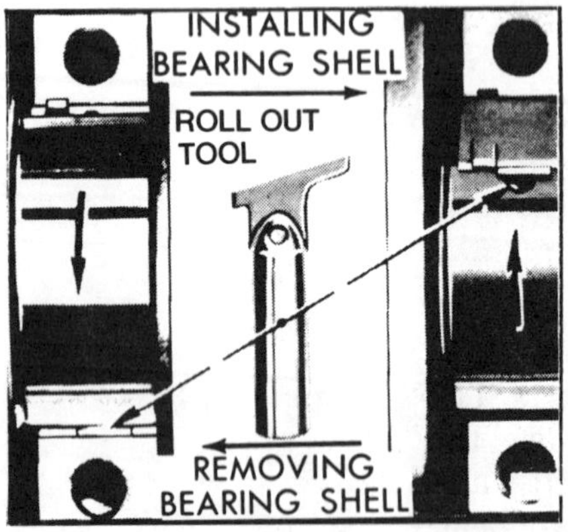

FIG 14-9 Roll-out tool and bearing removal directions. (Buick Motor Division, GMC)

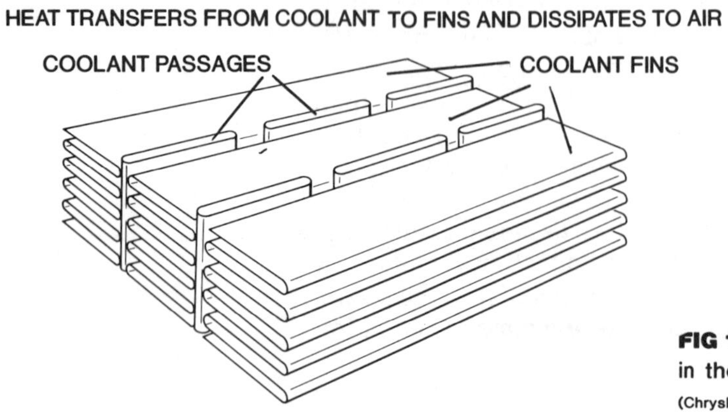

FIG 14-10 Heat transfer in the radiator core. (Chrysler Corporation)

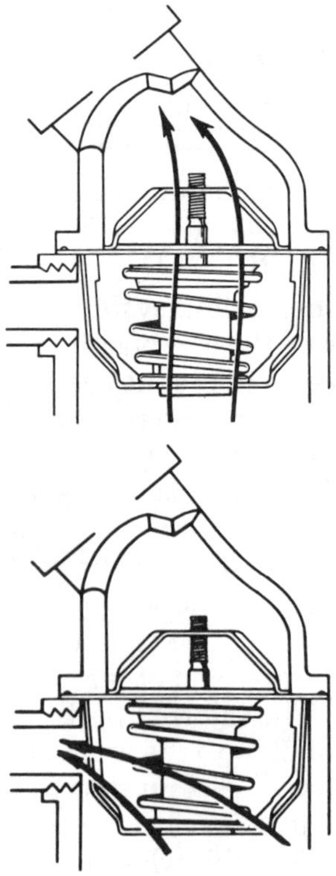

FIG 14-11 Thermostat position. (General Motors Corporation)

cap when the engine is stone cold, covering the grille area with a fender cover, and starting the engine. Watch for the flow of coolant. If the flow begins immediately, the thermostat is stuck open or missing. If the coolant begins to flow after three to five minutes, the thermostat is probably all right. The labor cost of removing and testing the thermostat is always more than the price of a replacement unit.

COOLANTS

Water has been used in cooling systems since the earliest days of the automobile. The disadvantages of water alone are that it has a relatively low boiling point, it lacks corrosion inhibitors and lubrication properties, and it freezes at 32° F. (0° C.).

A petroleum product known as ethylene glycol, with anticorrosion additives included can be mixed with equal parts of water to make up a good, all-around coolant (Fig. 14-13). A 50-50 solution of ethylene glycol will raise the boiling point of the coolant (with a pressure cap) by 20° F. (8° C.). A solution of more than 65% ethylene glycol is discouraged, for it will actually decrease freezing protection; pure glycol will freeze at –10° F. (–23° C.).

Along with anticorrosion agents, good quality coolants contain antifoam additives to maintain a stream of liquid coolant without air bubbles that tend to reduce the cooling effect.

Caution: Ethylene glycol coolants should be handled with care. They are poisonous and can cause injury to sensitive areas of the face or body and damage to paint and car finishes. Should this material be splashed on the skin or the vehicle's finished surfaces, flush immediately with clear water. Always wash hands after handling this coolant.

Although the coolants on the market today are called "permanent," they should not be used for longer than 24 months. After this length of time, additives are depleted which cannot be rejuvenated by adding more coolant.

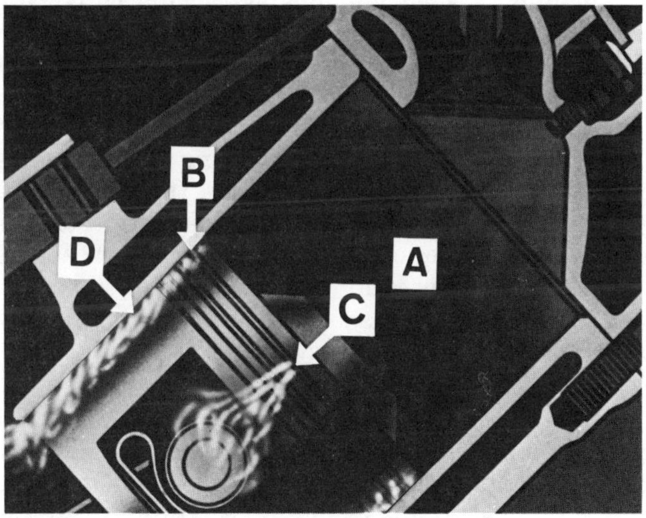

FIG 14-12 A. Blow-by begins with pressure in the cylinder.

B. PRESSURE LEAKS PAST THE RING GROOVES

C. RING END GAPS

D. PISTON

(Chevrolet Motor Division, GMC)

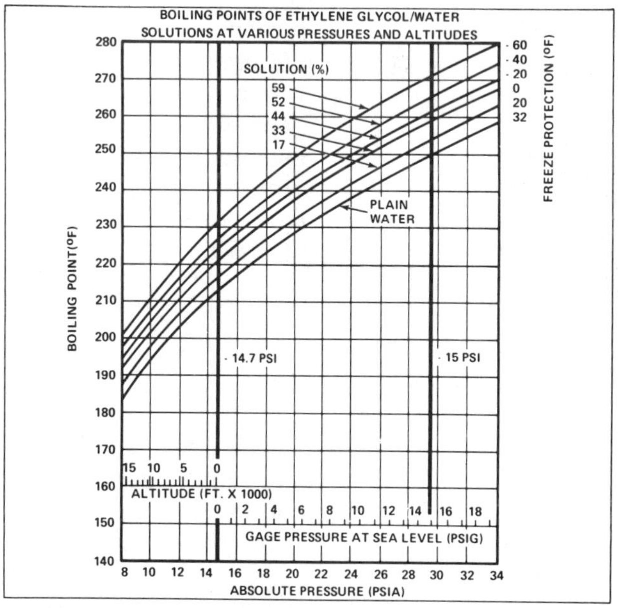

FIG 14-13 Boiling points of water-ethylene glycol solutions.

(Buick Motor Division, GMC)

CORROSION

Water solutions corrode cooling system metals when they are not protected by inhibitors (chemicals to stop the corrosive action). Loss of metal by corrosion may result in leakage or part failure. The formation of corrosion products on the surface of the metal reduces the heat transfer to the coolant in water jackets.

Corrosion is mostly the result of galvanic action (sometimes called electrolytic corrosion). Electric currents are produced between dissimilar metals in the cooling system (brass, iron, aluminum and even solder), and the coolant, a water base solution, serves as a conductor of the electricity. The galvanic action is further enhanced by the engine block, which serves as a grounding spot for the entire electrical system. Current flow of varying amounts is continually passing through the block and its cooling system to operate electrical devices in the vehicle.

Some corrosive action takes place as minor amounts of exhaust gases from the combustion chamber area penetrate the porous iron and gasket surfaces. Exhaust gas contains nitrous oxide, which, combined with hydrogen in the water, forms nitric acid (HNO_3). This acid is damaging to all parts of the cooling system. Most brand name coolants (antifreeze solutions) contain inhibitors to reduce or stop corrosion in the cooling system.

RUST AND MINERAL DEPOSITS

Water is the major, if not the only, component of the liquid in the cooling system. The basic makeup of water, two parts hydrogen and one part oxygen, encourages rusting (iron oxidation). Rusting shows up as the orangish color in the circulating liquid. Rusting action which continues for long periods has an insulating effect on the water jackets, which reduces cooling efficiency. The rust will also break loose from iron parts and block the flow of coolant by covering the tubes within the radiator.

Water also carries varying amounts of such minerals as calcium carbonate, which, when heated, will turn to lime (limestone) on the hotter areas within the water jackets (Fig. 14-14). This is the same material that builds up on the surfaces of a vessel used frequently to boil water. The damaging effect of the lime is that it insulates the hot areas from the coolant.

Lime and other minerals are normally removed during cooling system service with a two-step cleaner (Fig. 14-15). The first step involves adding the acid (oxalic) portion of the cleaner to a system with clean water from which the thermostat has been removed. The engine is operated for a period of time, and then the cleaner is drained. In the second step, an acid neutralizer is added to clean water and circulated through the system. Finally, the neutralizer is drained and fresh (preferably soft) water is added with a 50-50 mixture of coolant (antifreeze).

WATER PUMPS

Although water pumps circulate a combination of water and other coolant liquids, they are still referred to simply as water pumps. The pump is belt driven from the crankshaft at roughly the same rotational speed. The pump is positioned so it is always primed (filled with coolant), even when the radiator is only one-third full. The pumping action forces coolant through the water jackets in the block, head(s), and manifold.

Its major parts are the housing (iron or aluminum) shaft, shaft seal, bearing, and impeller (Fig. 14-16). The shaft drives the impeller so as to spin the coolant toward the outer edge of the impeller. The housing is formed to carry the coolant to one or more outlets in the housing. This design is referred to as centrifugal, since the coolant is fed to the center of the impeller and is thrown outward (centrifugally) toward the outlets.

The seal has the two-fold job of keeping coolant within the system and keeping air out of the circulating coolant. Pump seal failure causes coolant loss, which is easily seen as a drip or stream from the housing. If the seal is allowing air to enter the cooling system, it will usually go undetected. The best clues to air intake past the seal are rusting action seen in the coolant and overheating.

Note: Check other areas of air intake such as hoses or gaskets before making a pump change.

The pump bearing is designed to allow the shaft to turn freely within the housing. Bearing failure is easily detected by grasping each side of the fan blade, and making alternate front to rear movements. If the bearing is defective, the shaft will move within the pump housing. If the pump bearing is loose in its housing or the seal has failed, the pump must be replaced.

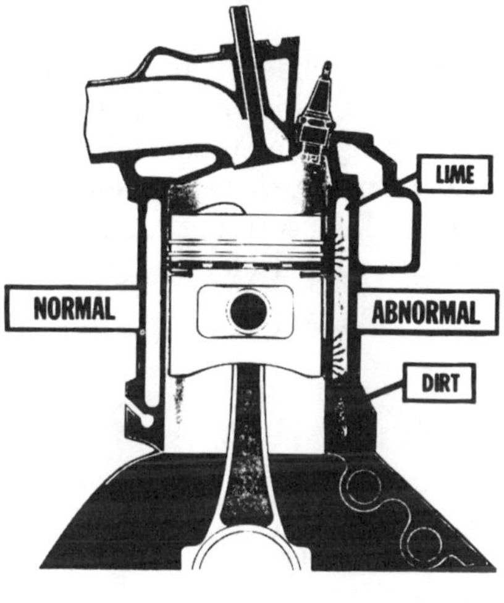

FIG 14-14 Friction heat can start with cooling system problems.

(Dana Corporation)

FIG 14-15 Two step cooling system cleaner.

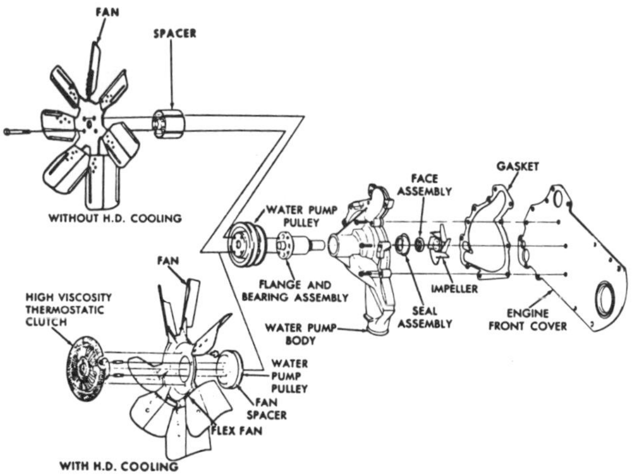

FIG 14-16 Water pump and fan assembly.

(Cadillac Motor Car Division, GMC)

RADIATORS

The radiator can best be described as a heat exchanger. Heated coolant is fed into the upper opening, and by the time it reaches the lower opening, much of its heat has been taken away.

There are two common types of radiators: the down-flow (Fig. 14-17) and the cross-flow (Fig. 14-18). The operating principles and materials of construction are the same for both. The cores (air flow area) of radiators have different designs because of differences in intended service, manufacturing capabilities, and costs.

Some radiators have a coolant recovery tank, which changes a system from pressurized to semi-closed pressurized. A hose between the radiator overflow outlet and the recovery tank catches coolant that might otherwise be lost on account of expansion of the liquid as it warms to operating temperatures. As the engine cools off, the expanded coolant is drawn back into the radiator.

The materials used in most modern radiators are brass, copper, or occasionally aluminum. Brass is used to make the tanks, tubes, fittings, and the fins of some radiators. Copper is used at times for fin metal because of its superior heat conduction characteristics. The most efficient radiators have copper fins soldered directly to the tubes. Louvered copper fins soldered to staggered tubes or all aluminum are the core designs often preferred by race car builders since high horsepower engines require the greater heat exchanging ability of copper and aluminum.

Vehicles equipped with automatic transmissions require some means of cooling the oil that circulates through them. The lower or outlet tank of the radiator usually houses this oil cooler. The oil is carried to and from the radiator by oil lines connected to the transmission.

The radiator and the rest of the cooling system can be checked for internal (inside) and external (outside) leakage with most cap testers. This involves attaching the tester to the filler neck of the radiator and pumping in a 20 psi (140 kPa) pressure. If the pressure begins to drop, check all external surfaces (inside the passenger compartment too) for coolant leakage. After making certain the tester is not leaking, it can be assumed that the leakage is internal. Refer to Internal Coolant Leaks in this unit for further testing.

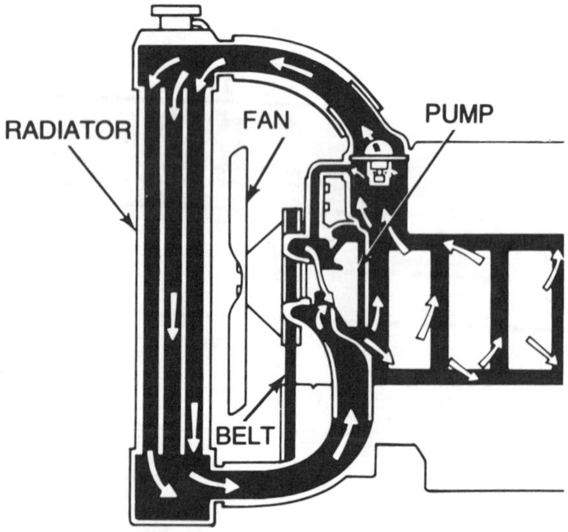

FIG 14-17 Down-flow radiator.
(Buick Motor Division, GMC)

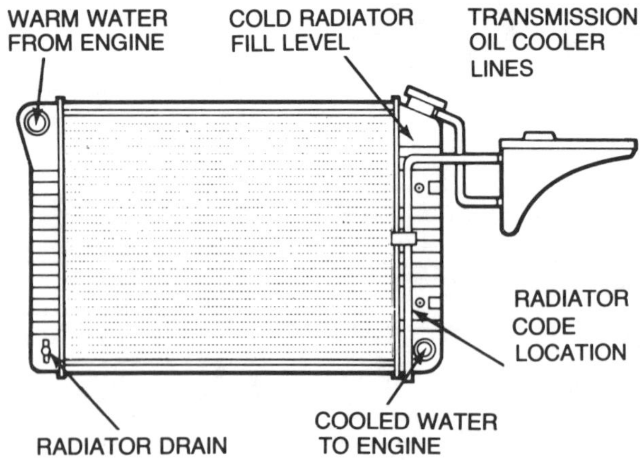

FIG 14-18 Cross-flow radiator with coolant recovery system.
(Chevrolet Motor Division, GMC)

RADIATOR SERVICE

The repairs made to a radiator should be done by a radiator specialist using equipment designed for that purpose. Repairs may range from soldering a slight leak to replacement of the entire core. *Rodding* is a term used to describe the general cleaning and minor repairs made to a radiator. It comes from the early automobile days when the round radiator tubes would become so plugged that a rod was forced through the tubes to unblock the restrictions.

The usual steps for cleaning and major repair of radiators are as follows:

1. Disassembly of the tanks from the headers
2. Thorough cleaning
3. Solder repairs or core replacement
4. Assembly of the top tank
5. Pressure test
6. Flow test
7. Painting surfaces (black only)

PRESSURE CAPS

The radiator pressure cap is one of the most important parts of the cooling system. It permits higher operating temperatures and reduces the loss of coolant, especially at periods of severe heat load. A radiator cap is rated by the pressure it takes to compress the main spring. Many caps in use today are rated at 15 pounds (105 kPa). This means the boiling point of the coolant is raised approximately 45° F. (7.7° C.) above what it would have been without a pressure cap.

The pressure valve of a cap is rubber faced to provide a good seal with the radiator filler neck. The pressure valve also contains a vacuum valve to prevent any undesirable vacuum buildup from occurring when the engine and its coolant temperature drop below normal operating level. There are two types of vacuum valves in use today: the light spring pressure and the weighted vacuum valve or surge type (Figs. 14-19, and 14-20). The spring pressure valve is closed at all times except when vacuum develops; the weighted valve closes only when there is a surge of vapor or coolant in the radiator.

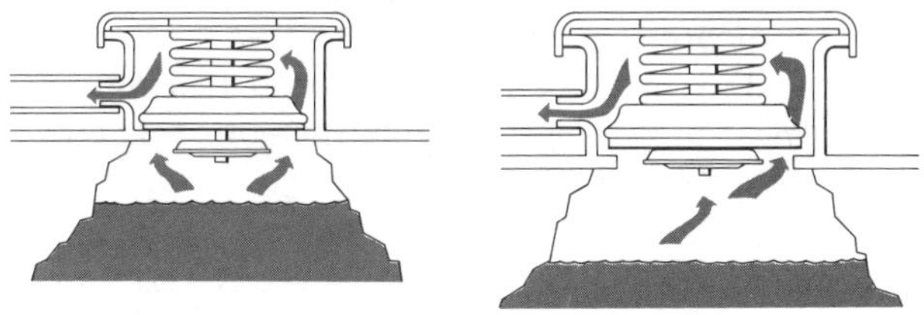

FIG 14-20 Action of the surge radiator cap. (Chrysler Corporation)

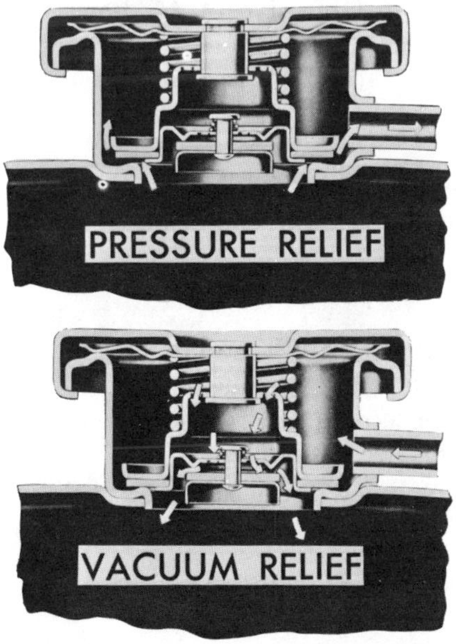

FIG 14-19 Spring pressure radiator cap. (Chevrolet Motor Division)

To discourage removal of caps on semi-closed (coolant recovery) systems, the cap labeled for use on this type of system (Fig. 14-21), has no gripping ears.

Although pressure caps require no maintenance, they should be inspected for cleanliness and operating pressure with a cap tester.

Caution: Do not remove the pressure cap from a hot radiator unless it is *absolutely necessary*. Removal will take away the pressure that provides a high boiling point, and the coolant will boil immediately, sometimes scalding the person removing the cap. If it must be removed, turn the cap counterclockwise part way to the safety stop. This will allow the steam and some of the coolant to escape through the overflow tube. When this stops, remove the cap by turning it the rest of the way.

FAN CLUTCHES

Automatic fan clutches are hydraulic devices to vary the speed of the fan in relation to the engine's temperature (Fig. 14-22). Fan clutches provide adequate cooling at reduced engine speeds while eliminating overcooling, excessive fan noise, and power loss at high speeds. Vehicles with air conditioning almost always have a fan clutch driving the fan blades.

Fan clutch diagnosis covers four general areas:

1. Noise
2. Silicone fluid leakage
3. Looseness
4. Engine overheating

Fan noise normally occurs during cold engine startup and continues until the silicone fluid is evenly distributed and when maximum cooling is needed. A fan clutch has failed when it never engages, especially during maximum cooling requirements; when it is engaged above 2500 rpm; and when there is a rough, grating feel as the fan is turned with the engine stopped.

Minor fluid leakage, seen as dust accumulation around the hub area only, is acceptable. Wetness (fluid) in that area means there is excessive leakage and the fan clutch should be replaced.

More than ¼ in. (6.35 mm) of lateral movement at the fan blade ends is considered excessive looseness and is grounds for replacing the fan clutch.

Excessive coolant temperature without engagement of the fan clutch usually indicates the clutch is defective.

Two methods of fan clutch testing are outlined in Objective 14-A1; however, most manufacturers have specific tests for fan clutches. Follow their recommendations if in doubt.

INTERNAL COOLANT LEAKS

If a leaking head gasket or a cracked engine block or head is a suspected cause of overheating, test for internal leakage in one of two ways: the Bloc-Chec method (Fig. 14-23) or the water pump disable method.

Bloc-Chec Method

1. Make certain that the coolant level in the radiator is at least 3 in. (7.62 mm) below the filler neck opening, so that coolant cannot enter the test unit when the engine is running. If the coolant level is too high, drain off coolant to about 3 in. (7.62 mm) below the radiator filler neck opening.
2. Unscrew the cap of the test unit and fill to the line with the special test fluid. Replace the cap securely.

Caution: Be sure the test unit is clean and free from oil and grease before filling.

3. Insert the rubber-covered fitting on the aspirator bulb firmly into the opening in the cap of the test unit.
4. Idle the engine for about 5 minutes. Then apply a load (transmission in Drive or in gear). Insert the rubber cone of the test unit firmly into the radiator filler opening.
5. Draw air from the radiator through the test unit by alternately squeezing the bulb for at least 2 minutes (unless the test fluid turns from blue to yellow before the 2 minutes). If the fluid turns to yellow, there is a combustion leak. If the fluid remains blue, there is no combustion leak.

Note: Keep in mind that air from the radiator, in the form of bubbles in the fluid, must pass through the test fluid.

If the test fluid turns yellow, indicating a combustion leak, it is necessary to isolate the leak. To locate the cylinder bank with the combustion leak,

1. Remove all the spark plug wires from one cylinder bank.
2. Run the engine on one cylinder bank for a minute or two to purge the cooling system of combustion gases from the previous test.
3. Retest with the Bloc-Chec testing unit.
4. If the fluid remains blue, the leak is in the cylinder bank that is not firing.
5. If the test fluid changes from blue to yellow, the leak is in the cylinder bank that is firing.

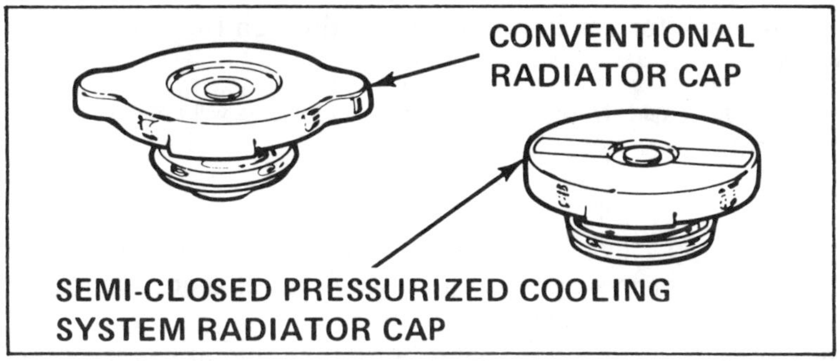

FIG 14-21 Radiator cap tops.
(Buick Motor Division, GMC)

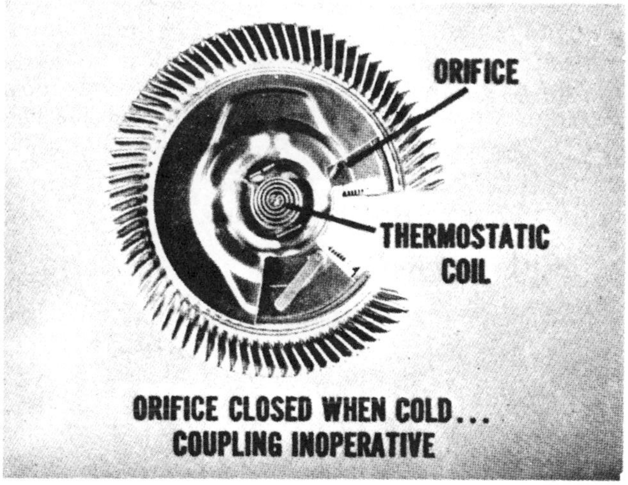

FIG 14-22 Thermostatic fan clutch.
(Chrysler Corporation)

FIG 14-23 Bloc-chek combustion leakage tester.
(Buick Motor Division, GMC)

Note: If a leak is found in the first bank, be sure to test the other cylinder bank in the same manner, for both cylinder banks may have combustion leaks.

To locate the cylinder or cylinders with the combustion leak,

1. Run the engine on the nonleaking bank.
2. Replace one spark plug wire in the nonfiring bank and recheck. If the test fluid remains blue, leave the spark plug wire in place and test the remaining cylinders in the same manner.
3. When the cylinder with the combustion leak is firing, the test fluid will turn from blue to yellow.

Note: There may be leakage from more than one cylinder. When a cylinder with a combustion leak is located, remove the spark plug wire from that cylinder and allow the engine to run for a minute to purge the cooling system from the previous test. Continue to test the remaining cylinders.

To isolate an internal leak in an in-line engine, remove one spark plug wire at a time and check each cylinder individually. A combustion leak may come and go or may occur only at certain engine temperatures. For this reason, it is necessary to test the engine at all temperature levels, starting from cold.

Pump Disable Method

1. Remove all drive belts that will turn the water pump.
2. Remove the thermostat and fill the engine with water until it is even with the top of the thermostat housing.
3. Start the engine and look for gas bubbles coming through the water at the thermostat openings.

Note: This check must be made quickly, otherwise small steam bubbles will appear which confuse the indications.

AIR ENTRAPMENT

When air is trapped in the upper cylinder head or manifold areas of the engine during the refilling operation, the system appears to be full, but there is extreme overheating after 3–4 minutes of operation.

Bleeding the air by leaving the "hot side" heater hose (manifold or water outlet) disconnected until the radiator is within 2 in. (50.8 mm) of the full mark will solve this problem.

LOWER HOSE COLLAPSE

During road speed operation or very fast idle, the lower hose (pump side) may collapse if the reinforcing wire is corroded away or pushed out of position. This will starve the pump for coolant, thereby decreasing its flow. Run the engine at high speed or squeeze the lower hose to feel the reinforcing wire (Fig. 14-24). Replace the hose or adjust the position of the wire to correct this problem.

EXHAUST SYSTEM RESTRICTION

Exhaust system restriction, resulting in poor performance, can occur in one or more of the following ways:

1. Collapsed exhaust pipe
2. Broken muffler baffles
3. Manifold heat control valve stick closed

Many exhaust pipes (manifold to muffler) are double walled. If such pipes are subjected to road splash (water, mud, ice, etc. splashing against them while they are hot) the inner walls may collapse (Fig. 14-25). This collapse may be total or in stages.

Muffler baffles and restrictors may break loose internally and cover the outlet opening. This slows the movement of hot exhaust gases leaving the engine.

Manifold heat control valves may stick in the closed position (Fig. 14-26). This too restricts the flow of hot exhaust gases.

To test for the first two causes of overheating and poor performance, disconnect the exhaust pipe from the manifold and operate the vehicle. Performance and cooling will return to normal if those types of exhaust restriction were the cause.

The manifold heat control valve can be checked manually by moving its counterweight. If it doesn't move, apply a special solvent for that purpose and try again. Should it still not move, tap the end of the shaft back and forth until it loosens; then try to rotate it between the open and closed position.

PUMP IMPELLER SEPARATION

Some water pumps fail when the impeller loosens from the shaft, reducing or stopping coolant circulation. This condition is difficult to detect in the early stages because the impeller does not always disengage from the shaft. This allows good cooling at times and little or no cooling at others.

Test by squeezing down on the upper hose by hand while accelerating the engine. A good flow force should be felt. If a flex hose is used that cannot be squeezed, remove the upper hose from the radiator. Momentarily start, then quickly accelerate, the engine. A good flow of coolant should come forth. If the coolant is to be reused, it will have to be caught in a container as it leaves the hose.

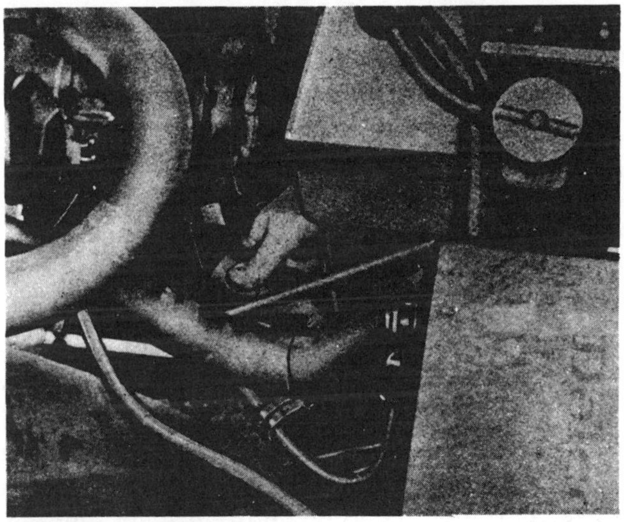

FIG 14-24 Checking for reinforcing wire in the lower radiator hose. (Buick Motor Division, GMC)

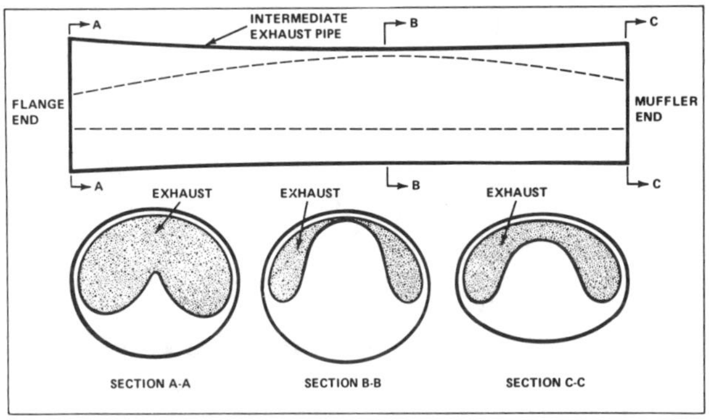

FIG 14-25 Exhaust pipe collapse. (Buick Motor Division, GMC)

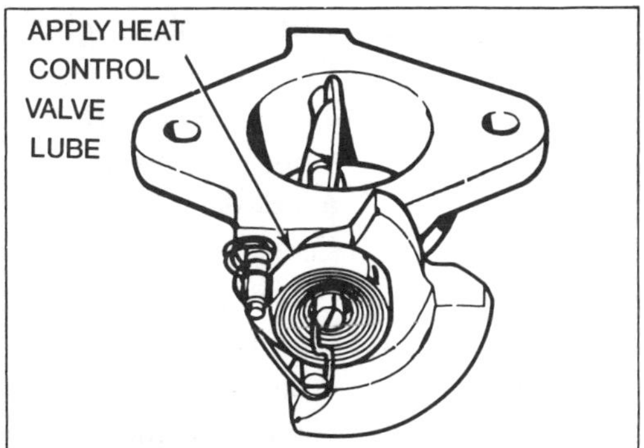

FIG 14-26 Manifold heat control valve. (Buick Motor Division, GMC)

OBJECTIVE 14-1 (manipulative)

Working as a team with one other student and with the aid of Job Report Sheet 14-A1, you will diagnose a liquid cooling system on an instructor-assigned automobile.

Time Recommended:

Twice the flat rate time allowed in *Motors, Chilton,* or *National Service Data* manuals. Instructor's estimate: 40 minutes.

Preparation:

1. Review pages 393 through 409.
2. Study pages 410 through 419.
3. Study manufacturer's or general repair manuals for the automobile being serviced.
4. Attend lectures and demonstrations on cooling system service.

Tools and Materials Needed:

1. Solution tester or hydrometer
2. Belt tension tester
3. Timing light
4. Vacuum hose diagram for system being checked

Procedure

Prechecks:

Test the operation of the warning lamp or temperature gauge if there is reason to feel the indicator is the only faulty component.

Thermostat replacement is required when there is overheating within 3–5 minutes of warmup time.

Summary of Checks

1. Check for obvious leaks in hoses, water pumps, etc. Check the condition of the radiator and heater hoses. Check the condition of the radiator front surface. Remove any foreign matter blocking the air flow.
2. Check coolant level.
3. Check coolant solution.
4. Pressure check the cooling system (radiator and pressure cap). For internal leakage test procedures, see pages 406–408.
5. Check for proper belt tension. Use a belt tension gauge and specifications, if available.
6. Check engine oil level and for coolant in oil.
7. Check fan clutch (if so equipped).
8. Check engine timing. Use a timing light and adjust distributor to specifications.
9. Check thermo-vacuum switch and hose routings (if so equipped). See manufacturer's shop manual.
10. Check transmission-controlled spark advance and hose routing (if so equipped). See manufacturer's shop manual.

Check:

1. Check for leaks and condition of hoses. Visually check the hoses for cracks, tight connections, deterioration, and other signs of damage. Squeeze the lower hose to check the pressure of the wire spring.
2. Check coolant level. If the coolant level is low, proceed to Diagnosis Chart, "Boiling or Coolant Loss" (Fig. 14-27).
3. Check coolant solution. If the coolant solution is below the specified protection, the boiling point is lower than the design level, which may be the cause of the cooling problem. A 50% solution is permissible. The Hydro-Spy (Duo-Chek) tester, or equivalent, is recommended for checking the coolant solution (Fig. 14-27). To use the tester, simply apply a few drops of coolant on the prism and look through the eyepiece. The scale inside the tester indicates the freeze protection of the antifreeze. Using the chart, the percent of ethylene glycol and boiling point can be determined.
4. Pressure check the cooling system.

Note: Before pressure checking the cooling system, make a thorough visual check of the entire system for any obvious leaks.

a. After the engine has cooled, carefully release cooling system pressure through the radiator cap.

Caution: The radiator pressure cap should not be removed while the engine is hot.

b. After the cooling system pressure is released, install the pressure cap adapter to the cap tester along with the cap (Fig. 14-28).
c. Pump the pressure tester to the specified amount of system pressure.
d. The cap should hold its pressure within 2 pounds for at least 10 seconds. If not, replace the cap.
e. Attach the pressure tester to the radiator filler neck (Fig. 14-29).
f. Pump the pressure tester to the rated system pressure shown on the cap. This point should be at or near the rated cap pressure.
g. Allow the system to hold a steady pressure for 2 minutes. If the system

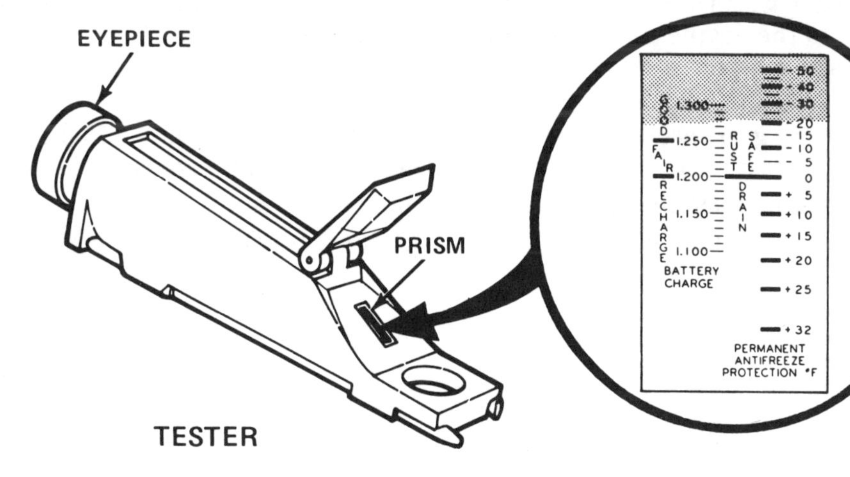

TESTER READING (°F)	COOLANT BOILING POINT (°F, Sea Level)*
− 50	228
− 34	226
− 15	223
0	220

* Non-Pressurized

FIG 14-27 Coolant solution tester.

(Buick Motor Division, GMC)

holds within 2 psi (14 kPa), the system is in good condition. While waiting to check the pressure, perform whatever other checks the time allows.

h. Should leak-down be greater than 2 psi (14 kPa) or should there be no pressure, check the following for leaks and repair as necessary:
   (1) Radiator cap and seating surface
   (2) All hose and hose connections
   (3) Water outlet gaskets
   (4) Thermo-vacuum switch at the manifold
   (5) Intake manifold bolts, and the front and rear of the manifold (see engines)
   (6) Water pump and mounting surface
   (7) Engine front cover
   (8) Radiator and filler neck, especially recovery tank joints
   (9) Head gaskets
   (10) Heater water control valve

Note: If the system appears to have internal leakage, see procedures in this unit.

5. Check for proper fan belt tension. Check fan belts to be sure they are tightened to specifications. Belts that are worn, smooth, frayed, or do not fit properly on the pulley may cause a cooling system problem by slipping.
6. Check engine oil level and coolant contamination. Be sure that the oil level is not above the full mark. High oil level can cause oil foaming and increased temperatures at higher engine speeds. An excessively low oil level can contribute to increased temperatures through increased friction. Also, check for coolant in the oil, which would indicate an internal leak. Insert the dip stick into the oil and quickly retrieve it. Touch the end of the stick to a *hot* manifold surface. If the oil contains coolant, it will spatter upon contact with the manifold. In severe coolant contamination, the oil will take on the same color as the coolant.
7. Check the operation of the fan clutch. The fan clutch can be checked in two ways:
   a. With the engine off, grasp one fan blade and spin it with a hard pull (Fig. 14-30). If it spins over five revolutions, the fan clutch is defective.
   b. Starting with the engine cold, run it at idle speed until the engine warms up. As the engine reaches operating temperature, the fan clutch should engage, creating a noticeable increase in fan noise. If fan noise does not increase, the fan clutch is defective.
8. Adjust timing to specifications, using manufacturer's procedures. Make remaining tests if engine is equipped with timing modification devices.
9. Check thermo-vacuum switch (TVS) and hose routings.
10. Check transmission-controlled spark (TCS) devices and vacuum hose routing.

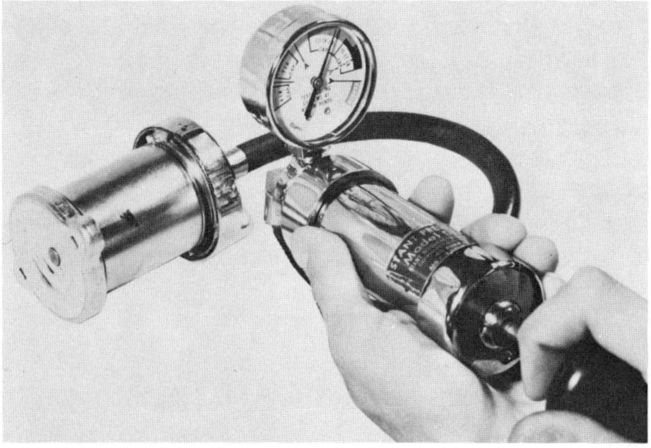

FIG 14-28 Testing the rated pressure of the radiator cap.
(American Motors Corporation)

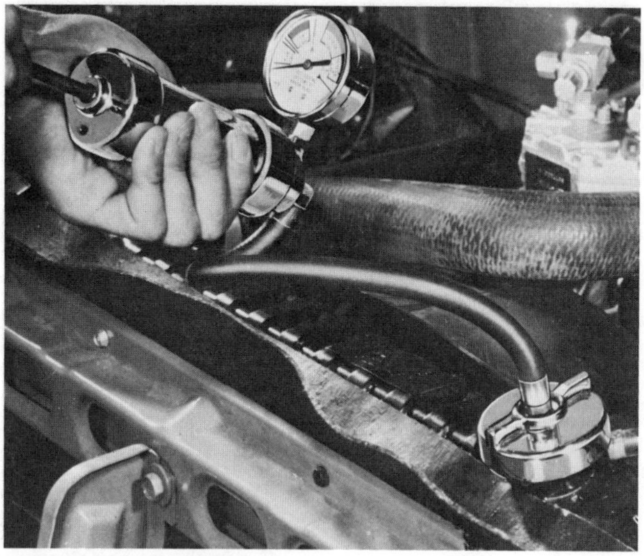

FIG 14-29 Pressure tester connected to the radiator.
(American Motors Corporation)

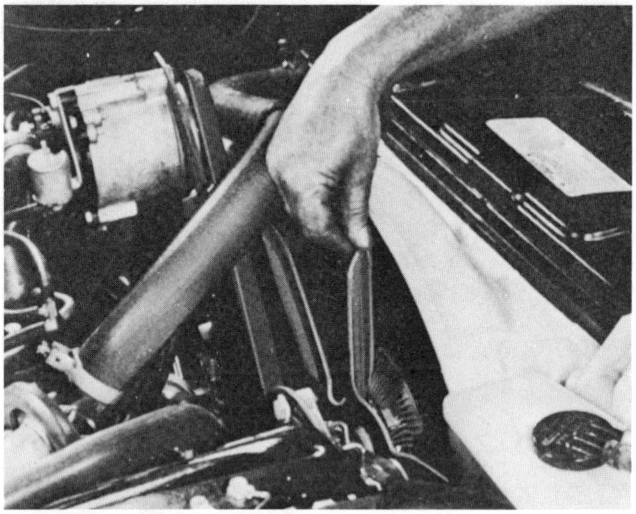

FIG 14-30 Checking the fan clutch.
(Buick Motor Division, GMC)

JOB REPORT SHEET 14-A1

COOLING SYSTEM DIAGNOSIS

Student Name ______________________ Date ____________

Make of Car ____________ Model ____________ Year ____________

Time Started ____________ Time Finished ____________ Total Time ____________

Flat Rate Time or Recommended Time ______________________

Specifications:

Cooling system pressure rating ______________________

Belt tension ____________ ____________ ____________

Ignition timing ____________° @ ____________ rpm

Checkup Questions:

1. What was the coolant protection level (degrees F. or C.)? ____________
2. How long did the cap hold its pressure? ____________
3. Was the passenger compartment checked for coolant leakage? ____________
4. How is the engine oil tested for contamination with coolant? ____________

DATE COMPLETED ____________ INSTRUCTOR'S SIGNATURE ____________

LUBRICATING SYSTEMS SERVICE

Lubricating systems are discussed in detail in Unit 10. This section is devoted to lubricating oil and lubricating system in-car service.

ENGINE OILS

The job of a modern engine oil is more than just to lubricate. It must

1. Clean interior engine surfaces
2. Cool engine parts, especially pistons and cylinder walls
3. Assist in sealing compression and power stroke gases
4. Reduce the tendency to foam, which provides marginal lubrication

Additives are included in blending modern oils because they assist in keeping the interior surfaces clean. Without detergent additives the oil has a greater tendency to sludge, which holds contaminates until the engine overheats. When overheating occurs, all the contaminates break loose, and bearing failure, among other things, usually occurs.

The interior of an engine becomes quite hot during normal operation. Oil is designed to assist the cooling system (air or liquid) in keeping engine temperature within safe levels. The maximum temperature level of most conventional petroleum engine oils is approximately 250° F. (121° C.). Most synthetic oils can reach 300° F. (149° C.) before temperature damage occurs.

No engine oil is right for all engines. The type of service the engines are asked to perform and the average air temperatures during that service are the two main reasons for selecting a certain oil. The engine manufacturer's recommendations should be followed, especially during warranty periods.

Engine oils are judged in two ways: the API (American Petroleum Institute) service classification and the SAE (Society of Automotive Engineers) grade and index (Fig. 14-31). The API service classifications now in use are as follows:

FIG 14-31 Engine oil service classifications and grade index.

SE—typical of engines in automobiles and light trucks beginning with 1972 and some 1971 models. Oil for this service provides all the necessary protection against oxidation, high temperature deposits, rust and corrosion above the level of SD oil.
SD—typical of engines beginning with 1968 model years. Oil of this classification provides adequate engine protection from rusting, high-low temperature deposits, and corrosion.
SC—typical of 1964–1967 engine models. Oil of this classification provides detergent-dispersant characteristics as well as protection against wear, rust, and corrosion.
SB—typical of engines operating in mild conditions where minimal compounding is required.
SA—typical of engines that require only a straight mineral oil during their operation.

The Society of Automotive Engineers has developed grading requirements for the flow characteristics (viscosity) of all engine oils. Conventional oils have a tendency to thicken when cold and thin out at high temperatures, while synthetic oils do not vary nearly as much. The SAE designs levels of flow that engine oils must meet to be classified into a particular grade. These grades, stamped on the top of an oil can, are wrongly referred to as the "weight." The number has nothing to do with oil weight, only flow ability at a given temperature. Automotive engines typically use either one of the four grades listed below or a multiviscosity grade:

SAE-10—used during winter months where temperatures are below freezing.
SAE-20—used where temperatures do not reach freezing in winter or where cooler temperatures prevail.
SAE-30—used in nearly all engines, especially during summer months.
SAE-40—used in engines with heavy load requirements during hot weather.

Multiviscosity oils are typically designated 5W-20, 10W-30, 10W-40, and 20W-50.

The *W* designation means the oil was tested at 0° F. (–18° C.) rather than 210° F. (99° C.), as is otherwise the case. For a multiviscosity oil such as 20W-50 the 20 grade test was successful at 0° F. (–18° C.), and the 50 test requirements were met at 210° F. (99° C). This indicates the 20W-50 oil has a wider range of flow than a single-grade oil.

Note: The grade numbers have no bearing on the quality of the oil.

SYNTHETIC OILS

Synthetic (man-made) oils are gradually replacing conventional oils. Their advantages are that they reduce friction (increasing engine and gasoline mileage), reduce oil consumption, and are not affected by temperature extremes. At present, they are more costly than regular oils; however, their benefits may outweigh the extra cost.

These oils have three distinct base materials. This means that oil made of one base *cannot* be added to an engine using a different oil base. The three bases are

1. Synthesized hydrocarbon
2. Diesters
3. Polyglycols

If the hydrocarbon base is chosen, conventional oil may be intermixed; however, this reduces its effectiveness.

OIL CONTAMINANTS

The oil used in an engine would last indefinitely if it were not contaminated by water, gasoline, carbon, heat, metal particles, and dirt. No filtering system or material has yet been devised to remove the following contaminants:

Water—Part of the combustion reaction produces water, some of which reaches the crankcase. Also, condensation (a vapor turning into a liquid) takes place when the engine is shut down warm, causing the temperature in the crankcase to drop. Much of the water is re-vaporized during the next operating period, but some combines with the oil to form sludge.

Gasoline—Since the piston rings and cylinder wall do not provide a perfect seal with one another, some fuel passes into the crankcase. This dilutes the oil, making it less effective and, in extreme situations, causing scoring of parts and bearing failure.

Carbon—During combustion, carbon is created, some of which eventually enters the crankcase, where it tends to score soft bearing materials.

Heat—Internal engine parts, especially pistons, operate at temperatures that vaporize (boil) the oil. When the oil makes contact with these hot surfaces, it is oxidized. Oxidation of the oil leaves carbon, which reduces its lubricating quality.

Metal particles—During engine operation, especially break-in periods, the wearing of metals rubbing against one another produces metal particles. These particles must be removed by draining the oil.

Dirt—As the engine operates, many different types of dirt enter the crankcase from outside sources, such as the air cleaner and crankcase ventilation system. Since they cannot be effectively filtered, the oil must be changed.

OIL DRAINING AND REFILLING

Oil change will be most complete if the pan plug is removed as soon as possible following engine operation at normal coolant temperatures. Contaminants are suspended within the oil while the engine is running; therefore, most of them will leave the crankcase if draining is done immediately upon shutdown.

While the oil is draining, watch for irregularities in color or thickness. Also smell a portion to see if it appears diluted with gasoline. Report any unusual conditions to your instructor. Service the filter if needed while the oil drains.

Replace the pan plug and tighten it securely. Add the recommended amount and type of oil. Check the dip stick for the proper oil level and start the engine. Watch the oil pressure light or gauge to be sure of normal oil pressure. Stop the engine, check for leakage, and add additional oil if necessary.

Name ______________________ Date ______________

DIRECTIONS:

Circle the best answer to each question.

1. Valve lift measurement
   I. requires a dial indicator.
   II. requires that all spark plugs be removed.
   A. I only B. II only C. Both I and II D. Neither I nor II

2. In-car camshaft removal
   I. from a V-8 engine requires removing at least one cylinder head.
   II. requires replacement of the cam bearings.
   A. I only B. II only C. Both I and II D. Neither I nor II

3. Valve spring replacement without removing the head
   I. involves removing the locks, spring retainer, and the valve.
   II. cannot be completed on OHV engines with dual springs.
   A. I only B. II only C. Both I and II D. Neither I nor II

4. Oil pan removal
   I. cannot be completed on engines with automatic transmissions.
   II. usually requires raising the engine slightly.
   A. I only B. II only C. Both I and II D. Neither I nor II

5. To replace the front seal (timing cover oil seal),
   I. the vibration damper must be pulled.
   II. drive the new seal in place above the old one.
   A. I only B. II only C. Both I and II D. Neither I nor II

6. Rear main seals
   I. do not require oil pan removal for replacement.
   II. are either neoprene or synthetic rubber.
   A. I only B. II only C. Both I and II D. Neither I nor II

7. Main bearing replacement with the engine in the vehicle
   I. should not be tried without removing the connecting rods.
   II. does not require oil clearance checks.
   A. I only B. II only C. Both I and II D. Neither I nor II

8. The engine's cooling system
   I. is designed to control the temperature of all engine parts.
   II. uses liquid and/or air to carry away the excess heat.
   A. I only B. II only C. Both I and II D. Neither I nor II

9. The coolant within the system
   I. flows only when the thermostat opens.
   II. should contain stop-leak additives.
   A. I only B. II only C. Both I and II D. Neither I nor II

10. A thermostat
    I. opens fully after 3–5 minutes of driving.
    II. will prevent overheating.
        A. I only   B. II only   C. Both I and II   D. Neither I nor II
11. Underheating occurs
    I. when the engine blow-by becomes excessive.
    II. most often as a result of a defective thermostat.
        A. I only   B. II only   C. Both I and II   D. Neither I nor II
12. Ethylene glycol coolant solutions
    I. are poisonous.
    II. are often a 50-50 mixture with water.
        A. I only   B. II only   C. Both I and II   D. Neither I nor II
13. Coolant contamination can be
    I. removed by draining and flushing the system.
    II. prevented by using a good quality coolant year round.
        A. I only   B. II only   C. Both I and II   D. Neither I nor II
14. Water pumps
    I. force cool liquid out of the top of the radiator into the engine.
    II. keep air out of the coolant.
        A. I only   B. II only   C. Both I and II   D. Neither I nor II
15. Radiators
    I. can be tested for leakage and flow by pressurizing to 20 psi (140 kPa).
    II. should be free of foreign matter in the fin and tube area.
        A. I only   B. II only   C. Both I and II   D. Neither I nor II
16. Radiator pressure caps
    I. lower the boiling point of the coolant to a safe level for all operation requirements.
    II. have vacuum valves to let cool air into the radiator after the engine warms up.
        A. I only   B. II only   C. Both I and II   D. Neither I nor II
17. All of the following are true *except*
    A. A coolant recovery type cap usually has no ears on its top.
    B. A coolant recovery system is a semi-closed system.
    C. Fan clutches are less noisy than regular fan blade systems.
    D. Fan clutches require more engine power to operate than regular fans.
18. Internal leakage
    I. has occurred when the test fluid in the Bloc-Chec turns blue.
    II. will usually show up as rusting at the core plugs.
        A. I only   B. II only   C. Both I and II   D. Neither I nor II
19. Lower hose collapse
    I. will take place when the engine is allowed to cool.
    II. is related to the air entrapment problem.
        A. I only   B. II only   C. Both I and II   D. Neither I nor II
20. Exhaust restriction can occur in all of the following ways *except*
    A. using dual exhaust pipes.
    B. exhaust pipe collapse.
    C. manifold heat control valve sticking in the closed position.
    D. internal muffler collapse.

DATE COMPLETED ______________ INSTRUCTOR'S SIGNATURE ______________________

unit 15

High Performance Considerations

This unit describes, in a very general manner, some of the operations performed by automotive technicians to produce a high performance engine.

The term *blueprinting* is often used to describe the clean, precision work needed to make an engine perform at its very best. Engine manufacturers attempt to produce engines to exact specifications; however, they also must turn out large numbers of engines by mass production. The result is that all parts have some size tolerance (amount of oversize or undersize). For example, a piston diameter may have a ± (plus or minus) tolerance of 0.0005 in. (0.0127mm). This means a 4.000 in. (101.600 mm) piston may measure anywhere between 3.9995 in. and 4.0005 in. (101.587 and 101.613 mm). This amount of tolerance is not critical, but remember that the cylinder bore has a size tolerance also.

The amount of tolerance in an engine part is calculated after a great deal of testing and consideration of past engine operation experience. For the average engine used to carry passengers, considerable tolerance is acceptable. For the race car engine that must produce high power output for long periods of time, very little size tolerance is acceptable.

Blueprinting means accuracy in measurement and machining, ample time to make all checks and assembly operations, and the ultimate in cleanliness to reduce chances of failure caused by foreign matter.

A WORD OF CAUTION

The delicate machining work described in this unit requires equipment with a high degree of accuracy. Even more important, the operator of the machinery (student or professional) must be dedicated to performing all jobs as precisely as possible.

When purchasing after-market high performance equipment or services, consult with people who are competent in this field. Discuss the intended use of your engine prior to beginning any work or making any purchases. As an example, there are many differences in camshafts for different performance uses.

Whether the engine is brand new or not makes little difference. The descriptions of work performed assume the parts have been *disassembled and thoroughly cleaned,* unless specifically mentioned. Also, all parts should be kept in their exact assembled order. This involves numbering or lettering all pieces with mating surfaces. Refer to the earlier units for disassembly and cleaning procedures.

CYLINDER BLOCK PREPARATION

Carefully remove all burrs and casting defects with a hand grinder. Smooth any casting irregularities in the same manner (Fig. 15-1). All bolt and stud holes should be cleaned with a bottoming tap to ensure full threads for the studs when installed.

Bolt holes should all be chamfered to remove burrs and to prevent the lead thread from pulling out between two mating surfaces. If the head surfaces are to be milled, as discussed later, chamfer after milling. This includes the oil drainage holes.

ALIGN BORING OR HONING

After an engine has been run, the cylinder block can change shape. Align boring ensures a straight support for the crankshaft, preventing any binding or friction-producing interference (Fig. 15-2).

Prior to align boring, grind a small amount of material from the block mating surface of each side of the main bearing caps (about 0.005 in.) to ensure perfect mating as well as to provide the proper main bearing crush after boring. Install the main caps and torque to specifications evenly and slowly. Follow directions for the setup and operation of the align boring or align honing equipment. If equipment is unavailable at your shop, carefully select a professional to complete this and other precision shop work.

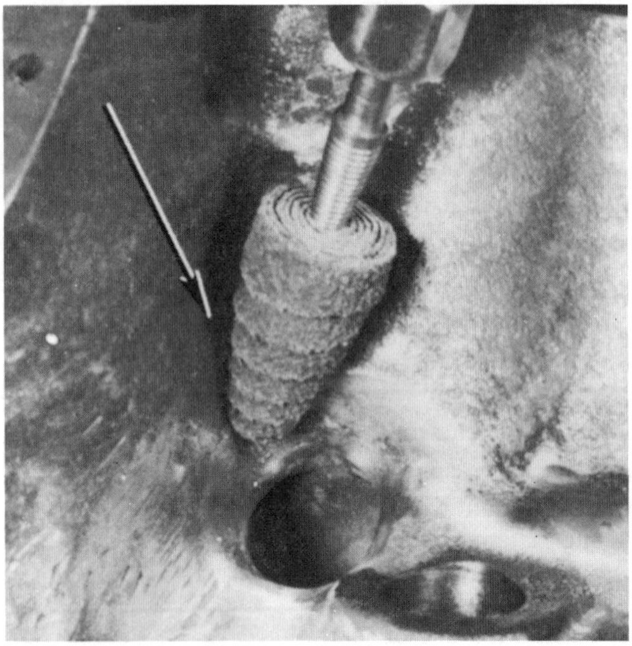

FIG 15-1 Deburring interior of block surfaces.
(Chevrolet Motor Division, GMC)

FIG 15-2 Align boring fixture mounted on the block.
(Dana Corporation)

CYLINDER BORING AND HONING

While the mains are torqued in place, bore and hone the cylinder (Fig. 15-3). The assembled torque for the main bearing caps is a factor in the "shape" of each cylinder. To ensure straight, round cylinders which are perpendicular to the crankshaft in an engine when buttoned up, always bore and hone with the main bearing caps in place and torqued to specifications.

Using the piston skirt diameter and the desired clearance, determine the finished cylinder size. Then, when boring, leave 0.002 in. to 0.003 in. stock for honing purposes. Hone with a rigid or flex hone to produce the desired diameter and the 30° to 45° cross-hatch pattern with a 20 to 30 rms or smoother finish. Use a deck plate (torque plate) to ensure straight cylinder walls (Fig. 15-4). A finished bore surface produced with a Sunnen CK-10 is considered by many engine builders to be the best (Fig. 15-5).

Lightly hone the valve lifter bores in the block with a brake cylinder hone to provide a clean, cross-hatched surface, but be careful not to enlarge the bores, for oil pressure and overhead oiling can be affected by too much clearance.

FIG 15-3 Boring cylinders to fit oversize pistons.
(Dana Corporation)

FIG 15-4 Honing cylinder walls with a deck plate (torque plate). (Chevrolet Motor Division, GMC)

FIG 15-5 Automatic honing machine.
(Sunnen Products)

Check the block deck for warpage. A light pass with a mill may be required to straighten the surfaces for proper sealing. A word of caution—on a V engine, be certain that both decks are machined to the exact same height and parallel to the crankshaft centerline, since a variation will later affect the combustion chamber volumes.

After all machining has been completed on the block, lightly chamfer the top of each cylinder opening to remove burrs and to provide an entering chamfer for the pistons and rings. Thoroughly clean the block, using light oil on the cylinder walls to remove honing particles and wiping clean with several clean rags. Soap and water should then be used to clean the entire block. Double check the bolt-hole chamfers as mentioned previously and reinspect the block for cracks and irregularities the machining operation may reveal.

Mask all machined surfaces and be certain that all oil and loose particles are removed before painting, since paint chips can plug oil passages or accumulate in the oil pump. Replace the block core plugs—threaded plugs and a sealer, if desired, may be used.

PREPARING THE CRANKSHAFT

The preparation of a crankshaft for high performance use may vary according to the intended usage as well as the owner's choice. However, there are a few basics that should be covered in preparation of any blueprinted engine. A crankshaft should be checked for straightness and crank-throw angularity. To correct angularity, the rod journals can be ground undersize with correct angular indexing. Journal oil holes should be chamfered or radiused with a wet stone. The journals should be radiused at the corners and then lightly polished with #400 sandpaper. The crankshaft should then be magnafluxed for cracks.

Check the main bearing clearances with Plastigage at this point. The crankshaft main journals can be ground undersize if additional clearance is needed. Check the protrusion of each bearing half for proper crush.

In performance engines, full 360° oiling is often desirable. Since, generally, only the top half of the main bearing is grooved, a cross-drilled crankshaft can be used to ensure a full oil supply to the main bearing.

Other optional treatments include hard-surfacing, such as Tufftriding; hard-chroming; and other processes that may be desired by the owner for the intended use. After any surface treatment, the crankshaft should again be checked for straightness.

CONNECTING ROD PREPARATION

In preparing a performance connecting rod that is durable, there are two basic areas to consider:

1. Specifications—length, size, and clearances
2. Stress relieving

The first point to consider under specifications is the center-to-center length of the rod pin end to the journal end. The desired length should be determined before any work is done. Then all machining should be done, holding this dimension equal for all rods.

One major factor in controlling the rod length is the piston-to-deck clearance. Assemble the piston, rod, and bearings on the crankshaft and check the depth of the piston head below the cylinder block deck. Generally, the manufacturer's minimum specified clearance is the goal. However, as a safety factor, be sure to allow a few thousandths for high-rpm rod stretch and piston rock. Rapping pistons against the cylinder head at high rpm can be expensive.

Remove the rod cap and take a light cut on the cap and/or rod with a rod and cap grinder (a few thousandths is sufficient). Then reassemble, torque, and hone to size (Fig. 15-6). Hone the rod pin end to size, if necessary, for desired clearance or to convert from press-fit to full-floating pins. Check rod alignment and straighten if necessary. The rod bore sides at the journal should be ground smooth to required clearance.

The connecting rod is a fatality-prone part of the reciprocating mass, so as many of the internal stresses and points of stress concentration should be removed as possible. Remove forging irregularities with a mill file and polish the rod lengthwise.

Again using the file, radius the corners, the area around the rod bolts, and polish. Internal stresses can be relieved by shot peening. After all of the above steps have been completed, the rods should be magnafluxed as further assurance of long life.

Use Plastigage again to check the rod bearing–crankshaft clearance.

PISTON AND PISTON RINGS

Continuing the blueprinting procedure, the hand-fit policy also applies to piston and rings. The piston-to-cylinder clearance has previously been covered, but now the piston must be fitted to the pin, the rings to the piston, and the rings to the cylinder.

Whether forged or cast aluminum pistons have been selected, the procedure is the same; only the clearances differ. The piston pin bores may be honed to the specified clearance. If converting from press-fit to full-floating, lock ring grooves must be cut before honing to size.

FIG 15-6 Reconditioning a connecting rod.
(Dana Corporation)

A light chamfer or radius of all sharp edges on the piston head will remove the localized hotspot tendencies, lessening the possibilities of preignition and detonation, critical in a high-compression, high-output performance engine.

The piston rings should be installed squarely in the cylinder to check end clearance. The head of a piston can be used to square the piston ring in the cylinder bore. Check the gap or end clearance with a feeler gauge. A file can be used to increase the gap, if necessary. As a rule of thumb, about 0.004 in. (0.107 mm) per inch of cylinder diameter is satisfactory, but never exceed 0.025 in. (0.635 mm) gap. Complete the piston ring fitting by checking the side clearance of the compression rings in the grooves, again using the feeler gauge. The side clearance should not exceed 0.005 in. Also check the ring groove depths. The piston ring should not protrude beyond the piston lands, nor should it be more than 0.005 in. below the smallest land. Sufficient groove depth avoids scuffed and/or broken rings as the piston expands at engine temperature.

ENGINE BALANCE

Now that all of the components have been selectively fitted and noted for cylinder location, the crankshaft, the rod bearings, the rods, the pistons, the pins, and the rings should be balanced. There may be variations in the weight of all components, particularly when some hand work has been done, such as polishing the rods. Calculations show that a 10-gram (about ⅓ ounce) weight differential becomes 60 pounds at 6000 rpm and 135 pounds at 9000 rpm. Beginning with the pistons, balance all of the reciprocating components. Some juggling can be done to find the combination of bearings, rods, pistons, and rings with the least weight variation throughout all cylinders. Then balance to the lightest assembly, removing the necessary material from the piston balancing area. A total weight spread of *less than one gram* should be attained if possible.

After the above assemblies have been balanced, the crankshaft should be balanced statically (at rest), then counterbalanced dynamically (in motion) for the determined reciprocating masses. The static balance will show heavy areas or the tendency for the crankshaft to roll to one point when unconfined and supported only by each end.

Dynamic balancing is generally done at low rpm (about 600 rpm). Bob weights are used to simulate the total rod-piston assembly weight (Fig. 15-7). Spinning the crankshaft on a balancer will indicate whether each counterweight should have material removed by drilling holes or added by welding a small metal plate to it (Fig. 15-8).

A complete, accurate balancing job will eliminate the out-of-balance forces that can tear an engine apart at high rpm or, at best, lower horsepower as the engine fights this condition. Balancing is not only good engine life insurance, but also a good investment for the performance to be gained by blueprinting.

Now that the basic block work has been completed, attention must be given to the cylinder head and the rest of the valve train components.

CYLINDER HEAD PREPARATION

Initial preparation of the cylinder head is much the same as for the cylinder block—disassemble, get it clean, and inspect for cracks and flaws. Check the head surface for straightness and determine the necessary surfacing required to clean it up. Check the condition of the valve seals and the valve guide bores. Remember that not only are the components and assembly to be blueprinted, but also the final combustion chamber volumes will be machined to the minimum specification and will be equal for both cylinder heads.

The valve guides should receive attention first. Install replaceable guides or oversize valves, knurl and hone, or use any other method, according to personal preference, to bring the guide bores to desired specifications. While reworking the guides, machine the guides for valve seals, allowing for adequate clearance between the seals and the valve spring and retainer when installed.

Do the valve job next. The valves should not be sunk into the head and should be seated with as little variation as possible from cylinder to cylinder. The valve stem tip protrusion above the valve spring seat should be the same for all valves. The intake valve seat width should be about 0.060 in. (1.524 mm), while the exhaust seat width should be in the neighborhood of 0.070 in. (1.778 mm to 2.032 mm) for better heat transfer. Any polishing of the combustion chamber should be done at this point (some competition classes don't allow this treatment). The polishing can be done with a hand die grinder. Ports may also be enlarged and polished if desired. If port work is done, be sure that the ports in the intake manifold match those in the cylinder head. Figure 15-9 shows a cylinder head on which the above work has been completed.

FIG 15-7 Bob weights attached to the connecting rod journals to simulate their weight while dynamically balancing the crankshaft. (Dana Corporation)

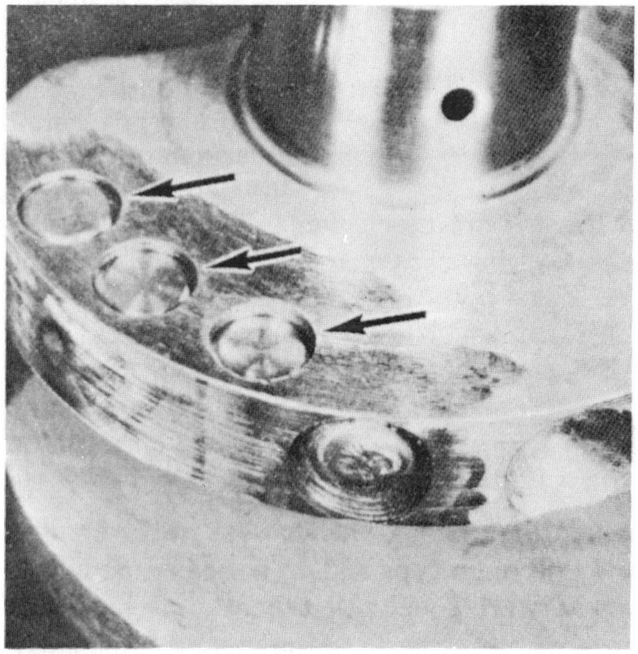

FIG 15-8 Holes drilled in the crankshaft counterweights to subtract weight for balancing purposes.
(Chevrolet Motor Division, GMC)

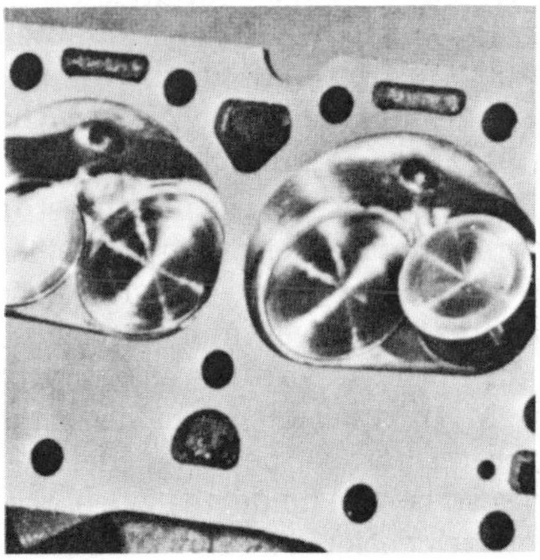

FIG 15-9 Refinished combustion chamber surfaces.
(Dana Corporation)

CC'ing

Now that the basic combustion chamber work has been done, all combustion chamber volumes should be equalized. The preferred method is to equalize all the combustion chamber volumes and then bring them to the minimum specification by milling the head surface.

The apparatus for measuring the volume in cubic centimetres consists of a standard chemical burette, a plexiglas plate, and light oil, solvent, or kerosene (Fig. 15-10). Food coloring can be added to show the liquid levels better.

Install the valves and the spark plugs. Use a light coating of grease to form a seal between the plate and the head surface. Pour the oil or solvent into the burette and measure the amount required to fill the combustion volume under the plate.

Measure both cylinder heads at the same time and record the volume. *Enlarge all combustion chambers to the largest one* on either head to equalize all eight chambers. The chambers can be enlarged by removing casting material from the chamber or by sinking the valves deeper. The first method is preferred. The volumes should be equalized within 0.1 to 0.2 cc.

Having recorded the existing volume and having determined the desired minimum, mill both heads the same amount to produce this desired minimum. Depending upon the amount of material to be removed, some material may have to be taken from the intake ports for correct intake manifold port alignment as installed.

Prior to the final assembly, check the installed valve spring height as shown in Figure 15-11 with the retainer in place. Check the valve spring tensions on a valve spring tester and determine whether shimming is required to give the valve spring the maximum tension specified as installed.

Install the valves, valve seals, shims, springs, retainer, and locks. On the rocker stud-type heads, use studs that are either threaded into the head or pinned to the head to prevent them from pulling out. Install the rocker arm assemblies. The cylinder head assembly is now complete.

Before further checking and assembly work, the crankshaft must be installed and torqued into place. After locating the main bearings, install the crankshaft and main caps. Torque the mains slowly and evenly in stages. After tightening each main cap each stage, turn the crankshaft by hand to be sure that it is still free and is not binding. It should turn freely by hand after all mains are fully torqued. If it does not, the trouble should be located before progressing further.

CAMSHAFTS AND VALVE TRAIN

Install the camshaft and valve lifters, coating both with a break-in lubricant that has an extreme pressure additive (90 weight gear lube or molykote, for example). Mount the timing gears and timing chain. Install a piston and connecting rod assembly in the #1 cylinder for checking purposes. Mount a degree wheel on the end of the crankshaft and a pointer on the cylinder block (Fig. 15-12).

Locating top dead center can be done with a dial indicator or by the positive stop method. In the positive stop method, mount a flat plate on the deck with an adjustable bolt in the center which will stop the piston travel near top dead center. Rotate the crankshaft until the piston hits the stop and note the degrees indicated by the pointer on the degree wheel. Rotate the crankshaft in the opposite direction until the piston hits the stop again and note the degrees. Midway between the two degree readings is TDC, so remove the stop, rotate the crankshaft to the correct setting, and then adjust the degree wheel and pointer to read zero. The degree wheel and pointer should not be moved for the remainder of the checks.

Figure 15-13 shows the method of checking the valve opening and closing points with relation to TDC. The dial indicator locates the valve opening point to determine whether the cam is advanced or retarded relative to the crankshaft rotation. Advanced or retarded conditions can be corrected by using offset keys or offset cam gear bushings to give the desired timing specifications. Worn cam gears or timing chains can alter the timing, so be sure they are in good condition.

PISTON-TO-VALVE CLEARANCE

Now that the cam has been timed, and while the piston and rod assembly is still in the #1 bore, check the piston-to-valve clearance. Lay two strips of modeling clay (about 1/8 in. or 3.175 mm thick) across the top of the piston—in the valve reliefs, if the piston has them. Install the cylinder head and pushrods and adjust the valve clearance. Turn the engine over by hand at least two revolutions. Remove the head and measure the thickness of the clay where the valves made impressions. It is necessary to check only one piston. About 0.090 in. to 0.100 in. (2.286 to 2.540 mm) minimum clearance is recommended. If it is less than that, the pistons should have valve reliefs cut for adequate clearance.

FIG 15-10 Cc'ing a combustion chamber to determine its volume. (Dana Corporation)

FIG 15-11 Measuring hhe installed height of a valve spring prior to installation. (Dana Corporation)

FIG 15-12 Using a dial indicator and degree wheel to determine camshaft index and TDC position. (Dana Corporation)

FIG 15-13 Dial indicator with a magnetic base used for determining lifter movements. (Dana Corporation)

FINAL ASSEMBLY

Install the pistons on the rods. If the piston pins are to be pressed into the rods, it is a good idea to heat the rod eye before pressing. This helps to avoid "shaving" material out of the rod during this step. Install the piston rings on the pistons with a suitable piston ring expander.

Slip a short piece of rubber hose over each rod bolt before installing the piston-rod assembly, for the bolt can easily nick a crankshaft journal. Having coated the rings, pistons, and bearings with a light film of engine oil, use a ring compressor to install each assembly. Install the rod caps and torque after each installation. Be sure that the rod caps are mated exactly with the rod and are not misaligned. Double check the rod side clearance at this point.

Install the cylinder heads, using the proper head gasket. Head gasket thickness and compressibility affect the compression ratio, so know the head gasket you are using. Torque the cylinder heads in place, using the manufacturer's recommended specifications and sequence. Install the pushrods and initially set the valves.

A cam with increased lift affects two more areas which need checking for adequate clearance. Rotate the engine by hand and check for valve spring "coil stacking" or "solid stacking" of the coils. Use a feeler gauge to ensure that there is clearance between coils. If the coils do go "solid," there will be severe damage to the valve train components. Check for adequate valve-seal-to-spring-retainer clearance with the valve open.

Secondly, on rocker stud heads, the rocker arm can bind on the stud at maximum valve opening. Bend a piece of 1/16 in. (0.794 mm) solder wire and insert one end between the stud and the valve side of the rocker arm slot. Rotate the engine until the valve opens and closes. If the rocker arm dents the solder, the slot may have to be relieved slightly with a hand grinder, since 0.060 in. (1.524 mm) minimum clearance is desired with the valve open.

Having made these final checks and adjustments, proceed with the complete buttoning up. A tip on the intake manifold gasket—be sure the gasket openings don't lap into the ports. If they do, they can be corrected by trimming.

The choice of equipment, special modifications, finishing, and surface treatments are all the choice of the builder. Every blueprinted engine reflects the individual touch. Use personal choice along with the manufacturers' service specifications, and take time to do a thoroughly clean and accurate job. The result is an engine that will give maximum performance and durability, a worthwhile and satisfying reward for the effort.

Glossary of Terms

Abrasive Any substance that cuts or grinds away another substance.

Additive A substance added to oil or gasoline to improve their function.

Adhesive A liquid material used to hold two or more things together.

After market A source of parts and materials that are applied to vehicles after assembly by the manufacturer.

Alloy A mixture of two or more metals.

Aluminum A metal used in the manufacture of engine parts, especially pistons.

API American Petroleum Institute.

ATDC After Top Dead Center.

Atomization Breaking a liquid (fuel) into fine particles or a mist.

Atmospheric pressure Pressure of the air, usually at sea level.

Babbit Bearing metal used in cam bearings.

Backlash The clearance or play between two parts, such as meshed gears.

Bank One-half of the cylinders of a vee or opposed engine.

Base circle The part of the cam with the smallest diameter, the side opposite the lobe.

BDC Bottom Dead Center, furthest downward travel of a piston.

Bearing An antifriction device.

Bearing clearance The difference in diameter between the bearing and the journal.

Bearing crush *See* Crush.

Bearing spread *See* Spread.

Bell housing Bell-shaped area between the engine and transmission which houses the flywheel and clutch or torque converter.

Bell mouth The oval-shaped wear typical of used valve guides.

Block *See* Cylinder block.

Blow-by Combustion gases that "blow by" the piston and its rings into the crankcase.

Bore The diameter of a cylinder; also used to describe the machining process of enlarging the cylinder by cutting with a boring bar.

Boring bar A device used to enlarge the cylinder diameter by cutting.

Break-in The period of engine operation following a reconditioning or the original operation of a new engine during which the parts wear together.

Broach To finish the surface of a metal by pushing or pulling a cutting or forming tool over or through it.

BTDC Before Top Dead Center.

Bushing A one-piece sleeve bearing pressed and expanded into an opening such as the small end of a connecting rod.

Cam follower A type of lifter used with OHC valve designs.

Camshaft The rotating shaft in the engine that opens the valves through the use of its cam lobes; driven at one-half the crankshaft speed by gears, belts, or sprockets and chains.

Catalytic converter An emission control device applied to the exhaust system to reduce the CO (carbon monoxide) and HC (unburned hydrocarbons).

Chamfer A surface that is 45° from two planes; to cut a sharp edge from a 90° corner such as a cylinder bore and deck surface.

Chassis The vehicle frame, suspension, and running gear—everything except the body.

CID Cubic inch displacement of a piston from BDC to TDC.

Clearances The difference in diameters of two parts or the amount of movement or play one piece makes against another.

Clutch The mechanism used with a standard transmission to engage and disengage the power train.

Clutch release bearing *See* Throwout bearing.

Compression pressure The pressure buildup resulting from pistons moving upward in the cylinder on the compression strokes.

Compression ratio A ratio of the volumes of a cylinder and combustion chamber with the piston at BDC compared to that with the piston at TDC.

Compression test The test performed to determine each cylinder's ability to pressurize the combustion chamber.

Connecting rod The internal engine part that connects the piston pin to the crankshaft (connecting rod journal).

Core The central portion of a casting mold used to displace the molten metal; the cylinder block core mold provides the passages for coolant flow when the cylinder block is cast.

Core plug Steel or brass plugs used to cover the openings where the core mold sand was removed; also called freeze plug, soft plug, welsh plug, expansion plug.

Crank pin *See* Connecting rod journal.

Crankshaft The largest rotating piece in the engine, which turns reciprocating motion into rotary motion.

Crush The slight extra length of an insert bearing that, when assembled, will assist in applying adequate contact pressure to the housing bore.

Cu in Cubic inch; the volume equivalent of a cube with six equal sides one inch square. Unit of volume measurement of cylinder size.

Cubic centimetre Cc A cube with 6 equal sides, each one being one centimetre square. Unit of volume measurement of cylinder size.

Deck Flat surface of a cylinder block where the head is mounted.

Degree A unit of circular measurement equal to 1/360 of a circle.

Detonation Abnormal combustion resulting in an audible knock, often caused by incorrect octane fuels or lugging of an engine.

Dowel A type of fastener used to align engine parts capable of being dis-assembled.

Draw file To smooth a surface by moving a file lengthwise across a surface while it is held perpendicular.

Dynamometer A complicated device used to measure the torque effort made by an engine or engine and vehicle; torque is then computed to horsepower.

EGR Exhaust gas recirculation.

Emission control A device or system designed to reduce the unwanted air pollutants from an engine or vehicle.

End gap *See* Ring End Gap.

End play The back and forth movement of a sliding part or shaft.

EP Extreme pressure, as related to lubricants.

Epoxy A chemical compound used to repair metal cracks or adhere materials together.

Ethylene glycol Chemical name for the base material used in most coolant and antifreeze solutions.

Expansion plug *See* Core plug.

Feeler gauge A strip or group of strips of a given thickness used to measure the distance between objects.

File A metal cutting tool used to remove minor amounts of metal.

Filter A device to trap contaminants in the fluids used in a vehicle.

Firing order The consecutive order in which the power strokes in an engine occur.

Flat rate An allotted time for journeyman mechanics to complete a particular job.

Floor jack A hydraulic lifting device on wheels; also called a service jack.

Flywheel A circular piece with enough mass (weight) to assist in smoothing out the power impulses of an engine.

Foot-pound The unit of energy or work to lift or pull 1 pound 1 foot.

Forged Made by forging.

Forging The process of shaping metal by hammering or pressing into the desired shape.

Four-stroke cycle The series of piston movements and valve openings of an internal combustion engine; the strokes in order are intake, compression, power, and exhaust.

Friction The resistance to movement of one object against another.

Full-floating pin One design of piston-to-connecting rod attachment; the pin has bearing surfaces in both the piston and the connecting rod.

Full-flow filter A filter and lubrication system in which the oil goes through the filter before it reaches the bearings.

Gallery Internal oil passages within the block.

Gas A state of matter not liquid or solid.

Gasket A sealing material of various types used to control gases and fluids.

Gasket sealer-adhesive Chemical compounds used to hold and seal liquids and gases.

Glaze The shiny surface of petroleum varnish that develops on a cylinder wall during engine operation.

Hacksaw A metal cutting tool with a hardened blade.

Harmonic balancer *See* Vibration damper.

Hemispherical combustion chamber (Hemi) One design of combustion chamber whose shape resembles a hemisphere.

High energy ignition system A type of breakerless ignition system applied to most 1975 and later GM engines.

Hoist Any device to lift a part or vehicle.

Hone A process of resizing and refinishing an opening to a given diameter; a tool used to perform the honing process.

Hydraulic lifter Part of the valve train used between the camshaft and push-rod of most modern engines.

id Inside diameter.

Impact wrench An air-operated or electrically-operated wrench for loosening and tightening fasteners.

Impeller The central rotating part of a water pump.

in Inch.

In-line The arrangement in which cylinders are placed front to rear.

Journal The rotating portion of a shaft that turns within a bearing.

Keeper The split lock device used with a retainer to hold valve springs.

Key A device used to prevent one part from turning against another.

Knock A variety of noises that develop in an engine where there is excessive clearance or abnormal combustion.

Knurl The forming process in which metal is expanded to reduce clearance in valve guides and on piston skirts.

kPa Kilopascal—a metric (SI) unit of pressure or stress measurement (1 kPa = 6.895 psi).

Litre A unit of metric volume measurement slightly greater than one U.S. quart.

Lobe The highest part of an eccentric such as a cam lobe.

Loctite Brand name for several thread-locking chemicals.

LPG Liquid petroleum gas—a petroleum gas used for engine operation.

Lugging Operating an engine at wide open throttle with no increase in engine speed; often occurs when an engine is overloaded in too high a gear.

Magna flux A brand name for a metal fracture detector.

Magnetic crack detection A process using magnetic energy with magnetic powder to detect metal fractures.

Manifold heat control valve A valve which creates slight back pressure on the hot exhaust gas to cause its movement toward the intake manifold.

Metre (m) A metric unit of linear (length) measurement equal to 39.37 inches.

Micrometer A precision measuring device that determines distances in fine units such as inches and millimetres.

Millimetre (mm) A metric unit of linear (length) measurement (1 000 mm = 1 m).

NO_x Chemical symbol for one part of the exhaust gases from a running engine.

Nose The area of the cam lobe producing the highest lift.

Octane rating A common method of judging a gasoline's ability to prevent abnormal combustion (knocking or pinging).

Off-road Any place beyond the normally traveled roads or highways, including race tracks.

OHC Overhead camshaft—an engine in which one or more camshafts are mounted on the head.

OHV Overhead valve—an engine whose valves are placed within the head.

Oil slinger A circular disc used to prevent excessive amounts of oil from touching an oil seal, often found on the crankshaft within the timing cover on an OHV engine.

Oscilloscope An electronic device which presents a picture of the input.

Oversize A term used to describe the increase in size of an opening necessary to correct clearance irregularities.

PCV Positive crankcase ventilation—a system used to recycle the blow-by gases from the crankcase.

Peen A process in which metal is showered with blows from glass beads, steel shot, hammers, etc. to clean and/or relieve internal stress.

Permatex Brand name of a variety of sealants and adhesives.

Pilot bearing bushing A bearing or bushing within the rear flange of a crankshaft used to support the input shaft of a transmission.

Piston displacement *See* CID.

Piston pin A hardened and plated steel sleeve used to joint the piston to the connecting rod, often called a wrist pin.

Piston ring compressor A tool used to compress the piston rings on the piston, allowing the assembly of the piston in the cylinder.

Piston rings Metal rings fitted to the grooves in the piston to control the combustion and crankcase oil; two types of piston rings are the compression and oil control.

Piston slap The noise heard when a piston reaches top dead center and rocks within the bore, caused by excessive clearance or improper piston installation on the rod.

Plastigage A wax string used to measure bearing clearances; its compressed width represents the distance between the bearing and the journal.

Pop-up A raised portion on a piston head.

Preignition Ignition of the air-fuel mixture before the spark jumps the electrodes at the spark plug.

Press fit One way two or more parts are joined together; mechanical or hydraulic press is often used for assembly and disassembly.

Pressure cap A spring-loaded cap used on a radiator to raise the boiling point of the coolant.

Pressure relief valve A valve with regulated spring pressure to bypass excessive pressure in the lubrication system.

Psi Pounds per square inch.

Puller A tool used to disassemble press-fit parts.

Radiator A device with finned passages used in the cooling system to remove heat from the liquid coolant.

Rails The thin part of steel rings used as part of the oil control (piston ring).

Ream A process using a multi-edge cutting tool (reamer) to size a hole.

Reciprocating motion Movement of an object back and forth or up and down.

Retainer A device to assist in holding an object in position.

Ring end gap The gap between the ends of a piston ring as it is placed within a cylinder.

Ring expander A tool used to expand a piston ring so it can be installed in the ring groove.

Ring ridge The upper unworn portion of a cylinder having no contact with the piston or its rings.

Ring ridge remover, Reamer A device to cut away the ring ridge until it is parallel with the upper wear area of the cylinder wall.

Rocker arm Part of the valve train, resembling a lever, that is used between the pushrod and the valve stem on an OHV engine to open the valves.

Rpm Revolutions per minute.

RTV Room temperature vulcanization—trade name for a silicone rubber adhesive sealant.

SAE Society of Automotive Engineers.

Score Advanced wear condition of scuffing where visible lines, grooves, or edges appear.

Scuff Wear in which one metal transfers to another in minor amounts.

Seal A flexible device designed to retain liquids within a given area; also used to keep contaminants out of an engine.

Service ratings Letter designations on lubricating oil containers indicating the type of service for which the oil was formulated, for example, SE, SD, SC, SB, SA.

Side clearance The distance between the sides of moving parts that do not serve as a load-carrying surface.

Silicone An organic base material used for lubricants, seals, and sealants; RTV base material.

Sludge A mixture of lubricating oil, water, and other crankcase contaminants usually formed into a mayonnaise-like material.

Smog An acronym derived from the words *smoke* and *fog* referring to undesirable air pollution formed from a variety of materials under sunlight; also called photochemical smog.

Soft plug *See* Core plug.

SOHC Single overhead cam—An engine valve design with one camshaft per cylinder head.

Spread Increased distance between two points beyond normal.

Spring Steel wire device wound into a coil to create tension or compression force.

Spring retainer Circular piece used to control the position and movement of a spring.

Strobe The length of piston travel within an engine; also, to increase the length of piston travel, as performed on some high performance engines.

Stud A headless fastener threaded on both heads.

Synthetic A man-made material closely resembling one of a natural origin.

Synthetic oil A man-made engine lubricating oil whose chemical structure is different from the common hydrocarbon oil.

Taper Difference in diameter (inside or outside) usually caused by wear.

Tappet *See* Lifter.

TDC Top dead center—Furthest movement of the piston toward the combustion chamber.

Thread chasing Using a tap to clean threads of foreign material.

Thread insert A device installed in a threaded hole to repair damaged threads.

Throating angle Valve seat grinding or cutting angle closest to the head ports; on a 45° angle seat, this would be 60°.

Throwout bearing The bearing mounted on the clutch release lever used to disengage the clutch.

Topping angle Valve seat grinding or cutting angle for narrowing seat width which is closest to the combustion chamber; on a 45° seat, this would be 30°.

Torque Turning or twisting effort.

Torque converter A fluid coupling used in an automatic transmission, located between the transmission and flywheel in the bell housing.

Torque specification The tightening effort required to bring two or more parts together with the proper tension.

Torque wrench A tool calibrated to measure the turning resistance of a fastener.

Undersize A term used to describe the diameter of a shaft that has been reground to less than standard size; undersize bearing inserts are thicker than standard to compensate for the metal that was ground away.

Vacuum Any pressure less than atmospheric.

Valve A device to control the flow of a fluid.

Valve clearance The operating distance between the rocker arm or cam follower and the camshaft.

Valve guide The opening in which the valve moves.

Valve lift The total amount of movement from closed to maximum open that the valve travels.

Valve overlap The period of camshaft rotation at the end of the exhaust stroke when both intake and exhaust valves are open.

Valve seat The surface in the head or block which contacts the valve face when it is closed.

Valve timing The relationship between camshaft and crankshaft movements.

Vernier A scale used with precision measuring instruments with alignment of lines and numbers.

Vernier caliper A measuring caliper with a vernier scale.

Vibration damper A device mounted on the nose (front) of the crankshaft to help absorb the vibration to the crankshaft caused by the power impulses of an operating engine.

VIN Vehicle Identification Number.

Viscosity index rating The ability of a liquid, usually lubricating oil, to flow; number on a can of motor oil below the letters *SAE*.

Vulcanization The process of curing (hardening) rubber.

Wire, mechanics Noninsulated soft steel wire used to maintain the position of a part or as a locking or pulling device.

Yoke (dummy) The transmission end of a drive shaft universal assembly which maintains the oil in the transmission case during its removal.

Index